Ultrasonic Nano/Microfabrication, Handling, and Driving

Ultrasonic nano/microfabrication, handling and driving is an emerging actuation technology, which utilizes ultrasonic vibration and the physical effects of ultrasonic vibration in fluids and solids to implement the fabrication, handling and driving of nano/micro scale objects. This book provides readers with the fundamentals, principles and characteristics of the ultrasonic devices for nano/micro fabrication, handling and driving, and design methods of the devices.

- Introduces fundamental concepts and offers examples of ultrasonic nano fabrication, including ultrasonic nano rolling, cutting and coating methods.
- Features a wealth of examples to illustrate the ultrasonic concentration and ultrasonic tweezers.
- Explains the principles of ultrasonic driving of gas molecules and demonstrates their applications in high-performance gas sensor systems and metal-air flow batteries.
- Teaches the principles of ultrasonic driving of microfluids and their applications in metal-air flow batteries and cooling of small solid heat sources.
- Provides examples for the finite element method (FEM) modeling and computation of ultrasonic devices for nano/micro fabrication, handling and driving.
- Summarizes the current and future trends in ultrasonic nano/microfabrication, handling, and driving.

This book shares the advances, methods and applications of ultrasonic micro/nano fabrication techniques for entry-level and advanced readers working on nano/microfabrication, gas sensing, biological sensing, metal-air batteries, electronic component cooling, and other related areas.

Emerging Materials and Technologies

Series Editor: Boris I. Kharissov

The *Emerging Materials and Technologies* series is devoted to highlighting publications centered on emerging advanced materials and novel technologies. Attention is paid to those newly discovered or applied materials with potential to solve pressing societal problems and improve quality of life, corresponding to environmental protection, medicine, communications, energy, transportation, advanced manufacturing and related areas.

The series takes into account that, under present strong demands for energy, material and cost savings, as well as heavy contamination problems and worldwide pandemic conditions, the area of emerging materials and related scalable technologies is a highly interdisciplinary field, with the need for researchers, professionals and academics across the spectrum of engineering and technological disciplines. The main objective of this book series is to attract more attention to these materials and technologies and invite conversation among the international R&D community.

Wastewater Treatment with the Fenton Process: Principles and Applications
Dominika Bury, Piotr Marcinowski, Jan Bogacki, Michal Jakubczak,
Agnieszka Maria Jastrzebska

Polymer Processing: Design, Printing and Applications
of Multi-Dimensional Techniques
Abhijit Bandyopadhyay and Rahul Chatterjee

Nanomaterials for Energy Applications
Edited by L. Syam Sundar, Shaik Feroz, and Faramarz Djavanroodi

Wastewater Treatment with the Fenton Process: Principles and Applications
Dominika Bury, Piotr Marcinowski, Jan Bogacki, Michal Jakubczak,
and Agnieszka Jastrzebska

Mechanical Behavior of Advanced Materials: Modeling and Simulation
Edited by Jia Li and Qihong Fang

Shape Memory Polymer Composites: Characterization and Modeling
Nilesh Tiwari and Kanif M. Markad

Impedance Spectroscopy and its Application in Biological Detection
Edited by Geeta Bhatt, Manoj Bhatt and Shantanu Bhattacharya

Nanofillers for Sustainable Applications
Edited by N.M Nurazzi, E. Bayraktar, M.N.F. Norrrahim,
H.A. Aisyah, N. Abdullah, and M.R.M. Asyraf

Chemistry of Dehydrogenation Reactions and its Applications
Edited by Syed Shahabuddin, Rama Gaur and Nandini Mukherjee

For more information about this series, please visit: www.routledge.com/Emerging-Materials-and-Technologies/book-series/CRCEMT

Ultrasonic Nano/ Microfabrication, Handling, and Driving

Junhui Hu

CRC Press
Taylor & Francis Group
Boca Raton London New York

CRC Press is an imprint of the
Taylor & Francis Group, an **informa** business

First edition published 2024
by CRC Press
6000 Broken Sound Parkway NW, Suite 300, Boca Raton, FL 33487-2742

and by CRC Press
4 Park Square, Milton Park, Abingdon, Oxon, OX14 4RN

CRC Press is an imprint of Taylor & Francis Group, LLC

ISBN: 978-1-032-51972-2 (hbk)
ISBN: 978-1-032-51973-9 (pbk)
ISBN: 978-1-003-40470-5 (ebk)

DOI: 10.1201/9781003404705

Typeset in Times LT Std
by KnowledgeWorks Global Ltd.

Contents

Preface

Ultrasonic nano/microfabrication, handling, and driving is an emerging technology, which utilizes ultrasonic vibration and the physical effects of ultrasonic vibration in fluid and solid to implement fabrication, handling and driving of nano/micro scale objects. It is of interest to a wide range of people, from the researchers and engineers who work on fabrication of nano/micro devices, syntheses of nigh-end nano materials, manipulations of biomedical samples and design of high-performance sensor systems, to the students who are entering the research field of nano/micro actuation as beginners. This book is written with the intention to provide the readers with the fundamentals and principles related to ultrasonic nano/microfabrication, handling, and driving, the detailed methods of designing the devices based on these principles and the characteristics of the devices. A lot of examples are given in this book, to illustrate how to design ultrasonic devices to implement the nano/micro fabrication, handling and driving, characteristics of the ultrasonic devices and the physical principles for these characteristics. All the examples in this book come from the research results of the author's research team.

Although the recent work of the author and other research teams has demonstrated that ultrasonic nano/microfabrication, handling, and driving has promising applications, there are very few reference books on this topic. The purpose of writing this book is to facilitate the related researchers and students and to contribute to the development and applications of ultrasonic nano/microfabrication, handling, and driving technology.

Ultrasonic nano/microfabrication, handling, and driving has a distinct interdisciplinary feature, and the researchers all over the world have contributed greatly to this technology. Although only the references closely related to the major issues are cited in this book, research work done by the other research teams in this area is equally important and invaluable.

This book is divided into eight chapters. Chapter 1 incorporates the fundamental concepts in ultrasonics, which are related to ultrasonic nano/microfabrication, handling, and driving, and an overall introduction of ultrasonic nano/microfabrication, handling, and driving. Chapter 2 gives and discusses examples for ultrasonic nano fabrication, including ultrasonic nano rolling, cutting and coating methods. Chapters 3 and 4 demonstrate the ultrasonic nano/micro handling technology with lots of examples. Chapter 3 describes and discusses the ultrasonic concentration or enrichment, and Chapter 4 the ultrasonic tweezers. Chapter 5 explains the principle of ultrasonic driving of gas molecules and demonstrates its applications in high-performance gas sensor systems and metal-air flow batteries. Chapter 6 explains the principle of ultrasonic driving of microfluid and demonstrates its applications in metal-air flow batteries and in the cooling of small solid heat sources. Chapter 7 gives more examples for FEM modeling and computation of ultrasonic devices for nano/micro fabrication, handling and driving. Chapter 8 summarizes the contents of this book, and elaborates on the development trends of the ultrasonic nano/microfabrication, handling, and driving technology.

My thanks go to my students who have contributed to the research work in this book, and the peers and colleagues who have supported the research activities of my team. I would very much like to acknowledge my gratitude to Dr. Qiang Tang, Dr. Pengzhan Liu, Dr. Xu Wang, Dr. Qingyang Liu, Dr. Gengchao Chen, Dr. Xiaomin Qi, Dr. Songfei Su, Dr. Xiaolong Lu, Dr. Hao Xue, Dr. Tianyu Zhang, Dr. Huiyu Huang, Mr. Yuchen Zhou, Mr. Huibin Ba, Dr. Zhao Luo, Mr. Jia Yin, Miss. Yumin Yang, Miss. Qiuxia Ma, Mr. Mu Wang, Mr. Junchao Che, Mr. Zongheng Xiang, Mr. Shihao Wei, Mr. Zihao Tang, Mr. Junchao Zhang, Mr. Kaibo Jia, Mr. Jiandong Xu, Mr. Shuo Zhang, Mr. Jiwei Li, Mr. Jianwen Zhou, Mr. Qi Zhan and Mr. Zhiyuan Zhu. for their academic contributions and hard work. I am also very grateful to Prof. Igor V. Minin, Prof. Oleg V. Minin, Prof. Tony Jun Huang and Prof. Takeshi Morita for the fruitful collaborations on ultrasonic micro/nano manipulations, and grateful to the National Natural Science Foundation of China for the funding support (Grant No. 11974183). Special thanks go to my wife Qun Yue and my parents for their understanding and kind support.

Junhui Hu
July 2023

Author Biography

Junhui Hu received his Ph.D. Degree from Tokyo Institute of Technology, Tokyo, Japan, in 1997, and B. E. and M. E. degrees in electrical engineering from Zhejiang University, Hangzhou, China, in 1986 and 1989, respectively. Currently, he works for Nanjing University of Aeronautics & Astronautics, China, as a full professor. His research interest is in ultrasonic sensors and actuators, ultrasonic nano fabrication, ultrasonic micro/nano/molecular manipulations, etc. He is a Chang-Jiang Distinguished Professor, China, and an IAAM Fellow.

He was an assistant and associate professor at Nanyang Technological University, Singapore, from 2001 to 2010, and an R&D engineer at Tokin, Japan, from 1997 to 1999. He authored and co-authored more than 300 publications, including more than 100 full research papers published in SCI journals, two books, one editorial review in an international journal and more than 60 disclosed/empowered China and Japan patents. He is the sole author of the monograph book *Ultrasonic Micro/Nano Manipulations: Principles and Examples* (2014, World Scientific). He has given more than 30 keynote/plenary/invited lectures at international conferences, and his research work has been highlighted by seven international scientific media. He served lots of international conferences as a Technical Program/Organizing/Scientific Committee member and was the chairman/honorary chair of seven international conferences. He won the Paper Prize from the Institute of Electronics, Information and Communication Engineers (Japan) as the first author in 1998, Also, he Scientist Medal of IAAM in 2023. and was awarded the title of valued reviewer by *Sensors and Actuators A: Physical* and by *Ultrasonics*. He was once supported by the Shuang-Chuang Project of Jiangsu Province, China, as a "Shuang-Chuang" expert. Presently, he is an editorial board member of four international journals, a board member of the Chinese Acoustical Society, as well as deputy director of expert committees on electronic information materials and devices and on Aerospace materials, Chinese National Think Tank for Materials and Devices.

1 Introduction

This chapter is to give the fundamentals of ultrasonics, which are essential to have a better understanding of the working mechanisms and characteristics of ultrasonic devices illustrated in this book and the research background of ultrasonic nano/microfabrication, handling, and driving. In the first section of this chapter, the fundamentals of ultrasonics that are related to ultrasonic nano/microfabrication, handling, and driving, are briefed. In the second section, the basic functions and working principles of ultrasonic nano/ microfabrication, handling, and driving methods are reviewed.

1.1 THE FUNDAMENTALS OF ULTRASONICS

Ultrasound, or ultrasonic wave, is a vibrational wave in solid, fluid and plasma with a frequency higher than 20 kHz, which cannot be heard by the human ear. Ultrasonics is the discipline that studies the generation, propagation, receiving, linear/nonlinear physical/chemical effects and various applications of ultrasound [1–3]. Ultrasound has wide applications in chemical engineering, environmental engineering, material engineering, biotechnology, medicine, electrical engineering, agriculture, etc.

1.1.1 GENERATION OF ULTRASOUND

1.1.1.1 Ultrasonic Transducers

Ultrasound can be generated by some animals, such as dolphins, whales and bats [4]. Bottlenose dolphins can generate ultrasound from their fatty foreheads in water for navigation and communication, and bats can generate ultrasound in a frequency range of 30–120 kHz to navigate and locate their prey. Ultrasound can also be generated artificially. The devices that are designed to generate ultrasound are termed ultrasonic transducers [4–6]. The early ultrasonic transducers are excited by electromagnetic force to generate ultrasonic waves. Presently, most of the ultrasonic transducers are excited by the piezoelectric effect, that is, the ultrasonic wave is generated by the same-frequency ultrasonic vibration of a piezoelectric component or a solid structure mechanically driven by piezoelectric component. The ultrasonic transducer excited by the vibration of a piezoelectric component is also termed a piezoelectric transducer.

Figure 1.1 shows the images of some commercial piezoelectric components. In ultrasonic nano/fabrication, handling and driving, one or multiple piezoelectric components are bonded onto the vibration excitation part of an ultrasonic device to mechanically excite the device.

Figures 1.2 and 1.3 show the structures and photos of two Langevin transducers (sandwich-type transducers), respectively. Figure 1.2(a) shows the structures of a traditional Langevin transducer working in the longitudinal vibration mode, in which two mass blocks (the front mass and back mass) are made of metal materials such as duralumin and stainless steel clamp two piezoelectric components

DOI: 10.1201/9781003404705-1

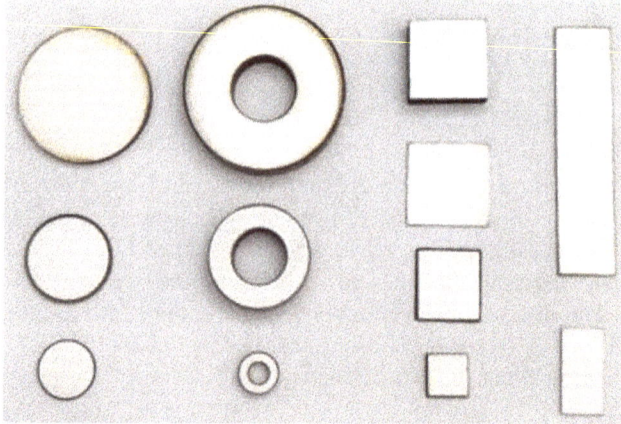

FIGURE 1.1   Photo of commercial piezoelectric components.

via a bolt structure. For the traditional Langevin transducer, the two piezoelectric components are polarized in the thickness direction, and their polarization
directions are opposite. One metal electrode is in between the two piezoelectric
components, and another one is in between one piezoelectric component and its
neighboring mass block (Fig. 1.2). The electrodes may be made of phosphor bronze,
copper covered with Ag and other metal materials. Figure 1.2(b) shows an optimized structure of the traditional Langevin transducer, in which the two cylindrical
terminal mass blocks are conically shaped with four supporting plates each. The
optimized structure brings in larger vibration velocity, lower temperature rise and
higher electroacoustic energy efficiency [7]. Apart from the sandwich-type transducer and piezoelectric components, the interdigital electrode transducer (IDT) is

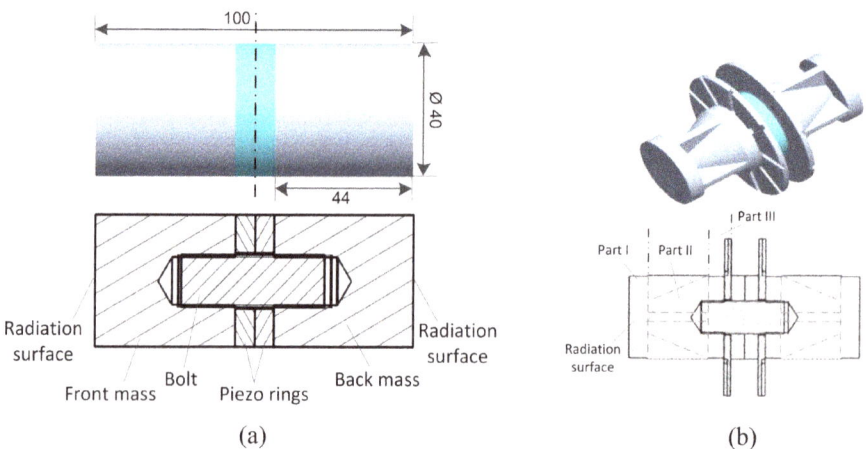

FIGURE 1.2   Configuration of (a) traditional Langevin transducer (TLT) and (b) mass-
optimized Langevin transducer (MOLT). The unit of the dimensions in (a) is mm. Reproduced
from Ref. [7] with permission from Elsevier.

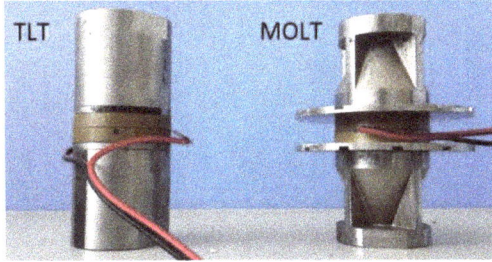

FIGURE 1.3   Photos of the transducers. Reproduced from Ref. [7] with permission from Elsevier.

another type of ultrasonic transducer, which has been widely employed to generate the surface acoustic wave (SAW) [8–10]. In ultrasonic nano/microfabrication, handling, and driving, one may use a piezoelectric component or IDT if there is a limitation on device volume and a sandwich-type transducer if power ultrasound is desired. For high-power ultrasonic applications, the ultrasonic transducers may also be excited by the vibration of magnetostrictive material [11]. However, in this case, the ultrasonic transducers usually have a larger size than the piezoelectric ones.

In practical applications, the maximum vibration of an ultrasonic transducer is limited by the temperature rise of the piezoelectric components. As the temperature rise increases, the resonance frequency of a transducer will become lower, and a frequency tracking circuit or operation is needed if one wants to keep the vibration constant. If the temperature rise is too large, even using the frequency tracking circuit cannot keep the vibration constant, and the transducer performance degrades.

1.1.1.2   Vibration Strength and Modes

One of the important characteristics of an ultrasonic device for nano/micro fabrication, handling and driving is the effect of vibration strength on the device's performance. The vibration strength may be represented by the vibration displacement u, velocity v and acceleration a at some point of an ultrasonic device. For most cases in nano/microfabrication, handling and driving, devices work at a single frequency. Thus, the amplitude of displacement u, velocity v and acceleration a of vibration with working frequency f have the following relationships:

$$v = \omega u; a = \omega v, \tag{1.1}$$

where ω is the angular frequency ($= 2\pi f$). For the root-mean-square (rms) values of them, there are similar relationships. To generate nano/microfabrication, handling and driving functions, the ultrasonic device must have sufficient vibration strength. This can be achieved by utilizing the resonance of the ultrasonic device.

The pattern of a vibration is termed vibration mode. Using a proper vibration mode of the device is another essential requirement for implementing ultrasonic nano/microfabrication, handling, and driving. Figure 1.4 shows the vibration modes of radiation surfaces of two transducers shown in Fig. 1.3, and the curve on the left side in Fig. 1.5 shows the flexural vibration mode of a micro manipulation probe in

TLT

MOLT

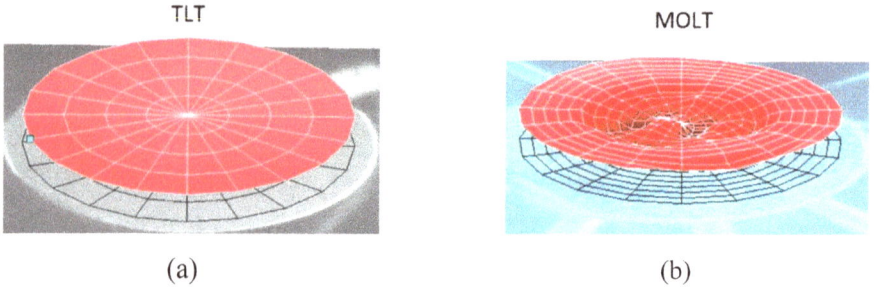

(a)   (b)

FIGURE 1.4   Vibration modes of the radiation surfaces of transducers shown in Fig. 1.3. (a) Perfect piston vibration mode. (b) Piston vibration mode with some irregularity at the central area. Reproduced from Ref. [7] with permission from Elsevier.

an ultrasonic micro tweezer [12]. The vibration mode in Fig. 1.4 was obtained by laser Doppler vibrometer (PSV-300F-B) and that in Fig. 1.5 by finite element method (FEM) computation (COMSOL Multiphysics).

1.1.1.3   Vibration Analyses

In design and optimization of the devices for ultrasonic nano/microfabrication, handling, and driving, vibration of the devices has to be analyzed to understand the vibration mode and strength under working conditions. The vibration can be analyzed by theoretical computation and measurement. In the vibration analyses, one has to understand two important concepts, that is, the out-of-plane and in-plane vibration velocities (or displacements). Either of them is the vibration velocity (or displacement) at a point on a vibrating surface. The out-of-plane one is normal to the vibrating surface, whereas the in-plane one is parallel to the vibrating surface.

In ultrasonic nano/microfabrication, handling, and driving, the devices usually have a rather complicated structure and consist of different materials, such as piezoelectric material and metal material. Thus, it is usually difficult to obtain the analytical solution of the device vibration. As a common practice, the FEM has been employed to compute the vibration mode and velocity of the devices. The commonly used FEM software packages include COMSOL Multiphysics and ANSYS. The

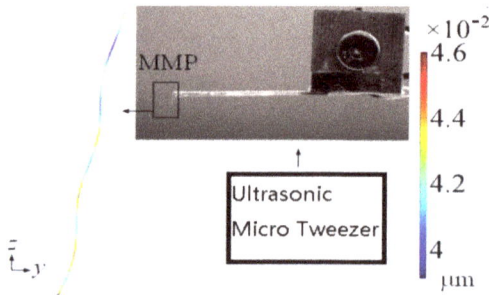

FIGURE 1.5   Flexural vibration mode of the micro manipulation probe in an ultrasonic micro tweezer. Reproduced from Ref. [12] with permission from Elsevier.

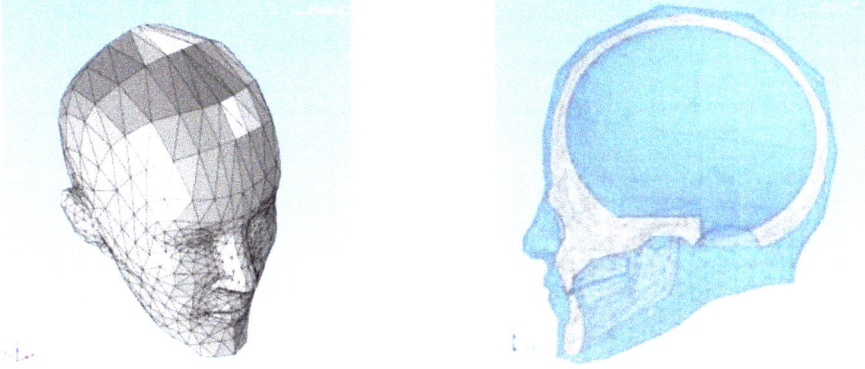

FIGURE 1.6 A 3D FEM model of the human head. Reproduced from Ref. [13] with permission from IEEE.

procedure of FEM computation may be roughly classified into four steps. In step I, a meshed FEM model is built; in step II, the model is solved numerically; in step III, the interested information in the solution is extracted and visualized; and in step IV, post-processing of the solution is carried out if it is necessary. Among these steps, step I is the most difficult, especially in cases where the solved area has a sophisticated shape. Fortunately, the higher version FEM software packages already have the function to import geometrical models generated by other software.

Figure 1.6 shows a 3D FEM model of the human head [13], which has a total element number of about 321,000 and minimum element size of 0.25 cm. The tetrahedral elements are used in the meshed model. Due to the complex shape of a human head, the geometrical model is created by the 3ds Max software first and then imported into COMSOL Multiphysics software. However, it must be pointed out that for most cases of ultrasonic nano/microfabrication, handling, and driving, functions in the commercial FEM software packages are sufficient to create the geometrical models.

If ultrasonic nano/microfabrication, handling, and driving are implemented in an ultrasonic field in a fluid, the vibration of the device is usually coupled with the ultrasonic field in the fluid. In this case, one has to take the coupling of the vibration and ultrasonic field into account in the building process of a FEM model.

The vibration can also be analyzed by measuring the vibration mode and velocity (or displacement) of surfaces of the devices. A laser Doppler vibrometer may be used in the measurement, which can not only measure the amplitudes of the x, y and z components of a measurement point but also give the phase relationships among them. A laser Doppler vibrometer can work with a scanning subsystem to measure the vibration velocity or displacement distribution, and the surface deformation. The vibration mode information can be obtained by analyzing the surface deformation, and the existence of traveling/standing waves may be analyzed by the measured phase distribution of vibration velocity or displacement on a surface [7, 14, 15].

However, it is still difficult to directly measure the vibration of a curved surface and liquid-solid interface, owing to the working principle of the existing laser

Doppler vibrometers. In such a case, the following method may be used to obtain the vibration velocity at point D on a surface, which is difficult to measure.

    I. Choosing point E on the surface of the device, where it is feasible to directly measure the out-of-plane or in-plane vibration velocity.

    II. Measuring the vibration velocity at point E.

    III. Computing the vibration velocity ratio of points E and D by the FEM.

    IV. Using the computed ratio and measured vibration velocity at point E to obtain vibration velocity at point D.

1.1.2 TRANSMISSION OF ULTRASOUND

Ultrasound generated by the transducer will transmit in the acoustic medium. During the transmission, the reflection, refraction and diffraction phenomena will happen, just like any other waves. Although the transmission of ultrasound involves lots of issues, this section only gives the concepts and processes that are closely related to the nano/micro fabrication, handling and driving.

1.1.2.1 Working Ultrasonic Fields

When an ultrasonic wave transmits in a fluidic acoustic medium such as air and water, an ultrasonic field will be generated in the medium. In the devices or equipment for ultrasonic nano/microfabrication, handling, and driving, the ultrasonic field usually has boundaries formed by solid substrate, liquid-air interface, etc. These boundaries can be classified as the acoustically hard boundary, acoustically soft boundary or acoustic impedance boundary, based on the ratio of specific acoustic impedance of the acoustic media on the two sides of a boundary. For example, the droplet-hard substrate interface may be treated as the acoustically hard boundary, and the droplet-air interface may be treated as the acoustically soft boundary. The out-of-plane vibration velocity is zero at the acoustically hard boundary, whereas the acoustic pressure is zero at the acoustically soft boundary. The boundary where both the out-of-plane vibration and acoustic pressure exist must be treated as an acoustic impedance boundary. For example, surfaces of biological tissues and cells may be treated as the acoustic impedance boundary.

In ultrasonic nano/microfabrication, handling, and driving, ultrasonic field in the vicinity of manipulated objects is termed working ultrasonic field (or working field). To analyze the working ultrasonic field, one has to pay attention to the acoustic boundary layer formed by the surface on a substrate or other solid object in the manipulation region, in which the viscosity of fluid medium has a very large effect on the ultrasonic field. The thickness of the acoustic boundary layer may be estimated by the following equation [1, 6]:

$$\delta = \sqrt{\frac{2\eta}{\omega \rho_0}}, \qquad (1.2)$$

where η and ρ_0 are the shear viscosity and density of the acoustic medium, respectively, and ω is the angular frequency of the acoustic field. The viscous boundary layer of an acoustic field is usually very thin, for example, it is 1.8 μm and 7.2 μm thick at 100 kHz and room temperature in water and air, respectively.

The working ultrasonic field may be irrotational and rotational. When the manipulated objects are beyond the acoustic boundary layer or the shear viscosity of acoustic medium is not important, it is irrotational, that is, the working ultrasonic field is governed by the wave equation [1, 5, 6]. When the manipulated objects are inside the acoustic boundary layer or the shear viscosity of acoustic medium is important, it is rotational, that is, vibration velocity of acoustic medium is the vector summation of longitudinal and rotational components [1, 6, 16] or the solution of the thermal viscosity acoustic field.

1.1.2.2 Governing Equations for Working Ultrasonic Fields

The commonly used parameters to describe an ultrasonic field include the scalar potential φ, vector potential Ψ, vibration velocity v and acoustic pressure p. The scalar potential φ and vector potential Ψ are two field variables to make field solving more convenient. The vibration velocity v is [16]

$$v = \nabla\varphi + \nabla \times \Psi, \tag{1.3}$$

in which $\nabla\varphi$ is the gradient of φ and $\nabla \times \Psi$ is the curl of Ψ. For the irrotational ultrasonic field such as the field outside the acoustic boundary,

$$v = \nabla\varphi. \tag{1.4}$$

The acoustic pressure p is the sound-induced pressure change in a fluidic acoustic medium, and thus, it is also termed acoustic pressure.

$$p \equiv P - P_0, \tag{1.5}$$

where P and P_0 stand for the total pressure and static pressure (the pressure when there is no sound wave), respectively. For the irrotational ultrasonic field, such as the field outside the acoustic boundary, the acoustic pressure is

$$p = -\rho_0\varphi_t + b\mu\nabla^2\varphi, \tag{1.6}$$

where ρ_0 is the static density of the fluidic acoustic medium, φ_t is the first partial derivative of φ with respect to time t, b is the viscosity number and μ is the shear viscosity. The Laplacian operator ∇^2 is defined by

$$\nabla^2 = \frac{\partial^2}{\partial x^2} + \frac{\partial^2}{\partial y^2} + \frac{\partial^2}{\partial z^2}. \tag{1.7}$$

The viscosity number b is

$$b = \frac{4}{3} + \frac{\mu_B}{\mu}, \tag{1.8}$$

where μ_B is the bulk viscosity.

For the irrotational ultrasonic field in inviscid fluid, such as the field outside the acoustic boundary in water and air, φ is solved from the following wave equation:

$$c_0^2 \nabla^2 \varphi = \varphi_{tt},\tag{1.9}$$

where φ_{tt} is the second derivative of φ with respect to time t and c_0 is the sound speed in acoustic medium. If the irrotational ultrasonic field is in a sticky fluid, the wave equation is in the following form:

$$\rho_0 \nabla \varphi_t + \nabla P - b\mu \nabla \left(\nabla^2 \varphi\right) = 0.\tag{1.10}$$

This equation is derived under the assumption of small-signal acoustics, that is, the sound-induced fluid density change $\delta\rho$ is assumed to be much less than ρ_0.

For a rotational ultrasonic field such as the field inside the acoustic boundary, the vibration velocity may be solved from the following equation [16]:

$$\rho\left(v_t - v \times \nabla \times v\right) + \nabla P + \mu \nabla \times \nabla \times v = 0,\tag{1.11}$$

where v_t is the partial derivative of vibration velocity with respect to time t. This equation forms the base of acoustic boundary theory.

As it is quite difficult to obtain the exact analytical solution of Eq. 1.11, the FEM may be used to solve the ultrasonic field inside and near the acoustic boundary. In this case, the rotational ultrasonic field can be solved by the thermo–viscosity module of the COMSOL Multiphysics software.

1.1.2.3   Energy Conversion and Transmission

If the rms values of the input voltage and current applied to an ultrasonic device are V_{rms} and I_{rms}, respectively, and the phase difference between the input voltage and current is θ, then the real electric power applied to the ultrasonic device (or input electric power) is

$$P_{in} = V_{rms} I_{rms} \cos\theta.\tag{1.12}$$

The vibration of the ultrasonic device will excite an ultrasonic field in the acoustic medium. If the output acoustic power of the ultrasonic device is P_a, then the electro-acoustic efficiency is

$$\eta_{ea} = P_a / P_{in}.\tag{1.13}$$

P_a may be measured by the so-called calorie method [7]. In this method, the temperature rise ΔT of the fluidic acoustic medium in time Δt is measured, and P_a is estimated by

$$P_a = c_s m \Delta T / \Delta t,\tag{1.14}$$

where c_s and m are the specific heat and mass of the fluidic acoustic medium, respectively. As the medium in the ultrasonic field is in elastic vibration, it has kinetic and potential energy. The kinetic energy density is [6]

$$E_k = \frac{1}{2}\rho_0 v^2. \tag{1.15}$$

The potential energy density is

$$E_p = p^2 / \left(2\rho_0 c_0^2\right). \tag{1.16}$$

The largeness and direction of acoustic intensity are defined by the large and direction of acoustic power flux per unit area, respectively. Thus, acoustic intensity can be used to represent the strength and direction of acoustic energy flow. It is usually denoted by I. When a sound wave in medium 1 is incident onto the interface between media 1 and 2, the reflection and transmission phenomena occur, that is, part of the sound power is reflected back into medium 1, and some transmits into medium 2. Denoting the intensity largeness of incident, reflected and transmission waves as I_i, I_r and I_t, respectively, the reflection and transmission coefficients R and T are defined by

$$R = I_r / I_i, \tag{1.17}$$

and

$$T = I_t / I_i, \tag{1.18}$$

respectively. For the plane wave and under the assumption that there is no energy loss at the interface, they are

$$R = \left(z_2 - z_1\right)^2 / \left(z_2 + z_1\right)^2, \tag{1.19}$$

and

$$T = 4z_2 z_1 / \left(z_2 + z_1\right)^2, \tag{1.20}$$

respectively, where z_1 and z_2 are the specific acoustic impedances of mediums 1 and 2, respectively [16]. The specific acoustic impedance z of a point in acoustic medium is the ratio of the acoustic pressure at the point to the vibration velocity at the same point. For the plane wave, there is

$$z = \rho_0 c_0. \tag{1.21}$$

Under the assumption that there is no energy loss at the interface, there is

$$R + T = 1. \tag{1.22}$$

Although the plane wave condition is not satisfied in most of the practical ultrasonic devices, the reflection and transmission coefficients R and T are still commonly used to estimate how much acoustic energy may transmit into the working ultrasonic field from a transducer and leak into the air from a droplet.

1.1.2.4   Traveling and Standing Waves

The acoustic traveling wave is a mechanical vibration wave progressing toward some particular direction in a solid, fluid or soft acoustic medium. In solid and soft acoustic mediums, the traveling wave is usually expressed by the vibration velocity or displacement of an arbitrary field point, whereas in fluidic acoustic medium, the traveling wave may be expressed by the acoustic pressure and vibration velocity/displacement. An important feature of the traveling wave is that the phase angle of the field parameters such as acoustic pressure changes linearly with the position of field point. For a single-frequency ultrasonic wave traveling along the z axis, its mathematical expression is

$$p = p_m \sin\left(\omega t \pm kz + \varphi_z\right), \tag{1.23}$$

where ω is the angular frequency (=$2\pi f$), k is the wave number $\left(= \dfrac{2\pi}{\lambda} = \dfrac{\omega}{c}\right)$ and the sign before kz indicates the traveling direction of the wave. The positive sign before kz means that the wave travels to the $-z$ direction, and the negative one means that the wave travels to the $+z$ direction. The amplitude p_m usually is not a constant for space in practical applications due to the acoustic energy loss caused by the acoustic bubbles, diffraction, relaxation, etc.

The acoustic standing wave is a stationary mechanical vibration wave in an acoustic medium. It may exist in solid, fluid and soft acoustic media. Examples of the standing wave include (but are not limited to): Ultrasonic vibration in a solid device working at or near a resonance frequency, resonant ultrasonic field between a flat radiation surface and reflector, and the vibration of resonant bubbles and droplets. For a single-frequency 3D ultrasonic standing wave in fluid, its mathematical expression is

$$p = p_m \sin\left(\omega t + \varphi_t\right)\sin\left(k_x x + \varphi_x\right)\sin\left(k_y y + \varphi_y\right)\sin\left(k_z z + \varphi_z\right). \tag{1.24}$$

In applications, a standing wave may be generated by the resonance of a solid vibrator and the superposition of two traveling waves with the same frequency and traveling in opposite directions. The latter often occurs in the gap between a flat solid radiation surface and a parallel reflector. An important feature of the standing wave is that the phase angle of the field parameters such as acoustic pressure does not change or only changes by 180°. The traveling and standing ultrasonic waves exist in bulk acoustic wave (BAW) and surface acoustic wave (SAW) forms, both of which have numerous applications in nano/micro fabrication, handling and driving.

In the working ultrasonic fields of practical devices, traveling and standing waves coexist in most of the cases. This may be caused by a reflecting surface with the specific acoustic impedance that is neither zero nor infinite. The percentage of the standing wave may be defined by the standing wave ratio (SWR) [17]

$$SWR = p_{ant}/p_{nod}, \tag{1.25}$$

where p_{ant} is the acoustic pressure at the antinode where the absolute value of acoustic pressure is the maximum, and p_{nod} is the acoustic pressure at the node where the absolute value of acoustic pressure is the minimum. *SWR* is used to represent the purity of a standing wave, that is, *SWR* is infinite for a pure standing wave and 1 for a pure traveling wave. Another definition of the percentage of standing waves is

$$SWR = \left(p_{ant} - p_{nod} \right) / \left(p_{ant} + p_{nod} \right). \tag{1.26}$$

In this case, *SWR* is 1 for a pure standing wave and 0 for a pure traveling wave.

1.1.2.5 Focused Ultrasound

In ultrasonic nano/microfabrication, handling, and driving, focused ultrasound is employed when a strong acoustic cavitation effect or a force potential well of acoustic field is needed or the working region that needs ultrasonic energy has a very small dimension. A strong acoustic cavitation effect may be used in the disruption of biological samples such as DNA and algae. A force potential well of acoustic field may be used in the capture of micro samples. In the gas molecular driving for high-performance gas sensing, only the gas molecules in the vicinity of the sensing surface need to be effectively driven by ultrasound. Thus, focused ultrasound may be used to raise the driving effect and increase the utilization rate of acoustic energy.

The focused ultrasound may be generated by an acoustic lens, concave acoustic radiation surface and other acoustic structures such as a flexurally vibrating metal case [18, 19]. Figure 1.7 shows the acoustic pressure distribution of a focused ultrasonic field with a working frequency of 49.1 kHz, generated by a bowl-shaped acoustic lens excited by a commercial sandwich-type transducer. The diameter of the concave radiation face is 28 mm. Controlling the size of the focal region is an important issue in many practical applications. The size of the focal region is dependent

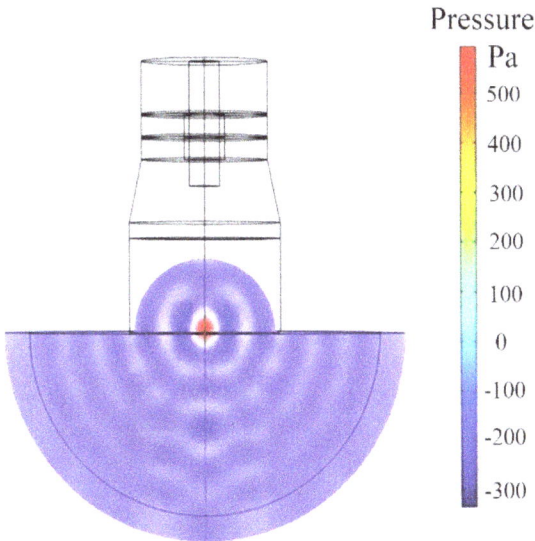

FIGURE 1.7 Acoustic pressure distribution of a focused ultrasonic field. Reproduced from Ref. [18] with permission from Elsevier.

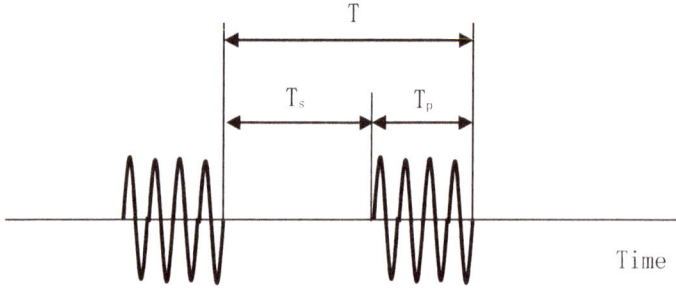

FIGURE 1.8   Pulsed ultrasound and its parameters.

on the ratio R_{rw} of the radiation surface size to the wavelength, and the focal region becomes smaller as ratio R_{rw} increases. In applications of focused ultrasound in nano/micro fabrication, handling and driving, one also has to pay attention to the strong flow generated by the ultrasonic field near the focal point, that is, the acoustic streaming. It may cool a solid surface at or near the focal point, and disturb the trapping process of acoustic tweezers.

1.1.2.6   Pulsed Ultrasound

Pulsed ultrasound, as shown in Fig. 1.8, is used in nano/micro fabrication, handling and driving, mainly for two reasons. One is that pulsed ultrasound with proper parameters can enhance the acoustic cavitation effect [20, 21]. Another is that pulsed ultrasound can decrease the temperature rise of the ultrasonic field and transducer. In terms of the first reason, it must be pointed out that the exact mechanism is still not fully understood yet, although there have been a lot of reports on the phenomena. In terms of the second reason, pulsed ultrasound technology has actually been applied in most of ultrasonic cleaning systems and ultrasonic disrupters for biological samples. In Fig. 1.8, T_p is the pulse width or on-time, T_s is the off-time and T is the pulse repetition period ($= T_p + T_s$). The pulse repetition frequency is $1/(T_p + T_s)$, and the pulse duty ratio is $T_p/(T_p + T_s)$.

In applications of pulsed ultrasound, one often has to control and calculate the time-averaged acoustic intensity $<I>$ or input electric power to the transducer $<P_{in}>$.

$$\langle I \rangle = \frac{1}{T} \int_{t=0}^{t=T} I(t)\, dt. \tag{1.27}$$

$<P_{in}>$ is calculated in a similar way. For a comparison of ultrasonic effect under different parameters of pulsed ultrasound, it is a common practice that $<P_{in}>$ is kept identical.

1.1.3   Physical Effects of Ultrasonic Waves

Ultrasonic nano/microfabrication, handling, and driving are implemented by utilization of various physical effects of ultrasound. Thus, it is necessary to understand the physical effect for better design and analyses of ultrasonic devices for nano/micro handling, fabrication or driving.

1.1.3.1 Acoustic Cavitation

When the total pressure in liquid is reduced to a value lower than the vapor pressure of the liquid by the negative half-cycles of a strong acoustic pressure, micro scale gas nuclei in the liquid can grow and oscillate with the ultrasonic wave. This phenomenon is known as acoustic cavitation [2, 6, 17], which has wide applications in chemical reaction catalysis, cleaning, emulsification, biological sample disruption, nano coating, etc. The bubbles in the acoustic cavitation may be roughly classified into two types. One is the stable bubble, which pulsates over a long-time interval and does not collapse. Another is the unstable or transient bubbles, which pulsates to some maximum size and then collapse violently. During the pulsation of the bubbles, the temperature and pressure inside a bubble can be up to 5000 K and 2000 atm, respectively. During the collapse, high-pressure shock waves are generated, which result in high-speed micro jet flows with a speed of up to several tens m/s. The acoustic cavitation effect is mainly caused by the transient bubbles, and the stable bubbles may hinder the transmission of ultrasonic waves and absorb the acoustic energy.

Factors affecting the acoustic cavitation effect include the acoustic pressure or intensity, working frequency, viscosity, surface tension, vapor pressure and temperature of the liquid. The acoustic pressure and working frequency play important roles in the acoustic cavitation. Increasing the acoustic pressure can enhance the cavitation effect. But the percentage of the stable bubbles, which hinders the transmission of ultrasonic waves, also increases as the acoustic pressure increases. Thus, the acoustic cavitation effect increases little with the acoustic pressure when the acoustic pressure is too strong. For the same acoustic pressure or intensity, increasing the working frequency weakens the acoustic cavitation effect. This is because as the working frequency increases, the duration of the rarefaction phase becomes shorter, which hinders the development of the transient bubbles.

An increase in the viscosity and surface tension of liquid means an intensified inter-molecular adhesive force, and thus hinders the acoustic cavitation effect. The effect of vapor pressure is similar to that of viscosity and surface tension. The temperature affects the acoustic cavitation effect through viscosity, surface tension and vapor pressure. At a higher temperature, it may be easier to generate the acoustic bubbles. However, the acoustic cavitation effect becomes weaker. Impurities in the liquid also affect the acoustic cavitation effect as they contain gas nuclei.

1.1.3.2 Surface Vibration Velocity of Solid Vibrators

Control of the surface vibration velocity of solid vibrators is important in ultrasonic nano/microfabrication, handling, and driving. It is well known that an elliptical and linear vibration velocity of the surface of a solid vibrator can be used to drive a rotor, solider or other macro scale movers. Such devices are termed ultrasonic motors [6, 22–24]. The elliptical and linear vibration loci can be generated by a traveling wave in the solid vibrator, or by a standing wave in the solid vibrator with specially designed driving feet. Recent researches have demonstrated that the elliptical and linear vibration velocity of a micro manipulation probe can excite various acoustic streaming eddies in a droplet and these acoustic streaming eddies can be used to capture nano/micro scale objects in liquid acoustic media [25–30], and they can also excite a micro tool to cut and roll nano/micro samples on a solid substrate [31, 32].

1.1.3.3   Acoustic Radiation Force

An object in a fluidic acoustic medium will experience the acoustic radiation force, which is caused by the non-uniformity of acoustic kinetic and potential energy densities on the object's surface and acoustic momentum transfer to the object. It can be calculated by [6, 33, 34]

$$F = \iint_{S_0} \left(\langle E_k \rangle - \langle E_p \rangle \right) n \, dS - \iint_{S_0} \rho_0 \langle v v_n \rangle \, dS, \tag{1.28}$$

where both integrals are for the whole surface S_0 of the object without sound, < > represents the time average over one time period, n is the outward normal unity vector of S_0, v is the vibration velocity of acoustic medium and v_n is the n directional component of v. The first and second terms on the right side are the contributions of acoustic energy density non-uniformity and momentum transfer, respectively. The acoustic radiation pressure given by the first term on the right side is always perpendicular to the object's surface and that given by the second term on the right side may contain a component in the direction parallel to the object's surface when there is v_n on the surface or the object is acoustically soft. When the object is acoustically hard, the second term on the right side disappears as v_n is zero on the surface. From Eq. (1.28), it is known that the acoustic radiation force exists even if the ultrasonic field is very weak.

The acoustic radiation force acting on a spherical object can also be estimated by the Gor'kov theory when the wave number k and spherical object's radius a satisfy $ka \ll 1$ [6, 35–38]. In the theory, the force potential of the acoustic field U is introduced, which is

$$U = V \left[-D \langle E_k \rangle + \left(1 - \gamma \right) \langle E_p \rangle \right], \tag{1.29}$$

where V is the volume of the spherical object, and D and γ are calculated by

$$D = \frac{3(\rho_s - \rho_f)}{2\rho_s + \rho_f}, \tag{1.30}$$

$$\gamma = \frac{\rho_f c_f^2}{\rho_s c_s^2}, \tag{1.31}$$

where ρ_s and ρ_f are the static densities of the spherical object and fluid, respectively, and c_s and c_f are the sound speed in the spherical object and fluid, respectively. The acoustic radiation force is

$$F = -\nabla U. \tag{1.32}$$

The above theory can be combined with the FEM to compute the acoustic radiation force acting on an object with an arbitrary shape in an arbitrary ultrasonic field [34].

1.1.3.4 Acoustic Streaming

Acoustic streaming is a kind of macroscopic or microscopic flow caused by a sound field, which exists in viscous fluid [6, 39–41]. It is well known that a sound wave in a fluid can drive a fluid element to vibrate near its equilibrium position back and forth. When the strength of a sound wave is proper, a time-independent flow will be generated in the sound field, which is called acoustic streaming. At the present stage, it is the major means that the ultrasonic nanomanipulation technology is using.

The conventional applications of acoustic streaming are in ultrasonic cleaning [2, 42], dispersion [2, 42, 43], mixing [2, 42, 44], non-contact driving [45–48], etc. The authors' research team found that acoustic streaming is a very useful means to manipulate nanoobjects. The mobile acoustic streaming field, generated by a micro-vibrating rod, can be used to suck, orientate, position and trap individual nanowires (NWs) in a noncontact or contact way, and to rotary drive a single NW at the water-substrate interface. The acoustic streaming field in a droplet on an ultrasonic stage can also be used to concentrate nano-objects in a controlled way.

In ultrasonic nano/microfabrication, handling, and driving, the acoustic streaming field is usually very complicated. Numerical calculation is an effective way to help us understand acoustic streaming thoroughly and accurately. Theoretically, an acoustic streaming field can be solved by viewing the acoustic field as a fluidic field, employing the Navier–Stokes equation, with the continuity equation and boundary conditions of the fluidic field, and filtering away the vibration component in the total fluidic velocity. This method consumes huge computational resources. A steady acoustic streaming field can be solved by a simplified method [49]. It is implemented by the FEM software COMSOL Multiphysics. In the method, the computation process consists of three steps. In the first step, the sound field is solved by the acoustic module. In the second step, vibration velocity and acoustic pressure of the sound field are used to calculate the spatial gradients of the Reynolds stress and mean pressure by the post-processing functions of the software, which generate the acoustic streaming. In the last step, the steady acoustic streaming is solved by the fluidic dynamics module, with proper boundary conditions.

Taking a 2D steady acoustic streaming field as an example, the simplified N–S equation for the acoustic streaming field is [39, 49]:

$$\rho_0(\bar{u}_i\, \partial \bar{u}_j / \partial x_i) = F_j - \partial \bar{p}_2 / \partial x_j + \eta \nabla^2 \bar{u}_j, \tag{1.33}$$

where $\bar{u}_i$ is acoustic streaming velocity, repeated suffix i and j represent x and y in the 2D model, respectively, ρ_0 is the fluid density in the undisturbed state, η is the shear viscosity of the fluid, and $\bar{u}_i$ is the time average of the second-order pressure. F_j is the spatial gradient of the Reynolds stress, which is

$$F_j = -\partial(\overline{\rho_0 u_i u_j}) / \partial x_i, \tag{1.34}$$

where u_i is the vibration velocity in the sound wave, and the bar signifies the mean value over one period. F_j acts on the fluid as a driving force of the acoustic streaming.

It must be pointed out that when the ultrasonic field is weak, acoustic streaming may not be generated as there is a frictional force between the fluidic acoustic medium and boundary and a cohesive force between the fluid molecules. The acoustic medium only starts to flow when the ultrasonic field is strong enough.

1.1.3.5   Ultrasonic Capillary Effect

The ultrasonic capillary effect is the phenomenon that the capillary effect is enhanced by ultrasound and usually happens within micro channels such as capillary tubes, micro/nano cavities in porous materials and gaps in a microfluidic system [2, 42]. To generate the ultrasonic capillary effect, the ultrasonic excitation may be applied to the inlet of a micro channel [2, 42, 50], or the solid wall of a microchannel [51–55]. The main application of the ultrasonic capillary effect in nano/micro fabrication, handling and driving is to drive the fluid to circulate in a fluidic system, and the speed of circuited fluid can be up to several cm/s.

According to the theory of the capillary effect, three forces affect the capillary effect's strength. The first one is the cohesive force among liquid molecules or the intermolecular bonding of the liquid. The second one is the adhesive force between the liquid and micro channel. The capillary effect appears only when the adhesive force between the liquid and micro channel is larger than the cohesive force among liquid molecules. When the cohesive force is stronger than the adhesive force, the capillary effect cannot happen, which is the case when mercury is in a glass capillary tube. Based on several experimental works, it is known that the acoustic cavitation effect has something to do with the ultrasonic enhancement of the capillary effect [2, 56].

The author's team proposed the following physical model to explain the ultrasonic capillary effect [6]. During the negative half circle of acoustic pressure, the negative acoustic pressure increases the distance among the molecules of liquid. Thus, during the negative half circle of acoustic pressure, the cohesive force among the molecules becomes weaker. During the positive half circle of acoustic pressure, the positive acoustic pressure compresses the molecules of liquid. But this does not increase the cohesive force so much, because the molecules will repel each other when they are very close, according to the classic model of molecular bonds. On the time average, the acoustic pressure makes the cohesive force among the molecules of the liquid weaker. This means that ultrasound can increase the difference between the adhesive and cohesive forces, which explains the ultrasonic capillary effect. Based on this model, acoustic bubbles are a sufficient condition for generation of the ultrasonic capillary effect, rather than a necessary condition. This model can explain the ultrasonic capillary effect better [51–55].

1.2   ULTRASONIC NANO/MICROFABRICATION, HANDLING, AND DRIVING

1.2.1   Ultrasonic Nanofabrication

Ultrasonic nanofabrication includes ultrasonic nanocutting [32], rolling [31], welding [57], coating [58], disruption [59], material fabrication [9, 60], etc. In this book, the "nano" in ultrasonic nanofabrication refers to the scale of fabricated objects rather than the scale of ultrasonic devices or the fabrication precision.

Ultrasonic nanocutting, rolling and welding are usually implemented by the linear or elliptical vibration of a tool ultrasonically excited [31, 32]. The tool may be made of tungsten micro needle, thin fiber glass or other elastic solid materials, depending on the mechanical properties of tooled samples. To transmit the ultrasonic vibration to the tool from an ultrasonic transducer, a vibration transmission needle which is usually made of metal may be employed. In the tooling, the sample such as a silver NW is on the surface of a solid substrate, and the tool is moved onto the sample by moving the ultrasonic device and pressed onto the sample to carry out the fabrication process. Different from ultrasonic nano handling, ultrasonic nanocutting, rolling and welding cannot work for fragile samples such as metal oxide materials, that is, ultrasonic nanocutting, rolling and welding have selectivity to the mechanical properties of tooled materials. This is because the ultrasonic vibration of the tool is directly applied to the tooled samples, and may cause the fragile samples to break into pieces. To implement the cutting, rolling and welding of a nano/micro sample on a solid substrate, the sample to be tooled must be fixed to the substrate. The water molecules in between the tooled sample and substrate can be employed to fix the nanoscale sample onto a solid substrate [32], which is convenient to use. This is mainly because the water molecules can expel the air in between the sample and substrate and cause a pressure onto the nanoscale sample.

In ultrasonic nanocutting [32], dynamic pressure generated by the normal (to the substrate) vibration of the tip of a probe is applied to the cut point of a NW on substrate, making the cut point thinner, and then the tangential vibration of the probe breaks the NW into two sections. In ultrasonic nanorolling [31], dynamic pressure generated by the normal vibration of the tool (a soft loop) is applied to a NW on a substrate, and the thickness of the whole NW is reduced greatly by rolling the micro tool back and forth on the substrate. The working principle of ultrasonic nano welding depends on the materials to be welded. Heat generated by ultrasonic vibration and melting of the materials at the interface is responsible for the welding effect in plastic-plastic welding, whereas vibration-enhanced atomic/molecular diffusion, mechanical rubbing effect and rapid plastic flow are the main mechanisms of metal-metal welding [2, 5, 42].

The ultrasonic nanocoating and disruption are implemented by the acoustic cavitation effect, and the ultrasonic nanomaterial fabrication may be realized by the acoustic cavitation effect or ultrasonic concentration methods [28, 61–63]. In ultrasonic nanocoating, nanoscale materials in a solution are ejected onto a flexible substrate such as the PDMS (Polydimethylsiloxane) substrate, by the shock waves and micro jets generated by the collapse of transient bubbles. In the ultrasonic nano disruption, nanoscale samples such as human DNA are cut into shorter or smaller pieces by the shock waves and micro jets generated by the collapse of transient bubbles, for further biochemical analyses or processing. The acoustic cavitation effect can be used in nano material fabrication, such as graphene material fabrication [64, 65]. Using the ultrasonic manipulation methods to fabricate nanomaterials is an emerging technique, which utilizes ultrasonic concentration, mixing and other ultrasonic physical effects to construct nanomaterials with sub-millimeter or larger size.

The ultrasonic nanofabrication functions described above have potential applications in advanced nanomaterial formation, high-performance nanosensor fabrication, biological sample treatment, etc.

1.2.2　Ultrasonic Nano/Micro Handling

Ultrasonic nano/micro handling refers to ultrasonic nano/micro concentration [28, 29, 61–63, 66], trapping [25, 27, 34, 36, 37], transportation [25, 67], release [38], sorting/separation [68–70], removal [71], etc. The "nano/micro" in ultrasonic nano/micro-handling refers to the scale of handled objects, rather than the scale of the ultrasonic devices.

The ultrasonic handling of nanoscale objects at a droplet–substrate interface or in a droplet is mainly implemented by the Stokes force acting on the handled objects and generated by acoustic streaming eddies, whereas that of micro-objects at a droplet–substrate interface or in a droplet is mainly implemented by the acoustic radiation force acting on the handled objects. The acoustic radiation force and acoustic streaming are generated by a properly controlled ultrasonic field in a droplet, which may be excited by a vibrating substrate or micromanipulation probe inserted into the droplet. Vibration of the substrate is in the form of BAW or SAW and that of the micromanipulation probe is usually in the elliptical or linear form, depending on the type of handling function.

The main features of ultrasonic nano/micro handling, compared to other physical handling methods, include little selectivity to the properties of handled materials, little heat damage to handled materials with the use of acoustic streaming, diversified manipulation functions and simple device structures.

The feature that ultrasonic micro handling has little selectivity to the properties of handled materials is explained as follows: As an ultrasonically transparent object does not exist in practice, the material property change of a practical object will not give rise to the disappearance of acoustic radiation force exerted on it. Thus, if some ultrasonic device has been proven to be effective in the handling of a micro object, then changing the material property of the handled object will not cause a failure of the same handling process, as long as the device can keep the pattern of working ultrasonic field constant and maintain its strength properly. The feature that ultrasonic nano handling has little selectivity to the properties of handled materials can be explained in a similar way, as the Stokes' force acting on a practical object, which is generated by the acoustic streaming eddies, is little affected by the material properties of the handled object.

The little heat generation in the utilization of acoustic streaming is because the vibration needed to generate the acoustic streaming eddies in ultrasonic nano handling is usually not too large, and moreover, the acoustic streaming eddies can flush away the heat generated by the vibrating manipulation probe or substrate.

The fundamental reason for the diversity of ultrasonic nano/micromanipulation functions lies in the diversity of physical effects of ultrasound, which include the acoustic radiation force, acoustic streaming, acoustic cavitation effect, ultrasonic frictional driving, Chladni effect, reverse Chladni effect, acoustically induced intermolecular force decrease (or acoustic capillary effect), etc.

In ultrasonic nano/micro handling, the main part of a probe-type device consists of an ultrasonic transducer (or piezoelectric component), a vibration transmission needle, a micro-manipulation probe and a solid substrate. The substrate-type devices based on SAW and BAW consist of a solid substrate and a layer of piezoelectric material with electrodes (IDEs or the whole electrodes), which is bonded onto the substrate. Thus, the probe-type and substrate-type devices have simple structures and are easy to be fabricated.

1.2.3 ULTRASONIC NANO/MICRO-DRIVING

Ultrasonic nano/micro-driving is the controlled drive of nano/micro-scale objects by ultrasonic vibration. The "nano/micro" in ultrasonic nano/micro-driving refers to the scale of driven objects rather than the scale of the ultrasonic devices. As the diameter of gas molecules is in the sub-nano range, controlled ultrasonic driving of gas molecules is also a kind of ultrasonic nano driving. In the existing researches, ultrasonically driven nanoscale objects include nanowires (NWs) [25], nanoparticles (NPs) [28], gas molecules [72–76], etc., whereas ultrasonically driven micro-scale objects include various solid micro-objects [77], micro-droplets [55] and liquid in micro-channels [54].

In controlled ultrasonic driving of NWs and NPs, the Stokes force, which is generated by acoustic streaming eddies, is the major driving force. The Stokes force acting on a sphere in fluid is

$$F_s = 6\pi\mu Rv, \tag{1.35}$$

where μ is the fluid shear viscosity, R is the radius of the spherical object and v is the flow velocity relative to the object. In controlled ultrasonic driving of gas molecules near a gas–solid interface, the driving force of gas molecules results from the acoustic pressure near and at the interface, based on experimental data analyses carried out by the author's team. However, how the acoustic pressure can generate such a driving force is still debatable.

In controlled ultrasonic driving of solid micro-objects, the acoustic radiation force is the major driving force, although the Stokes force resulting from acoustic streaming may affect the driving performance. In controlled ultrasonic driving of micro-droplets and liquid in micro-channels, the driving force is generated by the ultrasonic capillary effect, which results from the ultrasound-induced decrease of intermolecular forces [52–55].

Controlled ultrasonic driving of gas molecules near a gas-solid interface has been applied in ultrasonically catalyzed high-performance gas sensor systems and single-sensor electronic noses [72–76]. In ultrasonically catalyzed high-performance gas sensor systems, ultrasonic driving of gas molecules enhances the redox reactions between the target gases and oxygen species at the interface. In ultrasonically catalyzed single-sensor electronic noses, the dependency of ultrasonic driving capability on gas molecular mass and chemical property is utilized to discriminate gas species. In addition, controlled ultrasonic driving of liquid in micro channels has been applied in the saltwater Al-air flow battery to circulate the saltwater electrolyte [50]. Compared with the traditional saltwater Al-air flow battery with the electrolyte circulated by a mechanical pump, it has better discharge performance and less electric power consumption by the electrolyte driving system.

REFERENCES

1. P. M. Morse and K. U. Ingard, *Theoretical Acoustics*, (McGraw–Hill, New York, 1968).
2. O. V. Abramov, *High–Intensity Ultrasonics*, (Gordon and Breach Science Publishers, Singapore, 1998).
3. W. L. Nyborg, *Physical Acoustics*, in W. P. Mason and R. N. Thurston, Eds., (Academic Press, New York, 1966).
4. R. E. Berg and D. G. Stork, *The Physics of Sound*, *3rd ed.*, (Pearson Education, Inc., New Jersey, 2005), pp. 59–63.
5. D. Ensminger, *Ultrasonics: Fundamentals, Technology, Applications*, *2nd ed.*, (Marcek, Dekker, Inc., New York and Basel, 1988), Chap. 6.
6. J. Hu, *Ultrasonic Micro/Nano Manipulations: Principles and Examples*, (World Scientific, New Jersey, London, Singapore, 2014), pp. 15–19.
7. X. Lu, J. Hu, H. Peng and Y. Wang, "A new topological structure for the Langevin-type ultrasonic transducer," *Ultrasonics*, 75, pp. 1–8, 2017.
8. H. Ahmed, A. R. Rezk, J. J. Richardson, L. K. Macreadie, R. Babarao, E. L. H. Mayes, L. Lee and L. Y. Yeo, "Acoustomicrofluidic assembly of oriented and simultaneously activated metal–organic frameworks," *Nat. Commun.*, 10, p. 2282, 2019.
9. X. Y. Ding, P. Li, C. S. Lin, Z. S. Stratton, N. Nama, F. Guo, D. Slotcavage, X. L. Mao, J. J. Shi, F. Costanzo and T. J. Huang, "Surface acoustic wave microfluidics," *Lab Chip*, 13, pp. 3626–3649, 2013.
10. J. Friend and L. Yeo, "Microscale acoustofluidics: microfluidics driven via acoustics and ultrasonics," *Rev. Mod. Phys.*, 83, pp. 647–704, 2011.
11. S. Fang, Q. Zhang, H. Zhao, J. Yu and Y. Chu, "The design of rare-earth giant magnetostrictive ultrasonic transducer and experimental study on its application of ultrasonic surface strengthening," *Micromachines*, 9(3), p. 98, 2018.
12. Q. Liu, Q. Tang and J. Hu, "A new strategy to capture single biological micro particles at the interface between a water film and substrate by ultrasonic tweezers," *Ultrasonics*, 103, 106067, 2020.
13. F. Han and J. Hu, "Distribution and strength of sound in the human head", *Proceedings* of *2015 Symposium on Piezoelectric, Acoustic Waves and Device Applications (SPAWDA)*, Jinan, China, pp. 102–105, 2015.
14. J. Hu, X. Zhu, Y. Zhou and N. Li, "Principle of the rotation of small particles around a nodal point of strip in flexural vibration," *Sens. Actuator A Phys.*, 178, pp. 202–208, 2012.
15. X. Zhu and J. Hu, "Ultrasonic drive of small mechanical components on a tapered metal strip," *Ultrasonics*, 53, pp. 417–422, 2013.
16. D. T. Blackstock, *Fundamentals of Physical Acoustics*, (John Wiley & Sons, Inc., New York, 2000), pp. 74–77, Chaps. 3 and 4.
17. T. G. Leighton, *The Acoustic Bubble*, (Academic Press, San Diego, 1994), Chaps. 1 and 4.
18. Y. Hu, Z. Luo, Y. Zhou and J. Hu, "A focused ultrasound based cooling strategy for small solid heat sources," *Sens. Actuator A Phys.*, 331, 112932, 2021.
19. Z. Luo, Q. Tang, S. Su and J. Hu, "A high-performance structure for the bulk acoustic wave metal oxide semiconductor gas sensor," *Smart Mater. Struct.*, 28, 105015, 2019.
20. P. Ciuti, N. V. Dezhkunov, G. Iernetti and A. I. Kulak, "Cavitation phenomena in pulse modulated ultrasonic fields," *Ultrasonics*, 36, pp. 569–574, 1998.
21. W. Yang and Y. Zhou, "Effect of pulse repetition frequency of high-intensity focused ultrasound on in vitro thrombolysis," *Ultrason. Sonochem.*, 35, pp. 152–160, 2017.
22. X. Lu, J. Hu, L. Yang and C. Zhao, "A novel dual stator-ring rotary ultrasonic motor," *Sens. Actuator A Phys.*, 189, pp. 504–511, 2013.
23. X. Lu, Q. Zhang and J. Hu, "A linear piezoelectric actuator based solar panel cleaning system," *Energy*, 60, pp. 401–406, 2013.

24. X. Lu, J. Hu, S. Bhuyan and S. Li, "An ultrasonic contact-type position restoration mechanism," *Rev. Sci. Instrum.*, 85, 124901, 2014.
25. N. Li, J. Hu, H. Li, S. Bhuyan and Y. Zhou, "Mobile acoustic streaming based trapping and 3-dimensional transfer of a single nanowire," *Appl. Phys. Lett.*, 101(9), 093113, 2012.
26. X. Wang and J. Hu, "An ultrasonic manipulator with noncontact and contact-type nanowire trapping functions," *Sens. Actuator A Phys.*, 232, pp. 13–19, 2015.
27. G. Chen, N. Li and J. Hu, "Capture of individual micro metal wires in air by ultrasonic tweezers," *IEEE/ASME Trans. Mechatron.*, 20(6), pp. 3053–3059, 2015.
28. X. Qi, Q. Tang, P. Liu, I. V. Minin, O. V. Minin and J. Hu, "Controlled concentration and transportation of nanoparticles at the interface between a plain substrate and droplet," *Sens. Actuators B: Chem.*, 274, pp. 381–392, 2018.
29. Q. Liu, K. Chen, J. Hu and T. Morita, "An ultrasonic tweezer with multiple manipulation functions based on the double-parabolic-reflector wave-guided high-power ultrasonic transducer," *IEEE Trans. Ultrason. Ferroelectr. Freq. Control*, 67(11), pp. 2471–2474, 2020.
30. P. Liu, Q. Tang, S. Su and J. Hu, "Principle analysis for the micromanipulation probe-type ultrasonic nanomotor," *Sens. Actuator A Phys.*, 318, 112524, 2021.
31. X. Wang and J. Hu, "A flexible ultrasonic micro tool based AgNS fabrication process," *Appl. Nanosci.*, 8(6), pp. 1579–1586, 2018.
32. X. Wang and J. Hu, "Nanowire cutting by an ultrasonically vibrating micro tool," *Precis. Eng.*, 48, pp. 152–157, 2017.
33. T. Hasegawa, T. Kido, T. Iizuka and C. Matsuoka, "A general theory of Rayleigh and Langevin radiation pressures," *J. Acoust. Soc. Jpn. (E)*, 21(3), pp. 145–152, 2000.
34. Y. Liu, J. Hu and C. Zhao, "Dependence of acoustic trapping capability on the orientation and shape of particles," *IEEE Trans. Ultrason. Ferroelectr. Freq. Control*, 57(6), pp. 1443–1450, 2010.
35. L. P. Gor'kov, "On the forces on a small particle in an acoustical field in an ideal fluid," *Sov. Phys.–Dokl*, 6(9), pp. 773–775, 1962.
36. Y. Liu and J. Hu, "Trapping of particles by the leakage of a standing wave ultrasonic field," *J. Appl. Phys.*, 106(3), 034903, 2009.
37. J. Hu, J. B. Yang and J. Xu, "Ultrasonic trapping of small particles by sharp edges vibrating in a flexural mode," *Appl. Phys. Lett.*, 85(24), pp. 6042–6044, 2004.
38. X. Qi, P. Liu and J. Hu, "A low temperature-rise and facile manipulation method for single micro objects at the air-substrate interface," *J. Micromech. Microeng.*, 29, 105007, 2019.
39. J. Lighthill, *Waves in Fluids*, (Cambridge University Press, Cambridge, 1978), p. 329, 344–350.
40. J. Hu, K. Nakamura and S. Ueha, "Optimum operation conditions of an ultrasonic motor driving fluid directly," *Jpn. J. Appl. Phys.*, 35, pp. 3289–3294, 1996.
41. P. Liu, Z. Tian, K. Yang, T. D. Naquin, N. Hao, H. Huang, J. Chen, Q. Ma, H. Bachman, P. Zhang, X. Xu, J. Hu and T. J. Huang, "Acoustofluidic black holes for multifunctional in-droplet particle manipulation," *Sci. Adv.*, 8, eabm2592, 2022.
42. L. D. Rozenberg, *Physical Principles of Ultrasonic Technology*, (Plenum Press, New York, USA, 1973).
43. J. S. Taurozzi, V. A. Hackley and M. R. Wiesner, "Ultrasonic dispersion of nanoparticles for environmental, health and safety assessment – issues and recommendations," *Nanotoxicology*, 5(4), pp. 711–729, 2011.
44. Z. Luo, Q. Tang and J. Hu, "Effect of ultrasonic excitation on discharge performance of a zinc-air button battery," *Micromachines*, 12(7), p. 792, 2021.
45. J. Hu, K. Nakamura and S. Ueha, "An analysis of a noncontact ultrasonic motor with an ultrasonically levitated rotor,"*Ultrasonics*, 35(6), pp. 459–467, 1997.
46. J. Hu, K. C. Cha and K. C. Lim, "New type of linear ultrasonic actuator based on a plate-shaped vibrator with triangular grooves (Letter)," *IEEE Trans. Ultrason. Ferroelectr. Freq. Control*, 51(10), pp. 1206–1208, 2004.

47. Y. Hashimoto, Y. Koike and S. Ueha, "Transporting objects without contact using flexural traveling waves," *J. Acoust. Soc. Amer.*, 103, pp. 3230–3233, 1998.

48. J. Hu, K. Nakamura and S. Ueha, "A noncontact ultrasonic motor with the rotor levitated by axial acoustic viscous force," *Electron. Commun. Jpn. (Part III)*, 82(4), pp. 56–63, 1999.

49. Q. Tang and J. Hu, "Diversity of acoustic streaming in a rectangular acoustofluidic field," *Ultrasonics*, 58, pp. 27–34, 2015.

50. H. Huang, P. Liu, Q. Ma, Z. Tang, M. Wang and J. Hu, "Enabling a high-performance saltwater Al-air battery via ultrasonically driven electrolyte flow", *Ultrason. Sonochem.*, 106104, 2022, 10.1016/j.ultsonch.2022.106104,.

51. X. Zhang, Y. Zheng and J. Hu, "Sound controlled rotation of a cluster of small particles on an ultrasonically vibrating metal strip,"*Appl. Phys. Lett.*, 92, 024109, 2008.

52. J. Hu, H. Zhu, N. Li and C. S. Zhao, "Sound induced lobed pattern in aqueous suspension film of micro particles," *Sens. Actuator A Phys.*, 167(1), pp. 77–83, 2011.

53. J. Hu, N. Li and J. J. Zhou, "Controlled adsorption of droplets onto anti–nodes of an ultrasonically vibrating needle," *J. Appl. Phys.*, 110(5), 054901, 2011.

54. J. Hu, C. L. Tan and W. Y. Hu, "Ultrasonic microfluidic transportation based on a twisted bundle of thin metal wires," *Sens. Actuator A Phys.*, 135(2), pp. 811–817, 2007.

55. Z. Tan, S. Teo and J. Hu, "Ultrasonic generation and rotation of a small droplet at the tip of a hypodermic needle," *J. Appl. Phys.*, 104(10), 104902, 2008.

56. N. V. Dezhkunov and T. G. Leighton, "Study into correlation between the ultrasonic capillary effect and sonoluminescence," *J. Eng. Phys. Thermophys.*, 77(1), pp. 53–61, 2004.

57. C. Chen, L. Yan, E. S. Kong and Y. Zhang, "Ultrasonic nanowelding of carbon nanotubes to metal electrodes," *Nanotechnology*, 17, pp. 2192–2197, 2006.

58. H. Xue and J. Hu, "A liquid power-ultrasound based green fabrication process for flexible strain sensors at room temperature and normal pressure," *Sens. Actuator A Phys.*, 329, 112822, 2021.

59. P. Pilo, A. M. M. Tiley, C. Lawless, S. J. Karki, J. Burke and A. Feechan, "A rapid fungal DNA extraction method suitable for PCR screening fungal mutants, infected plant tissue and spore trap samples," *Physiol. Mol. Plant Pathol.*, 117, 101758, 2022.

60. A. Pucek-Kaczmarek, "Influence of process design on the preparation of solid lipid nanoparticles by an ultrasonic-nanoemulsification method," *Processes*, 9, p. 1265, 2021.

61. Q. Tang, X. Wang and J. Hu, "Nano concentration by acoustically generated complex spiral vortex field,"*Appl. Phys. Lett.*, 110, 104105, 2017.

62. Y. Zhou, J. Hu and S. Bhuyan, "Manipulations of silver nanowires in a droplet on low-frequency ultrasonic stage," *IEEE Trans. Ultrason. Ferroelectr. Freq. Control*, 60(3), pp. 622–629, 2013.

63. B. Yang and J. Hu, "Linear concentration of microscale samples under an ultrasonically vibrating needle in water on a substrate surface,"*Sens. Actuators B: Chem.*, 193, pp. 472–477, 2014.

64. X. Gu, Y. Zhao, K. Sun, C. L. Z. Vieira, Z. Jia and S. Huang, "Method of ultrasound-assisted liquid-phase exfoliation to prepare graphene," *Ultrason. Sonochem.*, 58 (12), 104630, 2019.

65. A. V. Tyurnina, I. Tzanakis, J. Morton, J. Mi, K. Porfyrakis, B. M. Maciejewska, N. Grobert and D. G. Eskin, "Ultrasonic exfoliation of graphene in water: A key parameter study," *Carbon*, 168, pp. 737–747, 2020.

66. P. Liu, Z. Tian, N. Hao, H. Bachman, P. Zhang, J. Hu and T. J. Huang, "Acoustofluidic multi-well plates for enrichment of micro/nano particles and cells,"*Lab Chip*, 20, pp. 3399–3409, 2020.

67. J. Hu, L. Ong, C. Yeo and Y. Liu, "Trapping, transportation and separation of small particles by an acoustic needle," *Sens. Actuator A Phys.*, 138, pp. 187–193, 2007.

68. X. Lu, A. Martin, F. Soto, P. Angsantikul, J. Li, C. Chen, Y. Liang, J. Hu, L. Zhang and J. Wang, "Parallel label-free isolation of cancer cells using arrays of acoustic microstreaming traps,"*Adv. Mater. Technol.*, 4(2), 1800374, 2018.

69. J. Hu, J. Yang, J. Xu and J. Du, "Extraction of biologic particles by pumping effect in a π-shaped ultrasonic actuator,"*Ultrasonics*, 45, pp. 15–21, 2006.
70. Y. Wang and J. Hu, "Ultrasonic removal of coarse and fine droplets in air," *Sep. Purif. Technol.*, 153, pp. 156–161, 2015.
71. P. Liu and J. Hu, "Controlled removal of micro/nanoscale particles in submillimeter-diameter area on a substrate,"*Rev. Sci. Instrum.*, 88, 105003, 2017.
72. S. Su and J. Hu, "Ultrasound assisted low-concentration VOC sensing,"*Sens. Actuators B: Chem.*, 254, pp. 1234–1241, 2018.
73. S. Su, P. Liu, Q. Tang and J. Hu, "Physical principle of enhancing the sensitivity of a metal oxide gas sensor using bulk acoustic waves,"*J. Appl. Phys.*, 124, 244902, 2018.
74. S. Su and J. Hu, "Gas identification by a single metal oxide semiconductor sensor assisted by ultrasound," *ACS Sens.*, 4, pp. 2491–2496, 2019.
75. S. Su, X. Qi, P. Liu and J. Hu, "Focused ultrasound assistance to the MOS gas sensor system," *IEEE Trans. Ultrason. Ferroelectr. Freq. Control*, 67(5), pp. 1009–1016, 2020.
76. T. Zhang, Y. Zhou, P. Liu and J. Hu, "A novel strategy to identify gases by a single catalytic combustible sensor working in its linear range,"*Sens. Actuators B: Chem.*, 321, 128514, 2020.
77. H. Ba and J. Hu, "Separation of small solid particles based on ultrasonic rotary," *Sep. Purif. Technol.*, 127, pp. 107–111, 2014.

2 Ultrasonic Nano Fabrication

Ultrasonic nano fabrication in this book refers to the nano device fabrication by an ultrasonic tool or field. In the first and second sections of this chapter, the principle, structure and characteristics of the ultrasonic devices for nano rolling and cutting are described, respectively [1, 2]. In the third section, the significance of exploring a new coating process for flexible nano sensors is explained. In the fourth section, a liquid power-ultrasound-based nano coating process, its application in flexible strain sensors and characteristics of the fabricated strain sensors are given in detail [3]. As another application example of liquid power-ultrasound-based nano coating process, fabrication method and characteristics of a high-performance flexible gas sensor with a nano sensing layer on the inner wall are demonstrated in the fifth section [4].

2.1 ULTRASONIC NANO ROLLING

Silver nano sheets (AgNSs) have promising potential applications in photonic products such as touch screens, photovoltaic cells, light-emitting diodes and nano sensors. The potential applications are motivated by their very good light transmission property, large surface-to-volume ratio and strong adsorbility to various gas molecules. So far, chemical methods including the photo-induced chemical reduction method, rapid reduction precipitation method, soft template method, thermal deposition and ultrasonic radiation-assisted method have been utilized to fabricate AgNSs. To boost practical applications of AgNSs, more convenient and greener AgNS fabrication methods are desired.

In this example [1], a flexible ultrasonic micro tool (MT)-based method to fabricate AgNSs by rolling silver nanowires (AgNWs) and Ag micro particles on a solid substrate in air is proposed and developed. A circular fiberglass ring with an inner diameter of 1–3 mm is utilized as the ultrasonic MT, and the fiberglass has a diameter of 5–15 μm. Experimental results show that the thickness of AgNSs produced by this method can be as thin as several ten nanometers, and the rolling effect is not sensitive to the preload when the MT vibration is sufficiently large. Analyses of the experimental results indicate that the thickness of fabricated AgNSs can be decreased further, if thinner AgNWs are used as the raw material, and the stiffness and vibration velocity of the MT are increased. This method is easy to be implemented, and the MT does not harm the solid substrate due to its circular and flexible structure.

2.1.1 DEVICE, EXPERIMENTAL SETUP AND ROLLING PROCESS

The ultrasonic device to roll AgNWs and Ag micro particles on a solid substrate in air, proposed in this work, is shown in Fig. 2.1(a). It consists of an MT, vibration transmission needle (VTN) and Langevin transducer. The MT is a circular fiberglass ring with

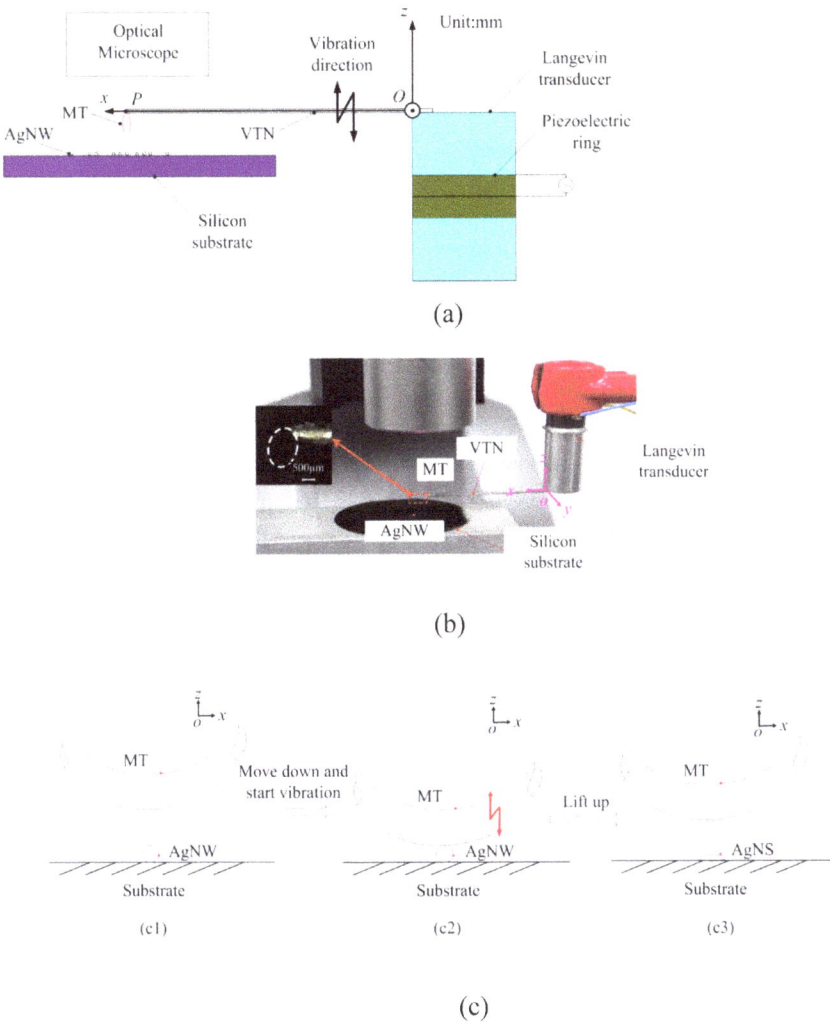

FIGURE 2.1 (a) Device structure. (b) Experimental setup. (c) Process of the ultrasonic nano rolling. Reproduced from Ref. [1] with permission from Springer Nature.

its top excited by the VTN's tip. The fiberglass used in the experiments has a diameter D_g of 5–15 μm, and the fiberglass ring has an inner diameter D_t of 1–3 mm. The VTN's root is excited by the Langevin transducer. The length of the VTN bonded onto the radiation face of the transducer is 12 mm, and its length outside the transducer is 47 mm. The device works at around 52.0 kHz, at which the transducer is in resonance. The design of the MT makes itself have a small stiffness and works as a soft spring, which results in uniform and stable contact between the MT and AgNWs during the rolling. Detailed dimensions and material properties of the device are listed in Table 2.1.

Figure 2.1(b) shows the experimental setup for the nano rolling process. Commercial AgNWs with a diameter D_s of 50–300 nm and lengths up to several ten microns, were

TABLE 2.1

Dimensions and Materials of the Device for Ultrasonic Nano Rolling

	Langevin Transducer	Vibration Transmission Needle	Micro Tool
Diameter	15 mm	1 mm	1–3 mm
Length	43 mm	47 mm	NA
Materials	Duralumin and piezoelectric rings	Stainless steel	Fiberglass

used in the experiments. The experiments were carried out under an optical microscope (VHX-1000E, Keyence). The ultrasonic transducer was mechanically fixed onto an X-Y-Z stage (LD125-LM-2, SELN) so that the MT could be moved in the x, y and z directions.

Figure 2.1(c) shows the rolling process of an AgNW on a Si substrate. First, the MT is moved to the location right above the AgNW. Then it is lowered until getting contact with the AgNW. After that, the MT is further lowered by several microns ($=d_0$) to gain sufficient contact with the AgNW. Then a driving voltage V of 5–100 V_{p-p} with a working frequency f of about 52.0 kHz is applied to the transducer, and the MT is moved along the substrate surface to roll the AgNWs. Finally, after sufficient time of rolling (10–30 s), the MT is lifted up and the ultrasonic vibration is switched off.

2.1.2 RESULTS AND DISCUSSION

Figure 2.2(a) shows the images of an individual AgNW before the ultrasonic rolling and fabricated AgNS after the rolling, recorded by an optical microscope. In the experiment, $V = 30$ V_{p-p}, $f = 52.0$ kHz, $d_0 = 5$ μm, $D_g = 15$ μm, $D_t = 2$ mm and $D_s = 300$ nm. It is seen that the length of the AgNW and AgNS has little difference. Thus, the cross-section area also has little change before and after the rolling. Based on this conclusion and the measured width W of an AgNS, the thickness of the AgNS can be deduced. Measured thickness t of the fabricated AgNSs is in the range of 50–100 nm. Figure 2.2(b) shows an AFM (Atomic Force Microscope) image of AgNS fabricated by the nano rolling process. It is seen that the AgNS thickness is quite uniform. Figure 2.2(c) shows the thickness of an AgNS, measured by an AFM. It also indicates that the AgNS thickness is quite uniform. The uniformity of the rolling effect is because of the flexibility of the MT.

To understand the working principle, vibration characteristics of the device were measured by a laser Doppler vibrometer (Polytec PSV-500), and the results are shown in Fig. 2.3. Figure 2.3(a) is the measured vibration magnitude at the VTN's tip versus working frequency. It is seen that the vibration magnitude has a peak at 52.0 kHz, which is also the resonance frequency of the transducer. Figure 2.3(b) shows the measured vibration velocity distribution along the VTN at 52.0 kHz. It is seen that the VTN vibrates flexurally and can magnify the vibration velocity of its root. Figure 2.3(c) shows the vibration displacement of the MT when the working frequency is 52.0 kHz, computed by the finite element method FEM (COMSOL MULTIPHYSICS). In the computation, the vibration velocity at the MT's top is 140 mm/s (0–p), $f = 52.0$ kHz, $D_g = 15$ μm and

(a)   (b)

(c)

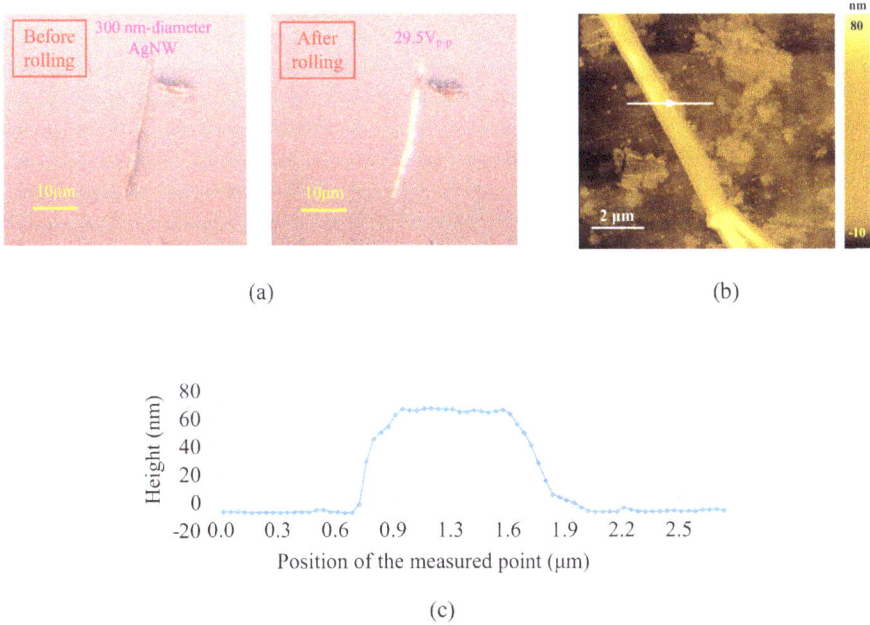

FIGURE 2.2 Effect of the ultrasonic nano rolling. (a) An AgNW before the rolling and AgNS fabricated by the rolling. (b) An AFM image of a rolled AgNW. (c) The AgNS thickness measured by AFM. 300 nm-diameter AgNWs were used in the experiments. Reproduced from Ref. [1] with permission from Springer Nature.

$D_t = 3$ mm. It is seen that the vibration of the VTN's tip does cause the MT to vibrate. Moreover, based on our FEM computation, it is known that the dynamic pressing force between the MT and AgNW along the z-direction is 4.76 µN (0–p) when the vibration velocity at the MT's top is 140 mm/s (0–p). Due to the good malleability of AgNWs (yield strength ≈ 35 MPa), the AgNWs are pressed into AgNSs under dynamic pressing force generated by the MT. The dynamic pressing force can be increased by increasing the MT stiffness in the pressing direction and by increasing the vibration velocity, as suggested by the vibration theory [5].

Spring constant k of the MT was computed by the FEM. The computed spring constant of the MT with an inner diameter of 3.0 mm, made of 15 µm-diameter fiberglass, was 0.073 N/m, which confirmed that the MT served as a soft spring during the rolling process [5]. It must be pointed out that although decreasing the stiffness of the MT can make the contact between the MT and AgNW more uniform and stable, it may result in a smaller dynamic pressing force between the MT and rolled sample, which makes the fabrication of thinner AgNSs difficult. Thus, spring constant of the MT must be optimized further if one wants to achieve thinner AgNSs by the method proposed in this work.

The dependency of the width and thickness of fabricated AgNSs on vibration velocity at the VTN's tip was investigated experimentally. Figure 2.4(a–d) show the measured AgNS width and ratio of the AgNS thickness t to AgNW diameter D_s versus vibration velocity, for AgNWs with diameter D_s of 50, 200 and 300 nm.

(a)

(b)

(c)

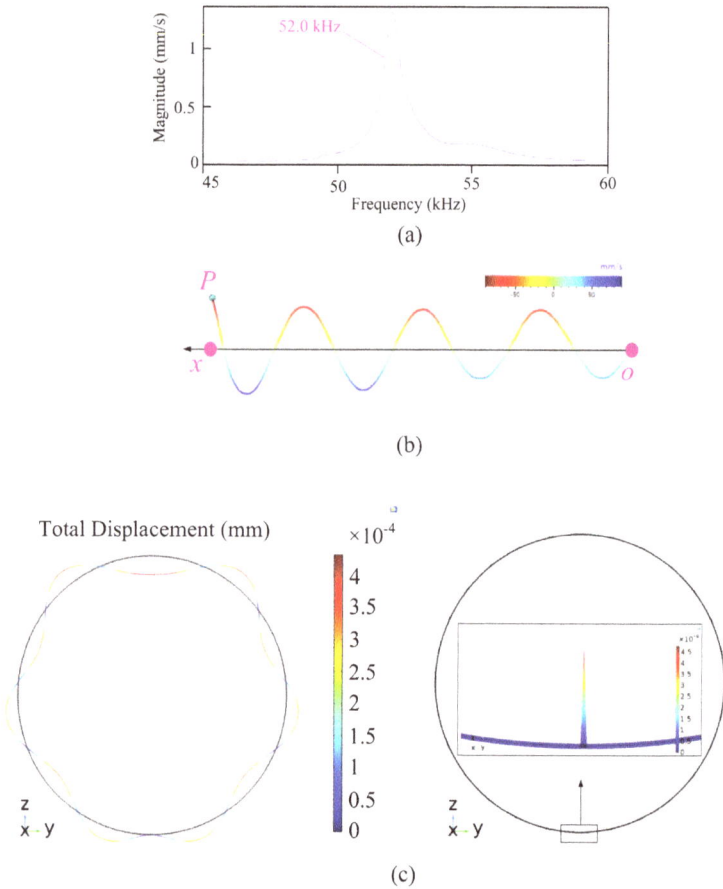

FIGURE 2.3 Vibration characteristics of the device. (a) Measured vibration velocity magnitude versus frequency at the micro tool's root. (b) Measured vibration distribution of the vibration transmission needle. (c) Computed vibration displacement of the micro tool, and the dynamic pressing force between the micro tool and rolled sample. Reproduced from Ref. [1] with permission from Springer Nature.

In the experiments, $d_0 = 5$ μm, $D_g = 15$ μm and $D_t = 2$ mm. From Fig. 2.4(a), it is seen that the AgNS width increases with the vibration increase, which indicates that AgNS thickness decreases with the vibration increase. This is because the dynamic pressing force between the MT and rolled nanoscale sample increases as the vibration velocity increases. From Fig. 2.4(a), it is also seen that the AgNS width does not increase when the vibration velocity is sufficiently large. The phenomenon indicates that the AgNS thickness does not decrease when the vibration velocity is sufficiently large. This is because as the width/thickness ratio of the AgNS increases, the AgNS stiffness in the thickness direction K increases rapidly, resulting in little deformation of the AgNS under the dynamic pressing force.

From Fig. 2.4(b), it is seen that the ratio of the AgNS thickness t to AgNW diameter D_s decreases as the vibration velocity increases and becomes little affected by the

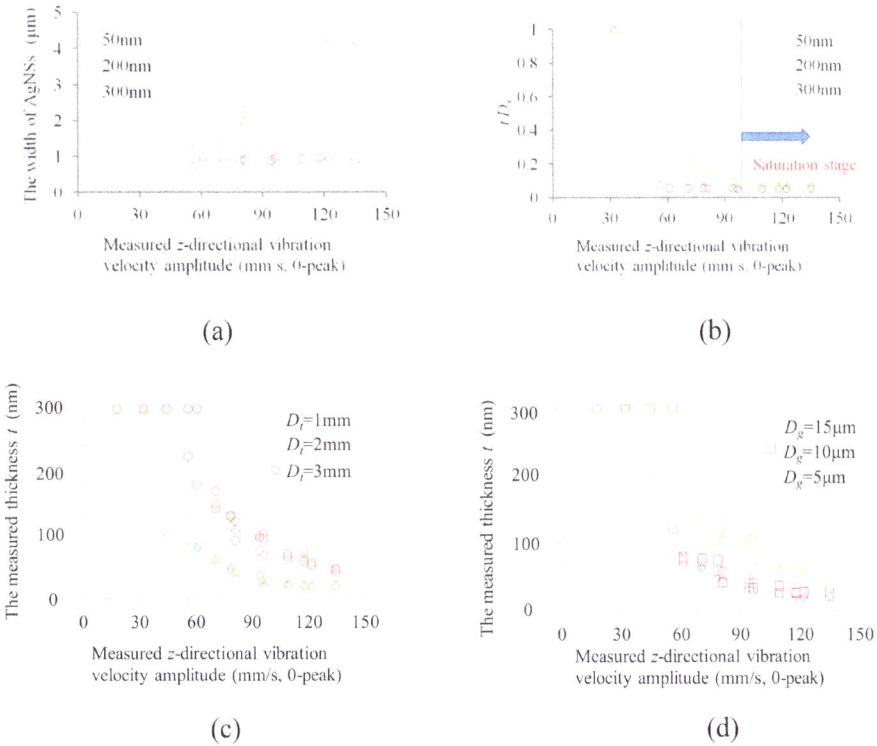

(a)

(b)

(c)

(d)

FIGURE 2.4 Measured dependency of nano rolling characteristics on vibration velocity of the micro tool's top. (a) Measured width of the AgNSs versus vibration velocity at different nanowire diameters. (b) The ratio of the AgNS thickness t to AgNW diameter Ds at different nanowire diameters. (c) The AgNS thickness versus vibration velocity at different micro tool diameter. (d) The AgNS thickness versus vibration velocity at different fiberglass diameter. Reproduced from Ref. [1] with permission from Springer Nature.

AgNW diameter when the vibration velocity is sufficiently large. The region in which t/D_s decreases little with the vibration increase is defined as the saturation stage of the nano rolling process. The former phenomenon is caused by the increase of the dynamic pressing force, and the latter results from a too large stiffness of the AgNS in the thickness direction. Also, it is seen that t/D_s at the saturation stage is not affected by the size of the rolled nanoscale sample (The measured t/D_s at the saturation stage = 0.05). Thus, thinner AgNSs can be fabricated if thinner AgNWs are used as the raw materials.

Figure 2.4(c) shows the measured effect of the MT's inner diameter D_t on the AgNS thickness. In the experiments, $d_0 = 5$ μm, $D_g = 15$ μm and $D_s = 300$ nm. It shows that the AgNS thickness is affected by D_t, and a smaller MT can cause thinner AgNSs. This is because a smaller MT has larger stiffness, generating a larger dynamic pressing force between the MT and rolled sample. Figure 2.4(d) shows the measured effect of the fiberglass diameter D_g on the AgNS thickness. In the experiments, $d_0 = 5$ μm, $D_t = 2$ mm and $D_s = 300$ nm. It shows that the AgNS thickness decreases as the fiberglass becomes thick. This is because the stiffness of the MT increases as the fiberglass becomes thicker.

The dynamic pressing force F_n and the preload F_0 between the MT and rolled sample were computed by FEM, with $d_0 = 5\ \mu m$, $D_g = 15\ \mu m$, $D_t = 3$ mm and a vibration velocity of 140 mm/s (0–p) at the micro tool's top. It was found that the zero-peak value of the dynamic pressing force F_n which was 4.8 μN, was much larger than the preload F_0 which was 0.37 μN. This explains the phenomenon that the preload in our experiments has little effect on the AgNS thickness in the saturation stage.

The device and fabrication process proposed in this work can also be used to roll Ag micro particles on the Si substrate into AgNSs. Figure 2.5(a) shows the 1 μm-diameter silver micro particles used as the raw material and AgNSs fabricated by the ultrasonic rolling, and Fig. 2.5(b) shows an AFM image of the AgNSs. In the experiments, $d_0 = 5\ \mu m$, $V = 25\ V_{p-p}$, $D_g = 10\ \mu m$ and $D_t = 2$ mm. The AFM measurement shows that the AgNS thickness is 400–500 nm, as shown in Fig. 2.5(c). Although the method is quite effective in rolling the AgNWs and Ag micro particles, our experiments show that it cannot be used to roll metal oxide nanowires (NWs) and particles, carbon micro fiber, Si micro wires, etc. For example, for a 60 nm-diameter ZnO NW shown in Fig. 2.6(a), it was experimentally found that no nano sheets could be generated by the rolling process. When the vibration velocity was large, the NW always broke during the rolling process, as shown in Fig. 2.6(b). When the vibration velocity was small, the NW shape had no change before and after the rolling. These phenomena are due to the poor malleability of oxide materials. It is predicted that the ultrasonic nano rolling method will be effective for the nano samples such as Au, Pt, Cu and Al, as they have better or similar malleability with Ag.

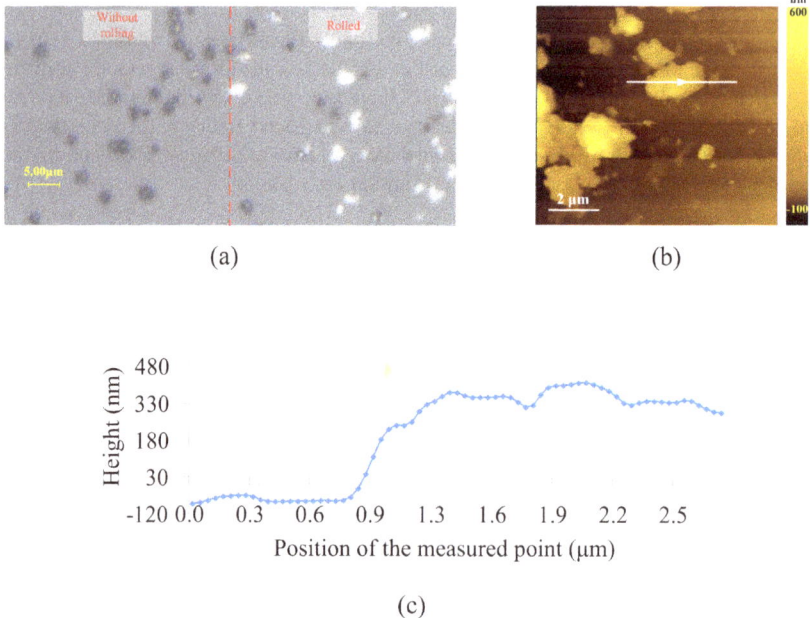

(a)   (b)

(c)

FIGURE 2.5   Ultrasonic rolling of 1 μm-diameter silver micro particles. (a) An optical microscopy image of the silver micro particles and the AgNSs fabricated by the rolling process. (b) An AFM image of the AgNSs. (c) Measured height of an AgNS. Reproduced from Ref. [1] with permission from Springer Nature.

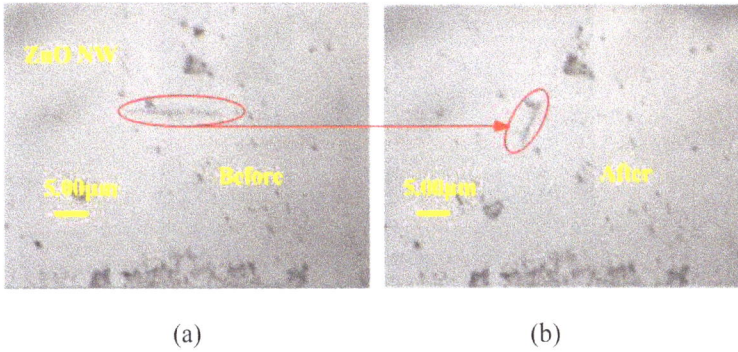

(a)   (b)

FIGURE 2.6   Effect of the ultrasonic rolling of 60 nm-diameter ZnO nano wires. (a) Before rolling. (b) After rolling. Reproduced from Ref. [1] with permission from Springer Nature.

2.1.3   SUMMARY

In this example, a flexible ultrasonic micro tool-based method is demonstrated, to roll commercial AgNWs and Ag micro particles on a solid substrate in air. The thickness of fabricated AgNSs can be as thin as several tens of nanometers, and the rolling effect is not sensitive to the preload when the vibration is sufficiently large. The thickness of fabricated AgNS can be decreased by using thinner AgNWs as the raw material, optimizing the MT structure and increasing the vibration velocity. Due to the utilization of the flexible MT, the rolling process does not harm the substrate, and the micro tool and processed samples have a stable and uniform contact, which raises the quality of the rolling process. The method provides a new and on-site way to fabricate AgNSs on a solid substrate, which can be used in the manufacture of photonic devices and nano sensors.

2.2   ULTRASONIC NANO CUTTING

Nanoscale materials can be used as the functional components of nano optoelectronic devices and bioelectronic sensors, and have large potential applications in high-performance photovoltaic panels, mechanical energy harvesters, high-performance batteries, super-capacitors, etc. Nano cutting is one of the important processes in nano fabrication and has potential applications in the testing of single nanoscale components and fabrication of nanoscale devices. To implement the nano cutting, researchers have proposed and investigated several methods with different working principles. They are based on principles such as mechanical breaking, electron beam and laser. Among them, the laser method has been most commonly used. It can cut individual NWs with a precise cutting position. However, the laser beam is harmful to biological samples owing to its intensive heat generation and causes the gasification of cut materials. The main features of the mechanical method include little heat damage to the samples and no gasification of cut materials. Electron beam cutting also has intensive heat generation.

In this example [2], we proposed and demonstrated a strategy to cut individual NWs by using the ultrasonic vibration of the tip of a micro cutting tool (MCT). The NW to

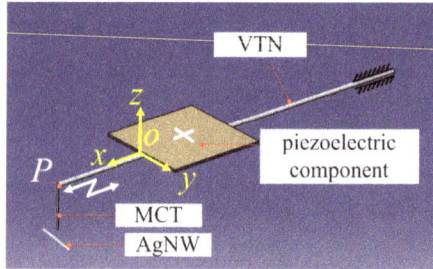

FIGURE 2.7 A schematic diagram of the experimental setup. Reproduced from Ref. [2] with permission from Elsevier.

cut was fixed on a substrate through the adhesion force between the substrate and the NW. Principle of the cutting, vibration velocity needed for the cutting and incision quality were investigated and analyzed. The main features of the method include no gasification of cut materials, real-time cutting under an optical microscope, and simple and light device structure.

2.2.1 DEVICE AND CUTTING PROCESS

Figure 2.7 shows the experimental setup for the ultrasonic cutting of a single AgNW. The device is made up of a piezoelectric component, vibration transmission needle (VTN) and an MCT. The VTN, made of steel, is bonded on the surface of the piezo-electric component, working as a wave guide to transmit vibration energy from the piezoelectric component to the MCT. The MCT is bonded to the tip of the VTN to generate a cutting force. The working frequency is around 96.9 kHz. We use a cone-shaped tungsten needle with a 2 μm-diameter tip as the MCT. A droplet of the nano suspension is dripped onto the silicon substrate. The cutting is carried out after the droplet dries off in air. During the cutting, the NW adheres to the substrate surface by water molecules between the NW and substrate, and the cutting process is implemented in air. Figure 2.8 is a photograph of the device. The entire process is carried

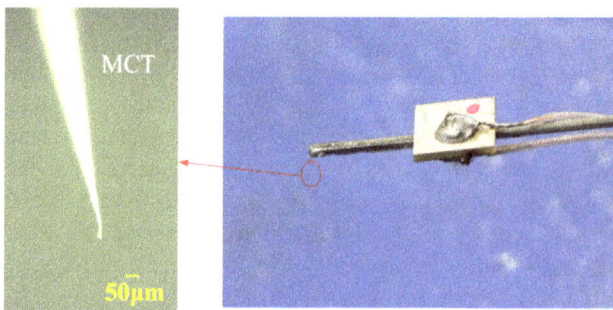

FIGURE 2.8 A photograph of the nanowire cutting device based on ultrasonic vibration. Reproduced from Ref. [2] with permission from Elsevier.

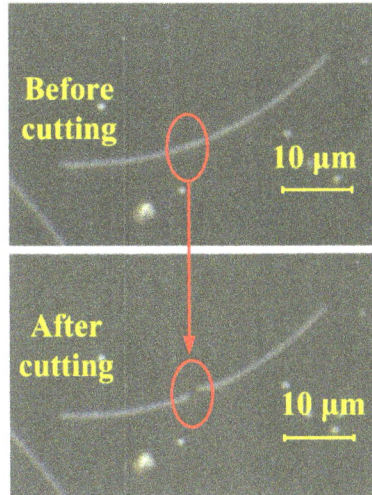

FIGURE 2.9   Cutting effect of linear vibration. The AgNW has a diameter of 120 nm, the cutting velocity amplitude is 143.4 mm/s and the working frequency is 96.9 kHz. Reproduced from Ref. [2] with permission from Elsevier.

out under an optical microscope (VHX-1000E, Keyence), which allows a real-time observation of the cutting process.

The cutting result is shown in Fig. 2.9, in which a single AgNW with a diameter of 120 nm is used, and the vibration velocity at the root of the MCT (point P) in the length direction of the VTN is 143.4 mm/s (0−p). It is observed that the AgNWs with a diameter from 50 to 400 nm can be cut at a working frequency of 96.9 kHz and driving voltage of 5–25 V_{p-p}. The cutting process is illustrated in Fig. 2.10. As shown in Fig. 2.10(b), the plane formed by the MCT and VTN is adjusted to make the main vibration component of the MCT (the x-directional vibration) approximately perpendicular to the AgNW, and the MCT is placed onto the AgNW during the cutting process. The purpose of keeping the AgNW perpendicular to the cutting velocity approximately is to decrease the incision size as possible as it can. Figure 2.10(d) shows two separated sections of the AgNW after the cutting process.

2.2.2   Cutting Principle

At 96.9 kHz (the working frequency) and 25 V_{p-p}, measured vibration components at point P are $V_x = 143.4 \angle{-51°}$, $V_y = 17.8 \angle 159°$ and $V_z = 9.1 \angle 128°$ mm/s, respectively. Thus $V_x \gg V_y$ and $V_x \gg V_z$ and point P (Fig. 2.7) vibrate linearly in the direction along the VTN (or the x-direction) during the cutting process.

The measured longitudinal (the x-directional) velocity distribution in the VTN between the piezoelectric plate and the VTN tip (point P) is shown in Fig. 2.11. In the measurement, the frequency and driving voltage were 96.9 kHz and 25 V_{p-p}, respectively. Furthermore, the vibration mode of the MCT was computed by the FEM, and the result is shown in Fig. 2.12. In the computation, the measured

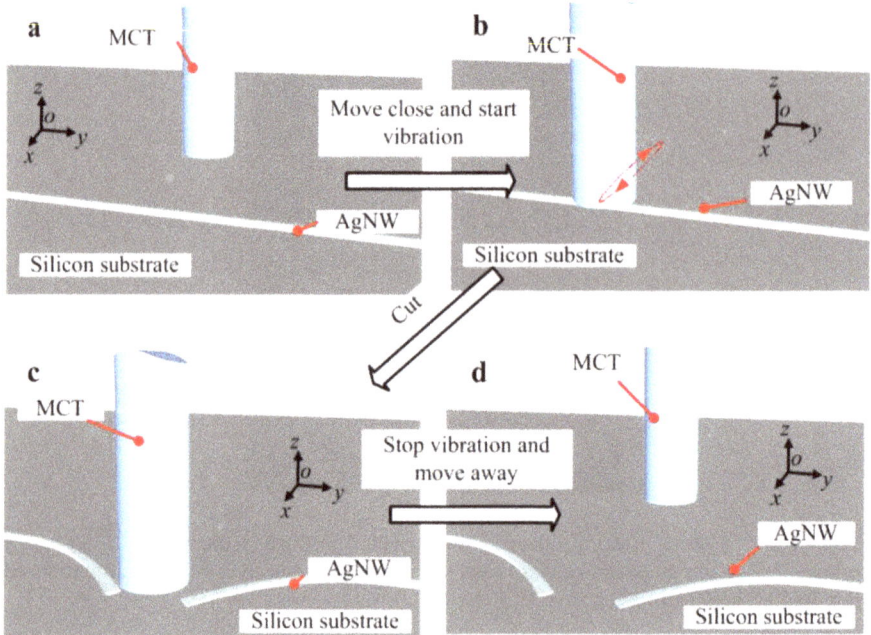

FIGURE 2.10 A schematic of the process of the ultrasonic nano cutting. Reproduced from Ref. [2] with permission from Elsevier.

vibration velocity components of point P at 96.9 kHz and 25 V_{p-p} were used as the excitation. It is found that the MCT vibrates flexurally, and its tip vibrates linearly. According to the computation, the MCT tip has a vibration amplitude 1.8 times larger than that of point P, which indicates that the MCT has a quite good vibration amplification function.

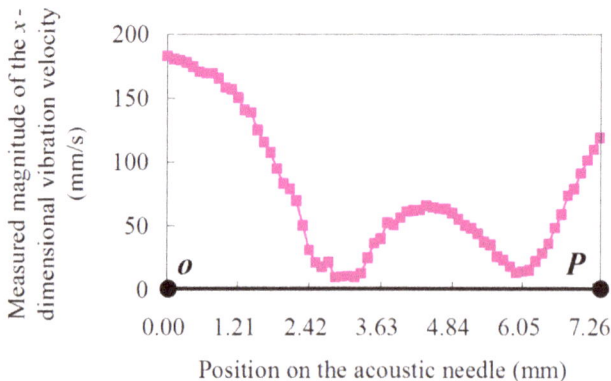

FIGURE 2.11 Vibration characteristics of the device. Measured distribution of the x-directional vibration velocity along the vibration transmission needle (VTN) between the piezoelectric plate and needle tip at a driving frequency of 96.9 kHz and voltage of 25 V_{p-p}. Reproduced from Ref. [2] with permission from Elsevier.

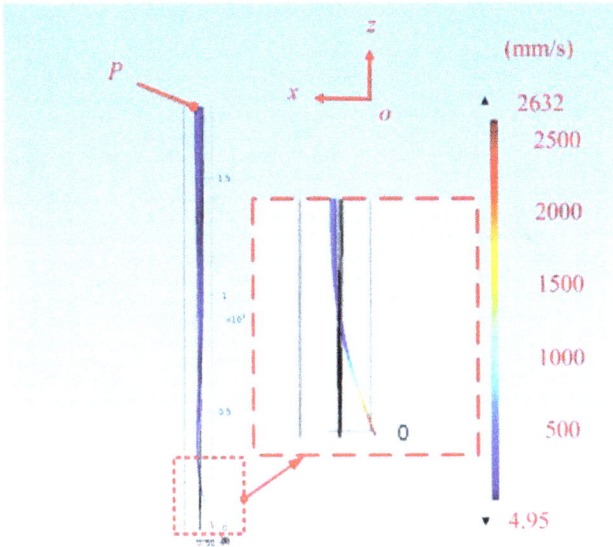

FIGURE 2.12 Calculated vibration mode pattern of the micro cutting tool (MCT) at a driving frequency of 96.9 kHz and voltage of 25 V_{p-p}. Reproduced from Ref. [2] with permission from Elsevier.

As the main vibration of the MCT (the x-directional vibration) is perpendicular to the AgNW approximately, the AgNW is applied with an x-directional force caused by the MCT's vibration. Due to this force, the AgNW in the incision section is stretched with a tensile stress. When the tensile stress reaches to the breaking point of AgNW, the AgNW is ruptured into two sections.

2.2.3 CUTTING CHARACTERISTICS

The incision morphology of a 90 nm-diameter AgNW cut at different x-directional vibration velocities was observed, and the results are shown in Fig. 2.13(a–e). In the experiments, the x-directional vibration at point P is 32.61, 64.87, 101.9, 143.4 and 210.5 mm/s, respectively. It is seen that the incision morphology depends on the vibration velocity. When the vibration velocity is small, the incision is not flat. When the vibration velocity is larger than a critical value ($\approx$101.9 mm/s), which is defined as the lower limit V_o of the optimum cutting velocity range, the incision becomes flat.

Figure 2.14(a–e) show the incision morphology of individual AgNWs with different diameters when the vibration velocity is kept constant ($V_x = 143.4\angle-51°$, $V_y = 17.8\angle159°$ and $V_z = 9.1\angle128°$ mm/s). It is seen that a too large NW diameter makes the incision not neat. As the NW diameter increases, the volume of the NW incision section V increases. This makes the NW incision section still connected to the rest part and the incision debris remaining at the incision, as shown in Fig. 2.14(d, e).

The vibration velocity below which the NW cannot be cut is defined as minimum cutting velocity V_m. It was measured for different diameters of the AgNW, and the

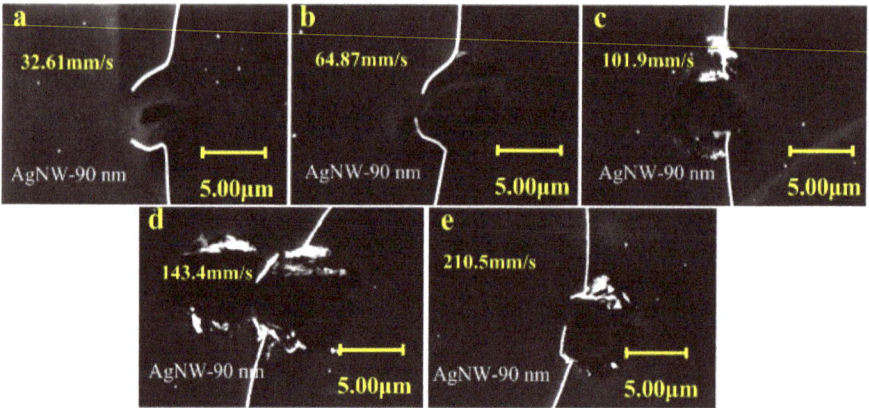

FIGURE 2.13 Incision morphology of the nano cutting based on the linear vibration of the micro cutting tool for different *x*-directional vibration velocities. The AgNW has a diameter of 90 nm. Reproduced from Ref. [2] with permission from Elsevier.

result is shown in Fig. 2.15. It is seen that the minimum cutting velocity increases with the AgNW diameter increase, and it is in the range of several ten millimeters per second.

Measured lower limit V_0 of the optimum cutting velocity range for the AgNWs with different diameters is shown in Fig. 2.16. It is seen that the lower limit of the optimum cutting velocity range increases with the AgNW diameter.

The use of ultrasonic vibration offers the opportunity to control the vibration trajectory of the MCT tip. We also investigated the incision morphology of the AgNWs cut by the MCT tip's elliptical vibration whose plane (the *xz* plane) is perpendicular to the substrate. In the experiment, the working frequency is 45.2 kHz, and $V_x = 28.4\angle72°$ mm/s, $V_y = 5.8\angle69°$ mm/s and $V_z = 38.5\angle158°$ mm/s. As V_y is

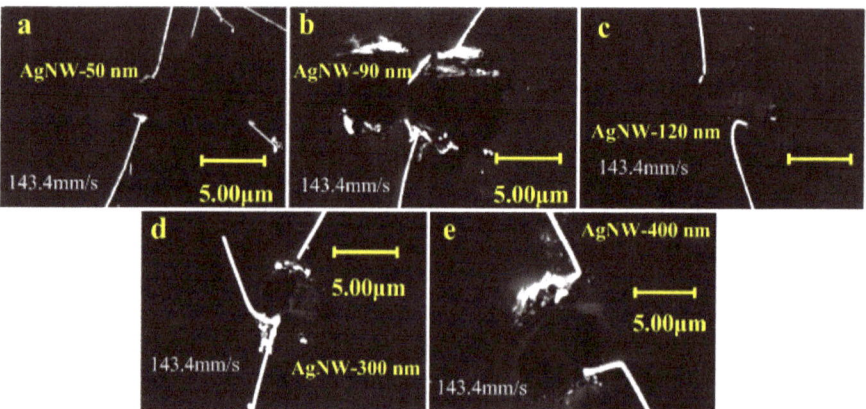

FIGURE 2.14 Incision morphology of the nano cutting based on the linear vibration of the micro cutting tool for individual nanowires with different diameters. The vibration velocity at point *P* is kept constant (143.4 mm/s). Reproduced from Ref. [2] with permission from Elsevier.

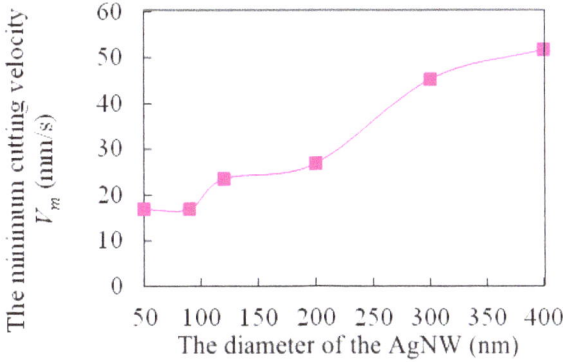

FIGURE 2.15   The minimum cutting velocity V_m versus the AgNW diameter. Reproduced from Ref. [2] with permission from Elsevier.

much less than V_x and V_z, the MCT tip's vibration is in the xz plane. As the phase difference between V_z and V_x is 86°, the vibration locus of the MCT is an ellipse. The incision of a 300 nm-diameter AgNW, which is caused by an elliptically vibrating MCT tip, is shown in Fig. 2.17. It shows that the AgNWs' ends at the incision take the shape of a flipper, which is caused by the z-directional vibration components of the MCT tip. When the MCT tip's elliptical vibration is used for cutting, a large z-directional cutting velocity needs to be avoided, because a large z-directional vibration velocity of the MCT makes the MCT tip bent beyond retrieve. Table 2.2 shows the measured minimum cutting velocity for two AgNWs with a 90 and 300 nm diameter, respectively, when the elliptical vibration is utilized.

Figure 2.18 shows the incision morphology of the AgNW cut by the MCT tip's elliptical vibration at different driving voltages when the device is in resonance (at about 45.2 kHz). When the driving voltage (or the vibration velocity) is not sufficient (Fig. 2.18), the AgNW cannot be cut into two sections and the incision takes the shape of a thin film, which is caused by the z-directional vibration component of

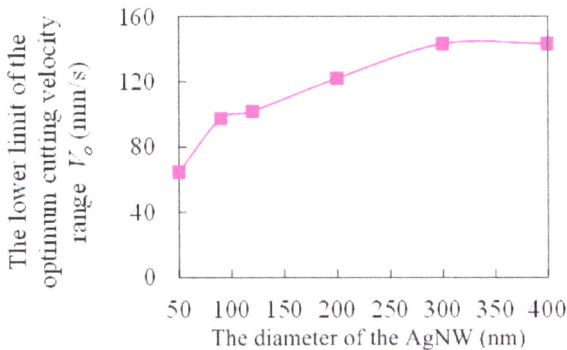

FIGURE 2.16   The lower limit V_o of the optimum cutting velocity range versus AgNW diameter. The working frequency is 96.9 kHz. Reproduced from Ref. [2] with permission from Elsevier.

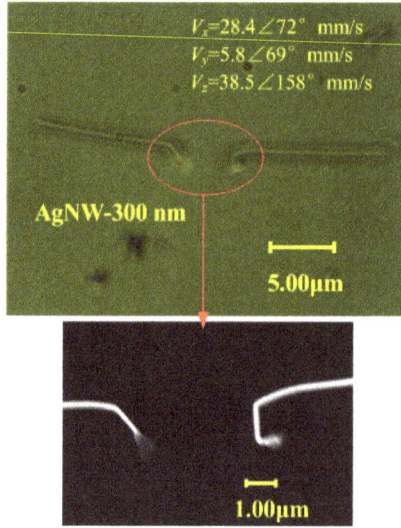

FIGURE 2.17 Incision morphology of the nano cutting based on the elliptical vibration of the micro cutting tool. The AgNW has a diameter of 300 nm, the driving voltage is 25 V_{p-p} and the working frequency is around 45.2 kHz. Reproduced from Ref. [2] with permission from Elsevier.

the MCT tip. The vibration velocities at the end point P of the VTN for the images shown in Fig. 2.18 are listed in Table 2.3. Figure 2.18 also indicates that the mechanism of the NW cutting based on the elliptical vibration is a bit different from that based on the linear vibration. In elliptical vibration cutting, the MCT squashes the incision section first and then breaks it by the horizontal vibration velocity.

2.2.4 SUMMARY

An NW cutting method, which uses the linear and elliptical ultrasonic vibration of the tip of an MCT and the adhesion force between the NW and substrate, is demonstrated and analyzed. It shows that both of the linear and elliptical vibration of the MCT can be used in the cutting of individual AgNWs, and the incision quality

TABLE 2.2

Measured Minimum Cutting Velocity When the Vibration Locus of the MCT is an Ellipse at around 45.2 kHz

	90 nm-Diameter AgNW	300 nm-Diameter AgNW
$V_{m,x}$ (mm/s)	17.8∠161°	24.9∠4°
$V_{m,y}$ (mm/s)	3.5∠154°	5.1∠5°
$V_{m,z}$ (mm/s)	19.2∠58°	33.8∠86°

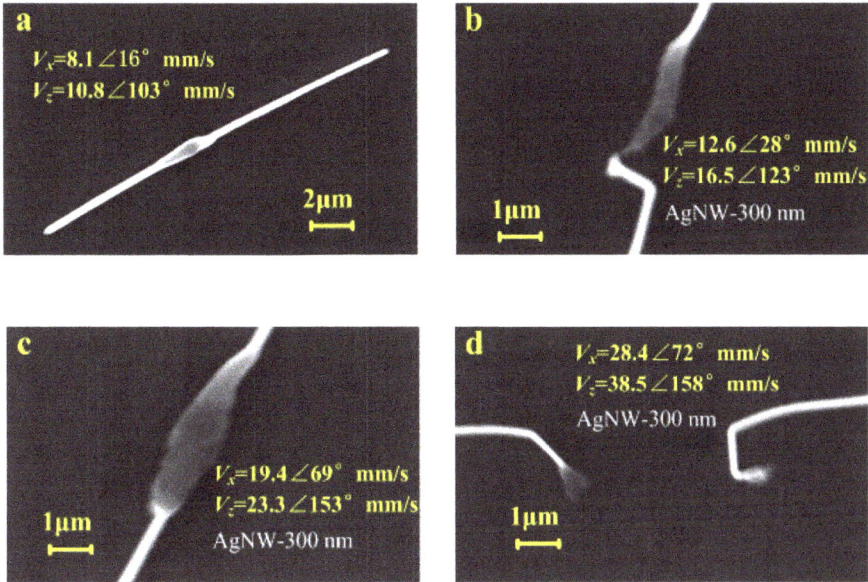

FIGURE 2.18 Incision morphology of the nano cutting based on the elliptical vibration of the micro cutting tool for different vibration velocities. The AgNW has a diameter AgNW of 300 nm and the working frequency is around 45.2 kHz. Reproduced from Ref. [2] with permission from Elsevier.

depends on the vibration velocity and NW diameter. For the cutting based on linear and elliptical vibration, the vibration velocity must be larger than a critical value to achieve a flat incision. For the nano cutting based on the elliptical vibration, the incision ends of AgNWs take the shape of a flipper. The working frequency is 96.9 and 45.2 kHz in the linear and elliptical cutting, respectively. To cut the individual AgNWs with a diameter from 50 to 400 nm, the MCT needs vibration velocity amplitude from 18 to 220 mm/s. The main features of the method include no gasification of cut materials, real-time cutting under an optical microscope and simple and light device structure.

TABLE 2.3

Measured Vibration Velocity at the Tip of the Vibration Transmission Needle for the Images Shown in Fig. 2.18

	V_x (mm/s)	V_y (mm/s)	V_z (mm/s)
a	8.1∠16°	2.2∠14°	10.8∠103°
b	12.6∠28°	2.9∠24°	16.5∠123°
c	19.4∠69°	4.2∠67°	23.3∠153°
d	28.4∠72°	5.8∠69°	38.5∠158°

2.3 EXPLORING NEW COATING PROCESS FOR FLEXIBLE NANO SENSORS

Flexible nano sensors such as strain and gas nano sensors are highly desirable because of their potential applications in health monitoring systems for human beings. The main part of these sensors is usually composed of a flexible substrate made of polymer materials and nano sensing material deposited on the substrate. The substrate may be formed by polydimethylsiloxane (PDMS), polyethylene terephthalate (PET), polyimide (PI), polyethylene (PE), polyurethane (PU), etc. The sensing layer is usually made of carbon nano materials (CNTs, GO, rGO, etc.) and other functional nano materials (metal oxide NPs, noble metal NPs, etc.), depending on the required sensing function. During applications, the nano sensors are attached to the human skin or clothes.

One of the key technologies for flexible nano sensor fabrication is how to coat nano materials onto the surface of a flexible substrate in an effective and reliable way. As the chemical vapor deposition (CVD) method and lithography cannot be used in the fabrication process of flexible nano sensors because the flexible substrate cannot stand the high temperature, the commonly employed fabrication methods in the flexible sensor fabrication include the sputtering deposition, drop casting, spin coating, spray coating, soak coating, inkjet printing, screen printing, electrospinning, transfer, etc. Different film fabrication methods have their own characteristics in the preparation process. For example, the sputtering deposition method uses accelerated particles to bombard the target material in a vacuum chamber and make the sputtered target material deposited onto the surface of a flexible substrate to form a film of nanoscale thickness. The uniformity and purity of the nano film prepared by this method are excellent, and the bonding strength between the film and flexible substrate is high. The flexible nano sensor prepared by this method has good performance uniformity and is easy to commercialize. In the efforts of film fabrication at room temperature and normal pressure, the spin-coating method is commonly used. It relies on the centrifugal force generated by the rotation of substrate, to spread the coating droplets on the substrate surface to fabricate the film. This method can produce functional film of controllable thickness, with a simple fabrication process. The ink-jet printing method can print functional ink onto a selected surface area of a flexible substrate through computer graphics control, with the advantages of convenient operation and excellent controllability of the film pattern. It can realize a rapid preparation of flexible nano sensors with arbitrary patterns. The traditional ultrasonic spraying method uses airflow to deposit the ultrasonically atomized nano material solution or solution on the surface of a solid or flexible substrate to form a film. It has the advantages of little selectivity to the nano material type, high utilization rate of raw nano material, etc. This method can deposit nano materials on various substrates in one step and has the advantages of low-cost and scaled-up fabrication.

Despite of lots of efforts to fabricate flexible nano sensors, green and effective new coating methods are still desired. For example, the sputter deposition method needs vacuum and high electric field for the deposition of the functional film. Thus, it has the disadvantages of complex equipment, high energy consumption, long fabrication time, etc., which greatly limits the large-scale preparation. Although the coating, printing and spraying methods are of mass production capability and can rapidly fabricate the sensors at room temperature and normal pressure, they have several drawbacks such as the use of large quantity of chemical reagents and the weak bonding between the film and substrate. Other drawbacks of these methods include: In the drop-casting method, the quality and performance consistency of the fabricated sensing films are poor; in the printing and spraying methods, there is a problem of nozzle clogging during the operation, which increases the cost of equipment maintenance; in the spin-coating method, the nano material utilization rate is low; in the ultrasonic spraying process, the micro droplets produced by ultrasonic atomization will cause nano pollution in the production environment; in the transfer method, it is still not easy to transfer the whole nano materials agglomerates onto the substrate.

As an endeavor to deal with the green fabrication challenge, the author's group proposed a liquid power ultrasound-based coating process for flexible nano sensors. This method uses the acoustic cavitation-induced high-speed jets and the acoustic streaming field in the nano solution or suspension to deposit the nano sensing materials onto a polymer substrate. The method has green features such as operation at room temperature and normal pressure, minimum use of organic solvent, low maintenance costs and recyclability of nano material in the solution.

2.4 ULTRASONIC FABRICATION OF FLEXIBLE STRAIN SENSORS

Flexible strain sensors with large stretchability, high sensitivity and good stability are highly desirable because of their potential applications in electronic skins and health monitoring systems. In this work [3], a liquid power-ultrasound based fabrication process for flexible strain sensors is presented. It uses the acoustic cavitation induced high-speed jets and the acoustic streaming field in the MWCNT (multi walled carbon nanotube)-water solution sonicated by power ultrasound of 19.9 kHz, to deposit the MWCNT nano materials onto a 200 μm-thick PDMS substrate. The prepared strain sensor has a wide sensing range (up to 420% strain), high sensitivity (gauge factor (GF) = 8.4/216.8) and excellent stability (>10000 cycles at 50% strain). The fabrication process is implemented at room temperature (25°C) and normal pressure (1 atm), and does not cause the atomization of nano solution. Also, it uses the minimum organic solvent (*N,N*-dimethylformamide (DMF) as the disperser), and the remained MWCNTs in the fabrication process can be recycled. Apart from these green features, the fabrication process is little selective to the nano materials. Therefore, the proposed technique can be used in the fabrication of flexible strain sensors, electronic skin and other flexible nano devices in an environment-friendly way.

2.4.1 EXPERIMENTAL METHODS

2.4.1.1 Experimental Setup

Figure 2.19(a) depicts the experimental setup for liquid power-ultrasound deposition proposed in this work. It mainly consists of a beaker with the MWCNT-water solution (0.026 wt%), a 200-μm-thick PDMS substrate placed at the bottom of the beaker and a 19.9 kHz ultrasonic transducer with a horn inserted into the solution. In the deposition, the ultrasonic transducer is electrically driven by a power amplifier, which receives the frequency signal from a function generator. The voltage and current applied to the ultrasonic transducer is measured and monitored by an oscilloscope.

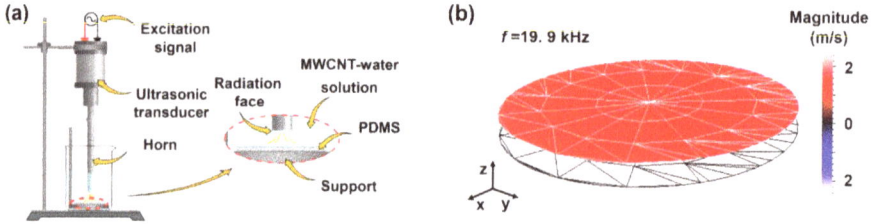

FIGURE 2.19 (a) Experimental setup for the ultrasonic deposition equipment. (b) Measured distribution of the z-directional (out-of-plane) vibration velocity of the radiation face at 19.9 kHz and 168 V_{p-p}. Reproduced from Ref. [3] with permission from Elsevier.

Vibration characteristics of the ultrasonic transducer were measured by a 3D laser Doppler vibrometer (PSV-300F, Polytec GmbH, Waldbronn, Germany). The measured distribution of vibration velocity of the radiation face at 19.9 kHz and 168 V_{p-p} is given in Fig. 2.19(b). The result shows the z-directional vibration velocity of the radiation face is quite uniform.

2.4.1.2 Fabrication Process of Flexible Strain Sensors

Figure 2.20 shows the fabrication process of flexible strain sensors, carried out in this work. It mainly includes the follows steps:

i. Weigh 100 mg MWCNTs (XFM22, Nanjing XFNANO Materials Tech Co., Ltd., China) without purification, and put it in a beaker.

ii. Add 200 ml of DI water (CJRO-20L/18, Nanjing Chaojing Technology Co., Ltd., China) and 200 ml of DMF (D112002-500 ml, Shanghai Aladdin Bio-Chem Technology Co., Ltd., China) into the beaker, and the mixed solution of MWCNTs is dispersed by ultrasonic cleaner (KQ3200, Kunshan Ultrasonic Instrument Co., Ltd., China) for 30 min, and then stands for 3 h.

iii. Pour 160 ml prepared MWCNT-water solution into a 250 ml beaker with a 200-µm-thick PDMS substrate (BD-KRN-200, Hangzhou Bald advanced Materials Technology Co., Ltd., China) placed at its bottom.

iv. Insert the horn of the ultrasonic transducer (HND-8AE-4520 Booster-D19, Hainertec (Suzhou) Co., Ltd., China) into the MWCNT-water solution, with the horn's end surface parallel to and 30 mm away from the PDMS substrate.

v. Drive the ultrasonic transducer at 19.9 kHz for 20 min and deposit the MWCNTs onto the PDMS substrate.

vi. Take the PDMS substrate with the deposited film out of the solution and let it dry naturally.

vii. Cut the PDMS substrate with the MWCNT film into several strips (5 mm × 40 mm) and connect the ends of each MWCNT film strip to external electrodes by silver paste (SPI# 05001-AB, USA).

viii. Seal the MWCNT film with a thin layer of liquid PDMS (BD-901, Hangzhou Bald Advanced Materials Technology Co., Ltd., China), pre-cure them at 60 °C for 20 min, ripen them at 80 °C for 30 min and finally form the strain sensor in sandwich structure to protect the deposited MWCNT film.

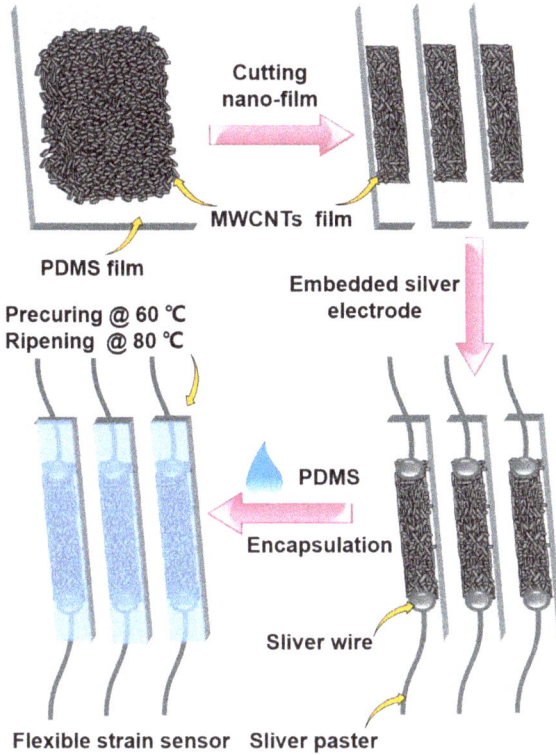

FIGURE 2.20 Liquid power-ultrasound-based fabrication process of flexible strain sensors. Reproduced from Ref. [3] with permission from Elsevier.

2.4.1.3   Characterization of Strain Sensors

Figure 2.21 depicts the composition of a test system for flexible strain sensors, constructed in this work. The morphologies and structures of the MWCNTs films were characterized by a field-emission scanning electron microscope (FE-SEM, JSM-7600f, Japan Electronics Corporation, Japan). The crystalline structure of MWCNTs before and after the liquid power-ultrasound deposition were analyzed by X-ray diffraction (XRD, PANalytical, Malvern, Netherlands) using Cu Kα radiation ($\lambda = 1.54$ Å) at 45 kV and 15 mA. The transmission spectrum was tested by UV/Vis/NIR spectrophotometer (Lambda 950, PerkinElmer, USA). A high-precision electronic universal testing machine (ZP110-15, Beijing Lianying Precision Machinery Technology Co., Ltd., China) controlled by a computer was used to stretch the strain sensor. The sensing response was recorded by an electrochemical workstation (CHI 760E, CH Instruments, Shanghai, China).

2.4.2   FINITE ELEMENT COMPUTATION

The ultrasonic field and steady-state acoustic streaming field were computed by the FEM, implemented by COMSOL Multiphysics 5.4 software (COMSOL, Inc., Stockholm,

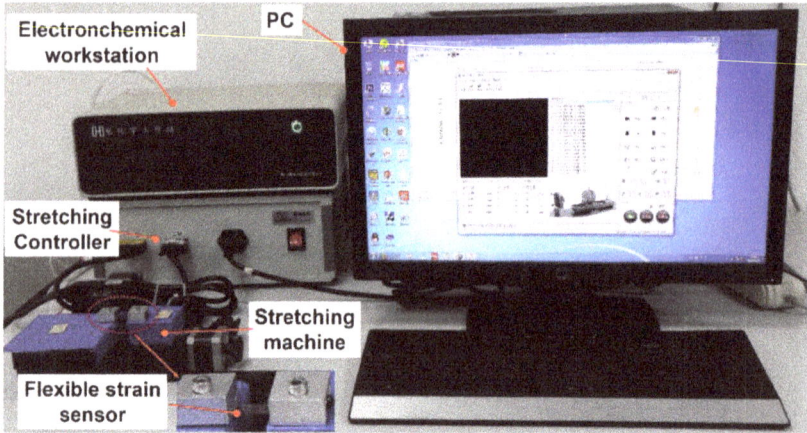

FIGURE 2.21   Testing system for flexible strain sensors fabricated in this work. Reproduced from Ref. [3] with permission from Elsevier.

Sweden). Figure 2.22(a) is a two-dimensional (2D) physical model for the FEM computation of the deposition system, and Fig. 2.22(b) shows a 2D meshed model for the FEM computation. To take a balance between the computational error and time, the mesh size of the acousto-fluidic field was not uniform and was smaller near the ultrasonic horn than in the rest region. The maximum mesh size was 1 mm, corresponding to about 1.3% of the wavelength of the ultrasonic field. The computational conditions were as follows: the vibration velocity of the radiation face was set to be uniform (=1.92 m s^{-1} (0–p)); the working frequency was 19.9 kHz. The computational process consisted of three steps. First, the sound field was solved by the acoustic module of the FEM software. Then, the vibration velocity and sound pressure were used to calculate the spatial gradients

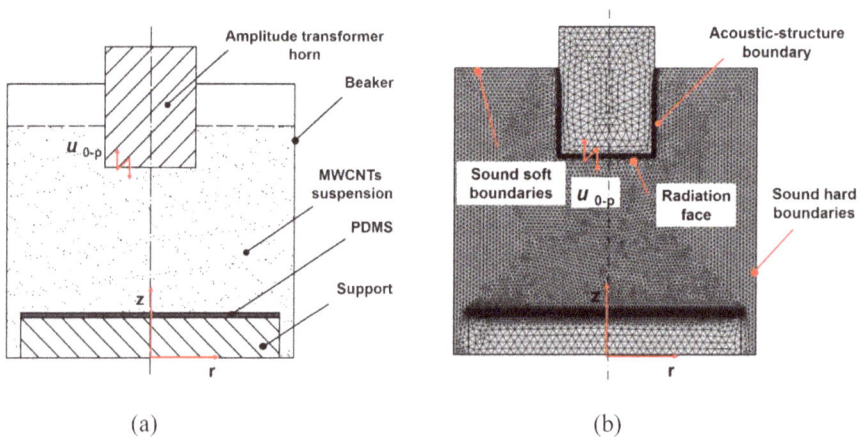

(a)   (b)

FIGURE 2.22   (a) a physical model for the liquid power-ultrasound deposition system. (b) A meshed model for the FEM computation. Reproduced from Ref. [3] with permission from Elsevier.

TABLE 2.4

Parameters of Ultrasonic System for the FEM Computation

Diameter of the Radiation Face (mm)	Height of the Support (mm)	Thickness of PDMS Substrate (μm)
19	5	200
Depth of the solution (mm)	Inner diameter of the beaker (mm)	Density of water ρ (kg/m³)
50	90	1000
Speed of sound in water c (m/s)	Shear viscosity of water η (Pa•s)	Volume-to-shear viscosity ratio in water $\dfrac{\eta'}{\eta}$
1500	0.001	2.1
Nonlinear parameter of water at room temperature $\dfrac{B}{A}$	Frequency of the vibration source f (kHz)	Out-of-plane vibration velocity of the radiate on face u (m/s) (0–p)
5	19.9	1.92

of the Reynolds' stress and second sound pressure by the post-processing functions of the software. Finally, the steady acoustic streaming was solved by the fluidic dynamics module of the same FEM software. The boundary conditions for the ultrasonic field were as follows: (i) The acceleration is continuous at the interfaces between the ultrasonic horn and liquid film. (ii) The boundary between the solution and air was sound soft (sound pressure $p = 0$). (iii) The boundaries between the water film and the inner surface of the beaker and substrate were sound hard ($\dfrac{\partial p}{\partial n} = 0$, where n denotes the unity vector of the boundary (see Fig. 2.22(b))). For the acoustic streaming field, all the boundaries were set to be no-slip. The material property constants and dimensions used in the computation are listed in Table 2.4.

2.4.3 RESULTS AND DISCUSSION

The bonding strength between the PDMS substrate and MWCNT film fabricated by the above process was qualitatively evaluated by the tape peeling test, with a comparison to that fabricated by the common drop-casting process, and the results are shown in Fig. 2.23(a–d). It was found that the MWCNT film fabricated by the common drop-casting process could be easily peeled, whereas MWCNT film fabricated by the above process could endure the peeling. Thus, the bonding strength between the PDMS substrate and MWCNT film fabricated by the liquid power ultrasonic deposition process is stronger than that fabricated by the common drop-casting process, as expected. Figure 2.23(e) shows the stretching of a fabricated sensor by a self-made stretching stage which can provide a maximum strain of 420%.

In order to investigate the in-depth mechanism of the liquid power-ultrasound deposition method, sound pressure and acoustic streaming in the ultrasonic field between the radiation face and PDMS substrate were computed by the 2D FEM.

The scotch peeling test 1:

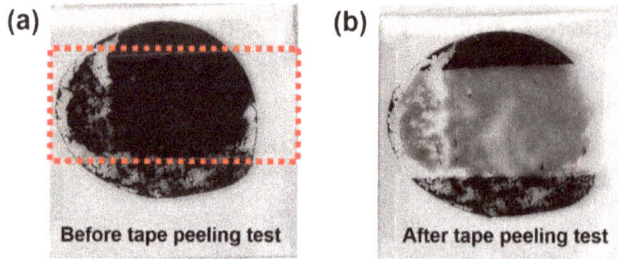

The scotch peeling test 2:

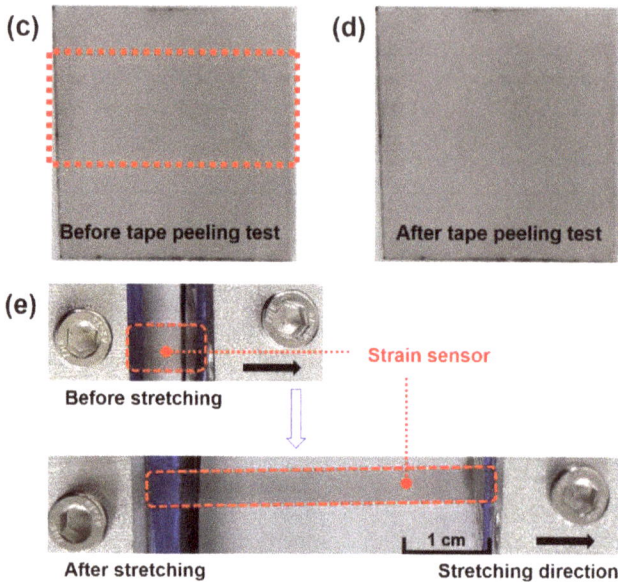

FIGURE 2.23 (a–d) The scotch tape peeling test of the bonding strength between the MWCNT film and PDMS substrate. The films in (a) and (b) were fabricated by the common drop coating method, and the films in (c) and (d) were fabricated by the liquid power ultrasonic deposition method. (e) Stretching of the MWCNT film strain sensor. The sensor in (e) is stretched with a strain of 420%. Reproduced from Ref. [3] with permission from Elsevier.

Figure 2.24(a, b) are the computed sound pressure distribution and acoustic streaming field when the frequency and out-of-plane vibration velocity of the radiation face are 19.9 kHz and 1.92 m s^{-1} (0–p), respectively, which are the working frequency and vibration velocity used in the experiments of this work. Figure 2.24a shows that there exists sound pressure on the PDMS surface, and it can reach about 1×10^5 Pa. This means that the ultrasonic cavitation could be caused by the sound pressure on the PDMS surface. The micro jet generated by the implosion of transient bubbles, which has a velocity of up to several tens meters per second, can eject the MWCNTs onto the PDMS surface, causing the deposition phenomenon. Figure 2.24(b) shows that

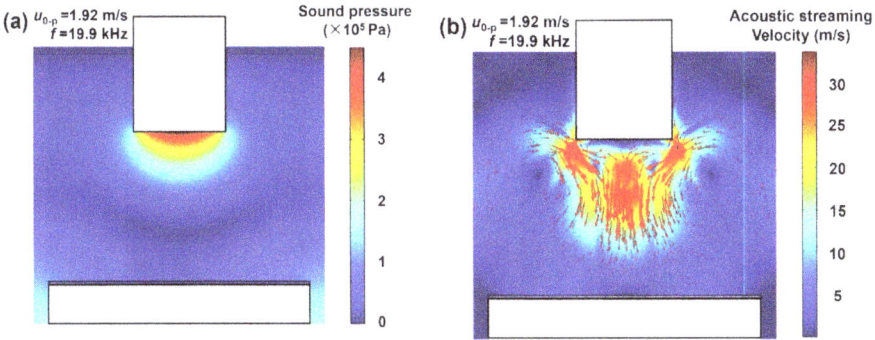

FIGURE 2.24 (a) Computed sound pressure field. (b) Computed acoustic streaming field. In the computation, the working frequency and out-of-plane vibration velocity of the radiation surface are 19.9 kHz and 1.92 m s^{-1} (0–p), respectively, and the distance between the radiation surface and substrate is 30 mm. Reproduced from Ref. [3] with permission from Elsevier.

the acoustic streaming flows toward the PDMS surface, indicating that the acoustic streaming could also contribute to the deposition phenomenon.

Figure 2.25(a, b) show the SEM images of the MWCNT film fabricated by the liquid power-ultrasound deposition method. The microscopic morphology of the MWCNT network is covered by a layer of impurities in the MWCNT material.

Figure 2.25(c) shows the presence of the MWCNT network in a coating crack. Thus, a network of MWCNTs should have formed, which is an essential condition to form a nano material film-based strain sensor. Figure 2.25(d) displays the XRD patterns of the MWCNT before and after the liquid power-ultrasound deposition. The diffraction peaks of the MWCNTs are also observed at about 25.89° (002) and 43.35° (100), which are ascribed to the interlayer space in the radial direction and in-plane graphitic structure of the CNT, respectively. The XRD patterns in Fig. 2.25(d) also indicate that the crystal structure of MWCNTs was not damaged by the deposition process.

Figure 2.26(a) shows a photograph of the fabricated MWCNT film (75 mm × 70 mm) with an A4 paper in the back. The characters on the A4 paper can be clearly seen, which indicates that the MWCNT film is transparent. Figure 2.26(b) shows the measured dependency of the transmittance on wavelength, which indicates that the transmittance can reach 80% in the visible range (with a wavelength range from 380 to 740 nm).

Figure 2.27(a) shows the measured current-voltage curves of the fabricated MWCNT film on the PDMS substrate. It is seen that the current flowing through the film is linearly proportional to the applied voltage and the slope of the lines changes with the strain. Thus, the fabricated MWCNT film can be utilized as a strain sensor. Figure 2.27(b) shows the measured sensing response versus strain of the sensors fabricated under three different vibration velocities (1.71 m s^{-1} (0–p), 1.83 m s^{-1} (0–p) and 1.92 m s^{-1} (0–p)) and the same sonication time (t_s = 20 min).

Here, the sensing response is defined as the ratio of the resistance change ($R-R_0$) to the baseline resistance (R_0). In the experiments, the strain sensor was stretched from zero strain to the maximum strain that our stretching equipment could provide, and the stretching rate was 1 mm s^{-1}. The reason for choosing the vibration velocities

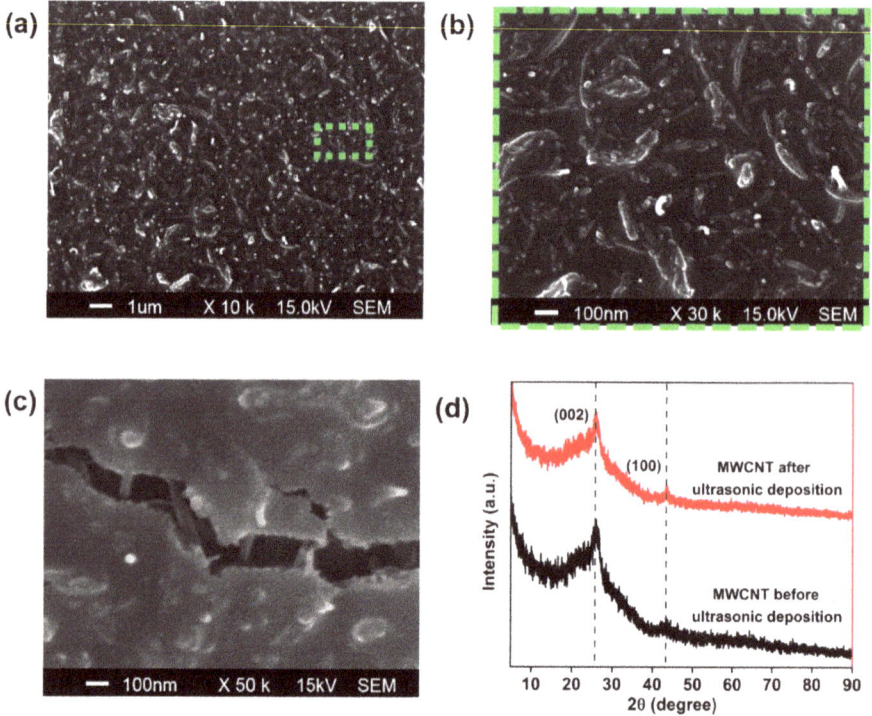

FIGURE 2.25 (a–c) SEM microstructures of the MWCNT film fabricated by the liquid power-ultrasound deposition method in MWCNT-water solution. (d) XRD spectra of the MWCNT before and after the liquid power-ultrasound deposition. The working frequency and out-of-plane vibration velocity of the end surface of the horn were 19.9 kHz and 1.92 m s^{-1} (0–p), respectively. Reproduced from Ref. [3] with permission from Elsevier.

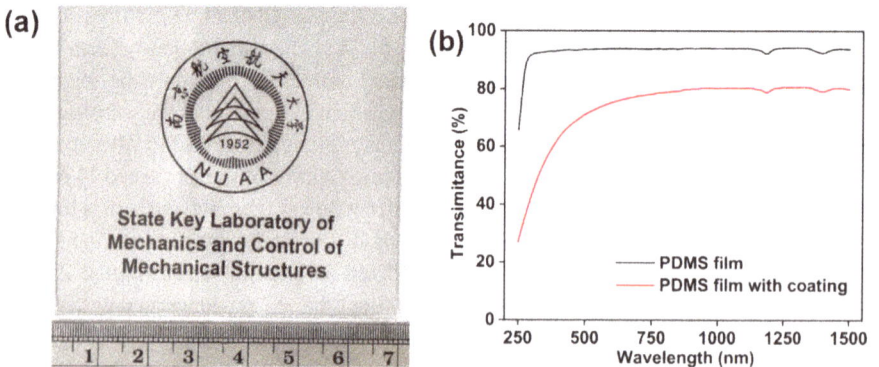

FIGURE 2.26 (a) Photograph of a 75 mm × 70 mm MWCNT film fabricated by the method. (b) Measured transmittance spectrum of the fabricated MWCNT film. The working frequency and out-of-plane vibration velocity of the end surface of the horn were 19.9 kHz and 1.92 m s^{-1} (0–p), respectively. Reproduced from Ref. [3] with permission from Elsevier.

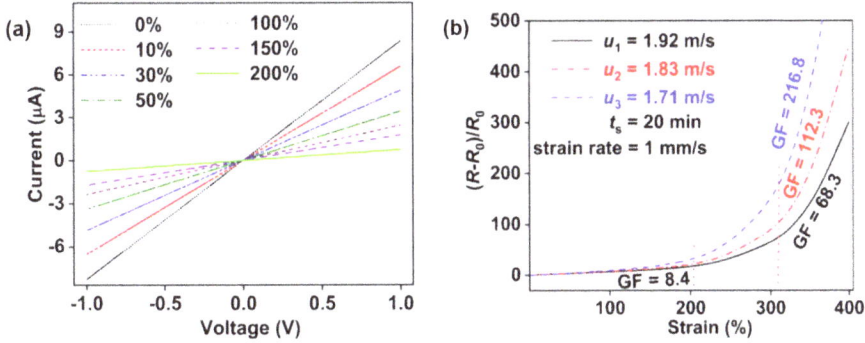

FIGURE 2.27   (a) Current-voltage relationship of the strain sensor at different levels of strain.
(b) Sensing response versus strain of the sensor fabricated under different vibration velocities
(0–p). Reproduced from Ref. [3] with permission from Elsevier.

of 1.71, 1.83 and 1.92 m s^{-1} was that too weak vibration caused bad quality of the
film due to a poor connection between MWCNTs and too strong vibration caused the
ultrasonic transducer unstable due to its high-temperature rise as the experimental
setup did not use a frequency tracking system.

It is seen that as the strain increases, the sensing response linearly increases first,
then nonlinearly increases and finally linearly increases again. The linearity in the
tensile strain range less than 210% is because the resistance increase in this range is
mainly caused by the increase of tunneling resistance and contact resistance between
MWCNTs. The linearity with a much larger slope in the tensile strain range higher
than 300% is because the resistance increase in this range is mainly caused by the
disconnection between MWCNTs and the rapid decrease of the conductive paths in
the MWCNT network. Between the above two regions, there is a transient region,
where the resistance response increases nonlinearly with the stretching strain, in
which the connection between MWCNTs starts to break and the conductive path
number starts to decrease. The GF, which is defined as the sensing response change
per unit strain change (or the slope of the sensing response-strain lines), is also given
in the figure. The GF was calculated to be 8.4 for strain range within 210% and
not less than 68.3 in the larger strain ranges, which are comparable to those of the
MWCNT-PDMS-based strain sensors in most reported work [6–12] (Table 2.5).
Moreover, it is seen that the vibration velocity affects the sensing response-strain
characteristic little at the strain range lower than 210%, and larger vibration velocity
causes lower sensing response (or sensitivity) when the strain is larger than 210%.
The former phenomenon is because the vibration velocity has little effect on the tunneling and contact resistance between the MWCNTs, and the latter one is because
the larger vibration velocity makes the decrease of the conductive path number in
the MWCNT network more difficult and the bonding between the MWCNT film
and PDMS substrate stronger. Although the strain sensor fabricated under 1.92 m s^{-1}
has a lower GF than that fabricated under 1.71 m s^{-1}, we cannot simply say that the
vibration condition of 1.92 m s^{-1} is inferior to that of 1.71 m s^{-1}, as the stability of the
strain sensor should also be examined in the performance assessment.

TABLE 2.5

Comparison between the MWCNT-Film Strain Sensor in This Work and the Carbon-Based Strain Sensors with the Same Sensing Mechanism (Resistive) Recently Reported

Work	Materials	Maximal Workable Strain Range	Average Gauge Factor	Durability	Linearity
This work	MWCNTs/ PDMS	420%	8.4 (0%–210%), 22.6 (210%–400%), 68.3 (320%–420%)	10,000 cycles at 50% strain	Piecewise linear to 420%
[7]	CNTs/PDMS	80%	234.9 (0%–40%), 1362.7 (40%–70%), 229.3 (70%–80%)	10,000 cycles at 8% strain	Piecewise linear to 80%
[8]	CNTs/PDMS	300%	0.91 (0%–100%), 3.25 (100%–255%), 13.1 (255%–300%)	2500 cycles at 200% strain	Piecewise linear to 300%
[9]	SWCNTs/ PDMS	280%	0.82 (0%–40%), 0.06 (60%–200%)	10,000 cycles at 200% strain	Piecewise linear to 280%
[10]	CNTs/PDMS	200%	9960 at 85%	5000 cycles at 10% strain	Piecewise linear to 85%
[11]	CNTs/PDMS	100%	87 (0%–40%), 6 (40%–100%)	1500 cycles at 20% strain	Piecewise linear to 100%
[12]	CNTs/PDMS	400%	24.3 (0%–150%), 437 (150%–260%)	10,000 cycles at 100%	Piecewise linear to 260%

In the following experiments, we carried out a pre-stretching process with a strain of 240% at a strain rate of 1 mm s^{-1} for 400 cycles, to achieve a monotonous relationship between the sensing response and strain in the experimental strain range. Figure 2.28(a) shows the measured dynamic responses at different strain values and a constant strain frequency (0.5 Hz). Figure 2.28(a) shows that the strain sensor has a very good repeatability. Also, the strain sensor can detect the minimum strain of

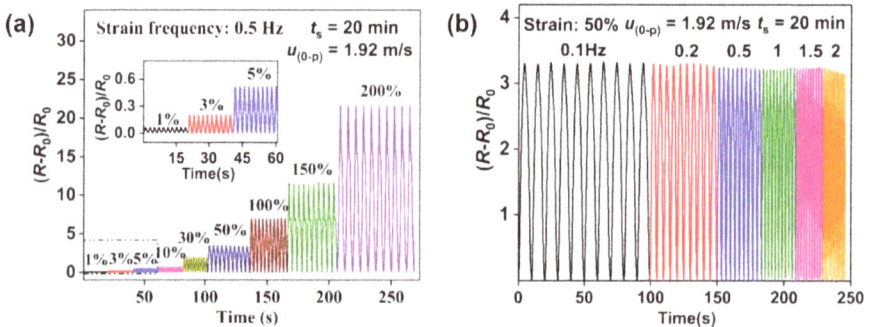

FIGURE 2.28 (a) Dynamic sensing responses at different strain levels and a constant strain frequency. (b) Dynamic sensing responses at different strain frequencies and a constant strain level. Reproduced from Ref. [3] with permission from Elsevier.

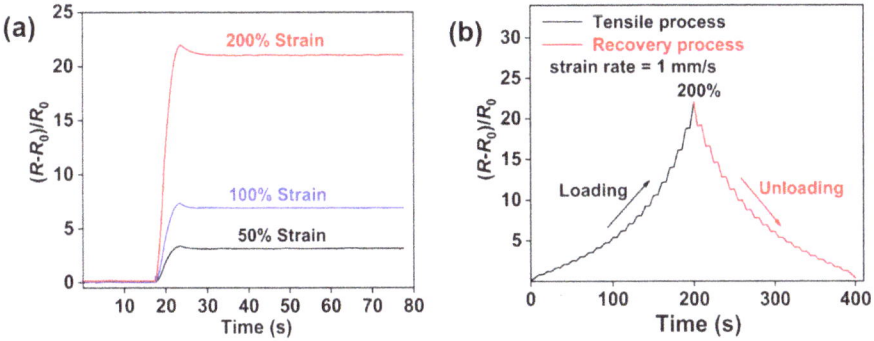

FIGURE 2.29 (a) Dynamic sensing responses of the MWCNT strain sensor under the step and hold tests. (b) Response of the strain sensor during the stepwise loading-hold and unloading-hold processes. Reproduced from Ref. [3] with permission from Elsevier.

1% that our experimental platform can provide. Figure 2.28(b) shows the measured dynamic responses at different strain frequencies and a constant strain (50%). It shows that the response is not sensitive to the strain frequency in the range of 0.1–2 Hz, which is the working frequency range of wearable devices to monitor human motion. Thus, the device is promising in the application of wearable sensor systems.

Figure 2.29(a) is the measured sensing response when the sensor is stretched to a strain of 50%, 100% and 200% with a strain rate of 25% s^{-1}. The creep recovery time of 50%, 100% and 200% strain is 3, 5 and 8s, respectively, which is smaller than the most reported values [6–12]. This also indicates that there is a good bonding between the MWCNT film and PDMS substrate in the liquid power ultrasound fabrication process.

Figure 2.29(b) shows the measured sensing response under stepwise strain loading and stepwise strain unloading. During the stepwise strain loading, every 5 s stretching (strain speed = 1 mm s^{-1}) is followed by a pause of 5 s and the cycle is repeated until the strain reached 200%. A similar way is repeated during the recovery process. The sensor exhibits very small overshoot in each step, which also results from a good bonding between the MWCNT film and PDMS substrate.

Stability is a very important issue in practical applications of strain sensors. Figure 2.30(a) shows the measured stability characteristic of the fabricated sensors under a cyclic loading-unloading of 50% strain at a rate of 0.5 Hz for 10,000 cycles. In the calculation of sensing response, we chose the sensor resistance at the no-load state at the 400th test cycle as the baseline resistance R_0. In the initial 400 strain cycles, the sensing response waveform has a rapidly downward shift, and the sensor becomes stable after the 400 cycles. In the initial 400 strain cycles, the cyclic loading-unloading process decreases the sensor resistance due to the newly built conductive paths in the MWCNT network. After about 400 cycles, the sensor resistance at the full-load and no-load states becomes stable because the conductive paths in the MWCNT network change little.

Figure 2.30(b) shows the measured change of the sensing response increase rate (Δ) during the long-term stability test under a cyclic loading-unloading of 50% strain at a rate of 0.5 Hz for 10,000 cycles. The strain sensors were fabricated under three

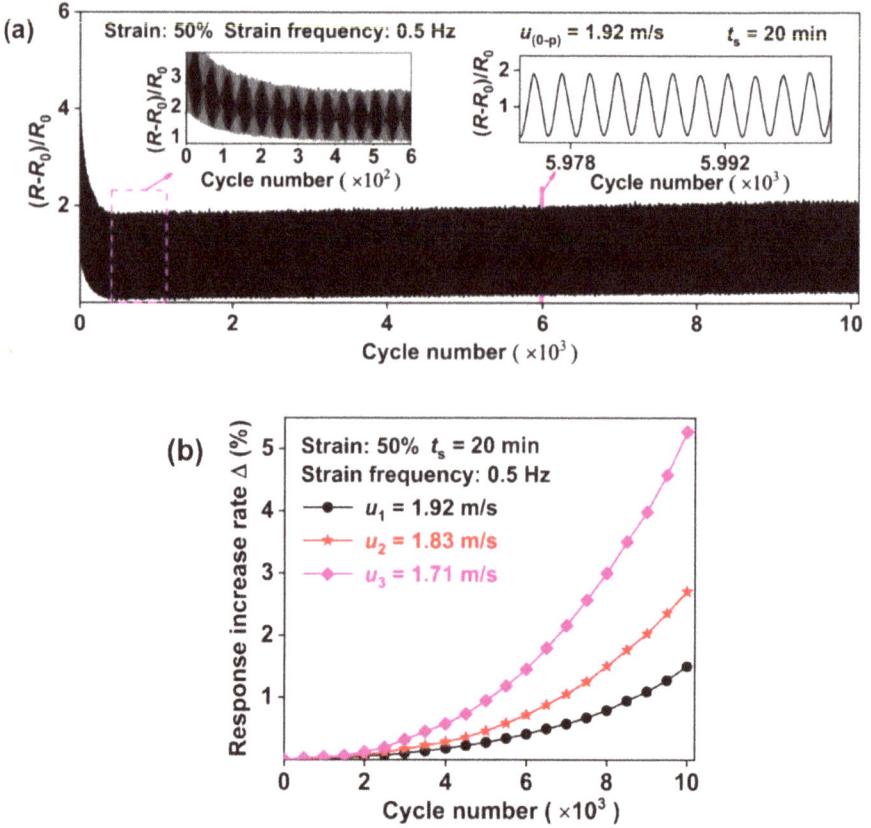

FIGURE 2.30 (a) Relative change in the sensor resistance under repeated loading-unloading of 50% strain with a strain frequency of 0.5 Hz for 10000 cycles. (b) The response increase rate with different vibration velocity (0–p) conditions in a long-term stability test. Reproduced from Ref. [3] with permission from Elsevier.

different vibration velocities (1.71 m s^{-1} (0–p), 1.83 m s^{-1} (0–p) and 1.92 m s^{-1} (0–p)) and the same sonication time (t_s = 20 min). The sensing response increase rate at the nth cyclic test is defined as

$$\Delta = \left(\gamma_n - \gamma_0\right)/\gamma_0, \qquad (2.1)$$

where γ_n is the sensing response at the nth cyclic test, γ_0 is the sensing response value at the beginning of the cyclic test (n = 1) and the measured baseline sensor resistance R_0 at each cycle was used in the calculation. It indicates that the sensor becomes more stable as the vibration velocity increases. This is because the bonding between the MWCTs and PDMS substrate becomes stronger as the vibration velocity increases. Based on the analyses of Figs. 2.27(b) and 2.30(b), it is known that large vibration should be employed in the fabrication of the flexible strain sensor that has to possess a long-term stability, and small vibration may be used for the one-off or disposable sensor to gain a better sensitivity.

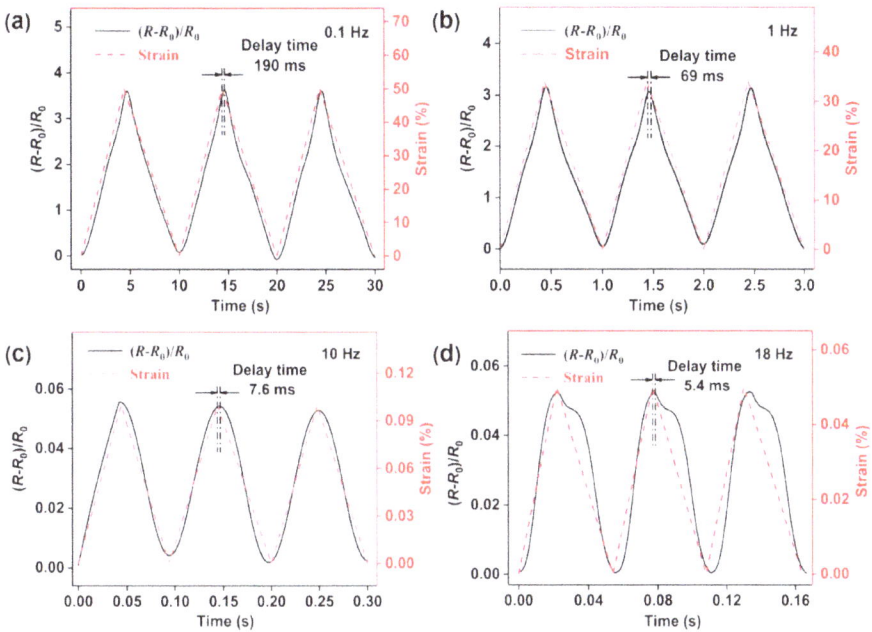

FIGURE 2.31 Measured delay characteristics at different strain frequencies. (a–d) Dynamic responses when the input strain changes with time in a triangular waveform, and the strain frequency is 0.1 Hz (a), 1 Hz (b), 10 Hz (c) and 18 Hz (d), respectively. Reproduced from Ref. [3] with permission from Elsevier.

Figure 2.31(a–d) are the measured dynamic responses when the strain changes with time in a triangular waveform and the frequency of the triangular waveform (strain frequency) is 0.1, 1, 10 and 18 Hz, respectively. Due to the viscoelasticity of the PDMS substrate and the interface between the MWCNT film and substrate, the dynamic sensing response has a delay to the input strain and a slight distortion of the output waveform occurs at larger strain frequency (>10 Hz). At a strain frequency of 0.1 Hz, the time delay is 190 ms and the time delay ratio is 1.9%. At a strain frequency of 18 Hz, the time delay is 5.4 ms and the time delay ratio is 9.7%. The good delay performance is because of a strong bonding between the MWCNT film and PDMS substrate.

In order to demonstrate the potential applications of the fabricated flexible strain sensors, the sensors were assembled onto various locations of the human body and the surface of other objects, and sensing responses to the motions were measured. Figure 2.6(a–d) show the ability of the sensors in the detection of various human motions, such as human joint movement, swallowing process, etc.

To monitor the bending degree of a finger in real-time, it was assembled onto a latex glove, as shown in Fig. 2.32(a). The measured response curves in this figure show that the finger-bending degree can be clearly reflected by the response curve. To monitor the bending of a wrist, the sensor was attached to a wrist guard, as shown in Fig. 2.32(b). Bending and recovery of the wrist can be clearly distinguished and monitored, with a very good repeatability. In order to demonstrate the ability of the sensor to monitor human tiny muscle movement during swallowing, the sensor was installed in the human throat. As shown in Fig. 2.32(c), the sensor can monitor the swallowing

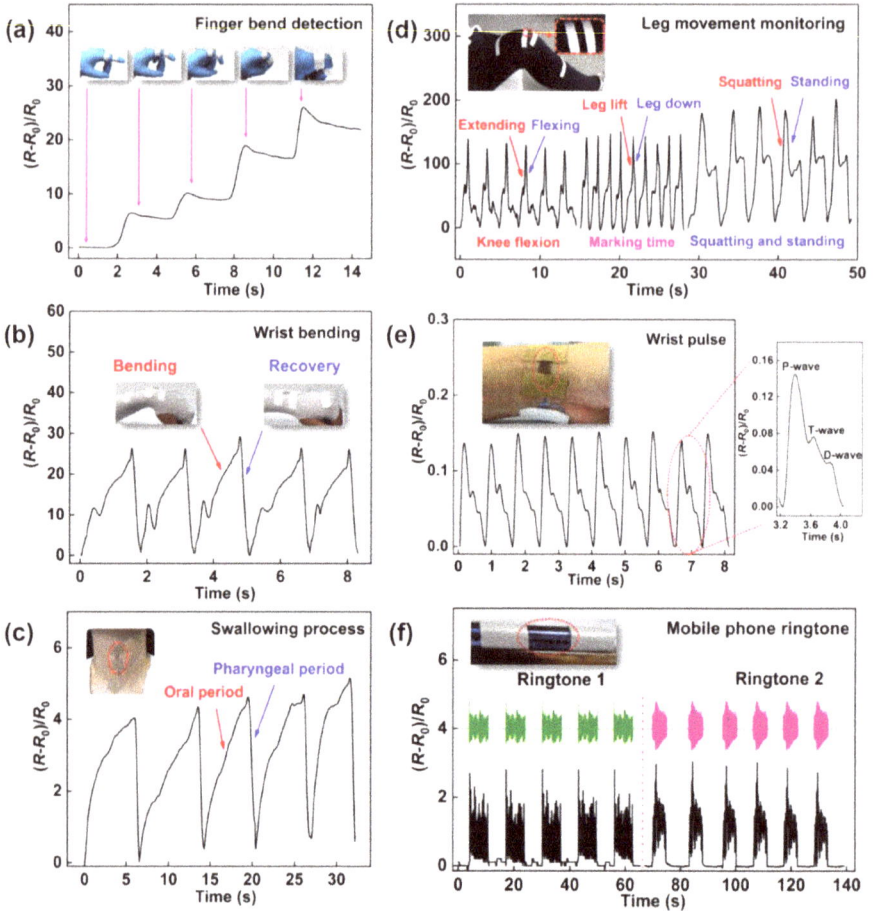

FIGURE 2.32 (a) Response of the sensor assembled on a disposable latex glove, to a finger bending. (b) Response of the sensor assembled on a sleeve, to a wrist bending. (c) Response of the sensor attached to the throat of a person, to the muscle movement during a swallowing action. (d) Responses of the sensor assembled on a stocking, to a knee flexion, marking time and squatting/standing, respectively. (e) Response of the sensor attached to the skin surface above the wrist artery, to a pulse. (f) Responses of the sensor attached on a mobile phone speaker, to two different ringtones. Reproduced from Ref. [3] with permission from Elsevier.

process in real-time. To detect the motions of leg movement, the sensor was integrated into a stocking, as shown in Fig. 2.32(d), which indicates that the sensor has potential application in medical training, home rehabilitation and sports monitoring.

Pulse is a key physiological indicator of the heart rate. Thus, it is of great significance to detect pulse signals by the flexible strain sensor. To demonstrate this capability of the sensor, the sensor was attached to the wrist skin, as shown in Fig. 2.32(e). The recorded dynamic response in the figure shows that this sensor is sensitive enough to give the percussion wave (P-wave), the tidal wave (T-wave) and the diastolic wave (D-wave) of the pulse. The pulse wave in this figure has a frequency of 75 beats min^{-1}, which is the value of a healthy adult in relaxation. To demonstrate the capability of the sensor in the pickup of other forms of subtle vibration, the speaker of a mobile phone was covered by the sensor and the response to two different ringtones was recorded, as shown in Fig. 2.32(f). It is seen that the dynamic response can pick up the main feature of the ringtone.

To address the issue of mass production with the fabrication method proposed in this work, one needs to redesign the tip of the horn of ultrasonic transducer to enlarge the area of the radiation face while keeping its out-of-plane vibration uniform on the radiation surface. In this case, the tip of the horn should be a metal block with sufficient thickness to avoid the occurrence of flexural vibration modes. Enlarging the area of the radiation face can increase the area of the MWCNT film on the PDMS substrate, which improves the fabrication efficiency. Another method to improve the fabrication efficiency is to move the radiation face in the liquid back and forth in the direction parallel to the substrate with a constant speed. The motion can be implemented by linearly driving the ultrasonic transducer assembled onto a motor system.

As the deposition of MWCNTs onto the PDMS substrate is caused by the acoustic cavitation and acoustic streaming near the substrate, it is understandable that the acoustic field parameters including the working frequency, distance between the radiation face and substrate, and vibration velocity of the radiation face, have an effect on the performance of the fabricated sensors. The working frequency of 19.9 kHz was chosen in this work because the power ultrasound was chosen for the investigation of fabrication effect. The distance between the radiation face and substrate was 30 mm in the above-stated experiments, which was a result of experimental searching. A too small gap may cause difficulty in the generation of acoustic streaming, and a too large gap may cause a large acoustic attenuation.

2.4.4 Conclusions

In this work, the author's group proposed a fabrication method for flexible strain sensors, which employs the power-ultrasound induced deposition of MWCNTs in liquid onto PDMS substrate. The fabricated sensor has large stretchability, high sensitivity and excellent stability, and the fabrication process has the merits of simple and low-cost equipment, easy operation and environment-friendliness which include the operation under room temperature and normal pressure, recyclability of the nano material, less organic solvent involved, etc. The analyses show that the fabrication process is based on the utilization of the acoustic cavitation and acoustic streaming field in the MWCNT-water solution, and employing proper vibration strength in the fabrication can improve the sensor performance such as stability, sensitivity, etc. Furthermore, it is experimentally demonstrated that the flexible strain sensors fabricated by the proposed method may be applied in the detection of pulse, swallowing, finger or wrist bending, knee flexion, squatting/standing and acoustic vibration. Thus, this work provides a promising and environment-friendly way to fabricate high-performance flexible strain sensors and other devices.

2.5   ULTRASONIC FABRICATION OF FLEXIBLE GAS SENSORS

The liquid-borne ultrasound can also be employed in the fabrication of flexible nano gas sensors. This section gives an example to demonstrate the detailed fabrication process and sensor performance. In the fabrication process, nano sensing material in solution is coated onto the inner wall of a flexible fine tube to form a gas sensing layer, and the tube with sensing coating is further constructed as a tubular gas sensor. Tubular SnO_2-rGO NH_3 sensors with this structure and fabricated by liquid-borne ultrasound have a measured lower detection limit of 1 ppm, which is lower than the lowest one (5 ppm) of reported NH_3 sensors based on the same sensing material. The sensing response only decreases about 3.6% for 1 ppm NH_3 after 2000 loading-unloading cycles at 40% strain and 0.5 Hz frequency. Apart from these merits, the inner wall sensing structure can protect the sensing layer of a flexible gas sensor from contaminants and touching.

2.5.1   EXPERIMENTAL SECTION

2.5.1.1   Fabrication of the Tubular Gas Sensor

The fabrication process of the tubular gas sensor proposed in this work, based on the liquid-borne ultrasound coating method, is shown in Fig. 2.33(a), and the detailed structure of the ultrasonic coating equipment is shown in Fig. 2.33(b). The fabrication

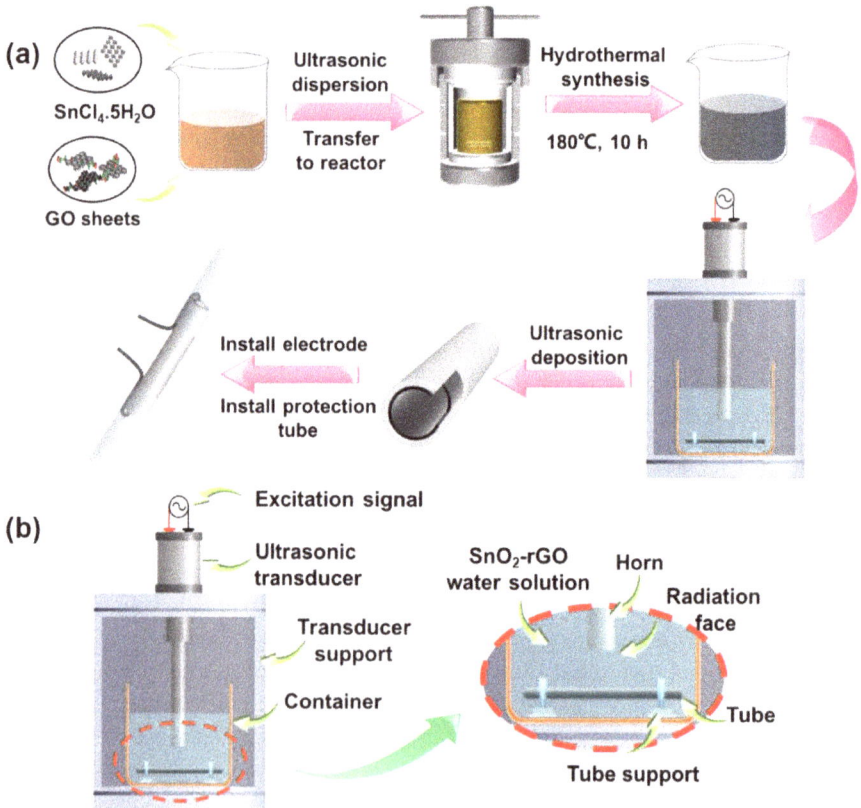

FIGURE 2.33   Schematics to demonstrate the liquid-borne ultrasound-based coating method for the fabrication of a flexible tubular gas sensor at room temperature and normal pressure. (a) Fabrication process of the flexible tubular gas sensor. (b) Experimental setup for the ultrasonic coating. Reproduced from Ref. [4] with permission from Elsevier.

process of the flexible tubular gas sensor shown in Fig. 2.33(a) mainly includes the following steps:

i. Use GO and $SnCl_4 \cdot 5H_2O$ as precursors; ultrasonically disperse 8 mg GO in 80 ml deionized water for 15 min to obtain a brown solution with dispersed GO flakes; add 0.2 g $SnCl_4 \cdot 5H_2O$ into the above solution and disperse again by ultrasound for 10 min.

ii. Transfer the uniformly dispersed brown solution into a 150 ml stainless-steel autoclave with Teflon lining, keep it at 180 °C in an oil bath for 10 h and then naturally cool it down to room temperature. Process the products by centrifugation, wash the products with deionized water several times and then disperse the precipitate in water for further use.

iii. Pour SnO_2-rGO nanohybrid prepared with 0.6g $SnCl_4 \cdot 5H_2O$ into a container, and then add 200 ml deionized water and 200 ml DMF. Disperse them by ultrasound for 10 min to form a uniformly distributed nano solution.

iv. Fix the silicone rubber tube with a support at the bottom of the container and ensure that there is a proper gap between the silicone rubber tube and the container's bottom.

v. Insert the horn of the ultrasonic transducer into the nano solution, with the radiation surface parallel to the container bottom and 37 mm away from the central axis of the tube.

vi. Drive the ultrasonic transducer at 19.88 kHz and 192 $V_{p\text{-}p}$ (in resonance) for 25 min to deposit the SnO_2-rGO on to the surface of the tube.

vii. Take out the coated tube from the nano solution, rinse with a water stream (about 1 m s^{-1}) to flush away the nano material loosely lying on the coating surface and then dry the coating with a hair dryer for the subsequent sensor fabrication process.

viii. Apply the sliver paste onto the inner surface coating at both ends of the tube to form the electrodes, connect lead wires to the electrodes and install a wire protection tube to complete the fabrication. The plane-shaped gas sensor on PDMS film was fabricated by a similar fabrication process.

2.5.1.2 Gas Sensing Measurement

To evaluate the sensing response of the tubular gas sensor to different target gases, sensing characteristics of the tubular gas sensor were tested with a laboratory-made gas test system, as shown in Fig. 2.34. One end of the sensor was connected to the outlet of a peristaltic pump through a flexible tube and the other end to a long tube for exhausting gas, and the flow rate of the carrier gas was controlled by the pump.

The change of the gas sensor's resistance caused by the target gas was measured by an electrochemical workstation in real-time and under 1 V measurement voltage. The response time is the time for the sensor response to reach 90% of the steady response, and the recovery time is that for the sensor response to drop to 90% of the steady response in ambient air. The sensing response γ was obtained by the relative resistance change when exposed to a target gas with controlled concentrations at a constant measurement voltage of 1 V.

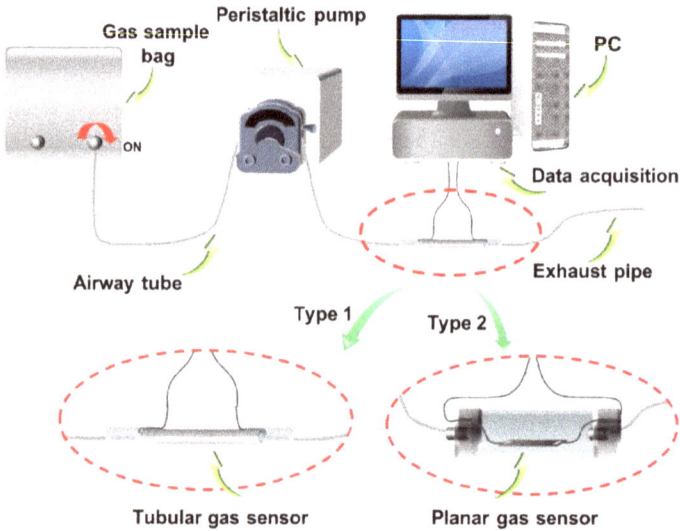

FIGURE 2.34   The test system for the flexible gas sensors fabricated in this work. Reproduced from Ref. [4] with permission from Elsevier.

$$\gamma = \frac{R_a - R_g}{R_g} \times 100\%, \tag{2.2}$$

where R_a and R_g are the resistance of the sensors in the air and target gas, respectively. In the experiments, the ambient temperature and humidity were 18 ± 2 °C and $54 \pm 3\%$ RH, respectively, except for the long-term stability experiments. The ambient temperature and humidity were monitored by a commercialized sensor (Guangzhou Aosong Electronics Co., Ltd, China, AHT20).

2.5.1.3   Characterization

The compositions and crystal structure of the samples were recorded by XRD (PANalytical Malvern) with Cu Kα radiation ($\lambda = 1.54$ Å) in the 2θ range from 10° to 90° at 45 kV and 15 mA. Raman spectra were measured by a J-YT64000 Raman spectrometer from 400 to 3500 cm^{-1} with 514.5 nm wavelength incident laser light. The transmission electron microscopy (TEM) and high-resolution transmission electron microscopy (HRTEM) images of SnO$_2$-rGO nanohybrid were recorded on a TEM system (FEI Tecnai G2 F20, FEI), and the elemental mapping was conducted by the TEM with an energy-dispersive spectroscopy (EDS). The sensing coating morphology of the tubular gas sensor was recorded by FE-SEM (Regulus 8100, HITACHI), and the optical pictures of the tubular gas sensor were measured by a Nikon camera (D7200).

2.5.2   RESULTS AND DISCUSSION

Photos of the sensor samples with a sensing layer on the inner wall, coated by the liquid-borne ultrasound, are shown in Fig. 2.35(a). The samples have different inner

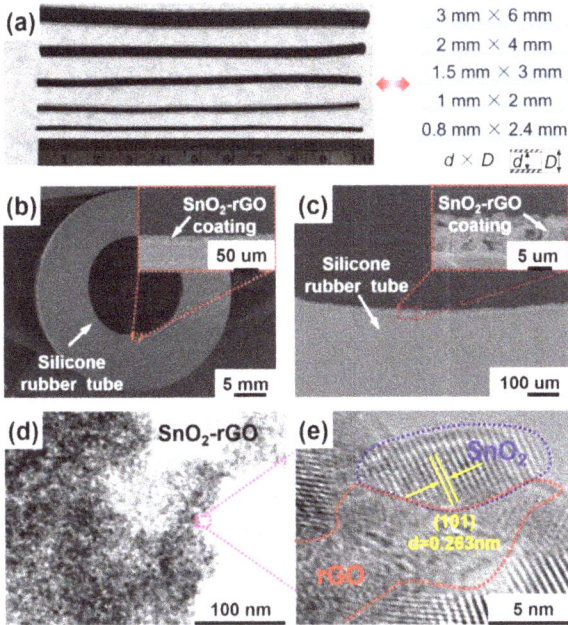

FIGURE 2.35 Morphology and microstructure of the SnO_2-rGO sensing layer. (a) Optical images of the fabricated tubular gas sensors with different inner diameters coated with SnO_2-rGO. (b,c) FESEM images of the cross-section of the SnO_2-rGO sensing layer with different magnifications. (d) TEM image and (e) HRTEM image of the SnO_2-rGO nanohybrid. Reproduced from Ref. [4] with permission from Elsevier.

and outer diameters, from 0.8 to 3 mm. Figure 2.35(b) shows a cross-sectional image of the samples observed by a field emission scan electron microscopy (FESEM). It can be seen that the nano materials were deposited on the inner wall of the tube, forming a coating layer with uniform thickness. The partially enlarged view of the tube cross-section in Fig. 2.35(c) shows that the SnO_2-rGO nanohybrid forms an interconnected network with some roughness on its surface, which is helpful to increase the actual sensing area. From the low-magnification TEM image shown in Fig. 2.35(d), it is seen that the SnO_2 NPs are uniformly and densely distributed on the rGO surface. The HRTEM image of nanocrystals in Fig. 2.35(e) further confirms the existence of SnO_2 NPs (observed D-spacing of (101) plane of SnO_2 = 0.26 nm) and shows that the particle size of SnO_2 NPs in SnO_2-rGO nanohybrid is less than 10 nm. The corrugation of rGO in Fig. 2.35(e) indicates the formation of hybridized structures between the SnO_2 NPs and rGO, which is necessary in gas sensing. After that, to confirm the material properties of the SnO_2-rGO nanohybrid, the prepared nanohybrid was analyzed by XRD, Raman and EDS, respectively. The detailed material properties of the nanohybrid are described in Fig. 2.36.

In order to investigate the mechanism of coating process, the sound pressure distribution inside the tube was computed by the FEM, as shown in Fig. 2.37. Figure 2.37(a, b) are the computed sound pressure fields in the y–z section (passing the tube axis) and x–z

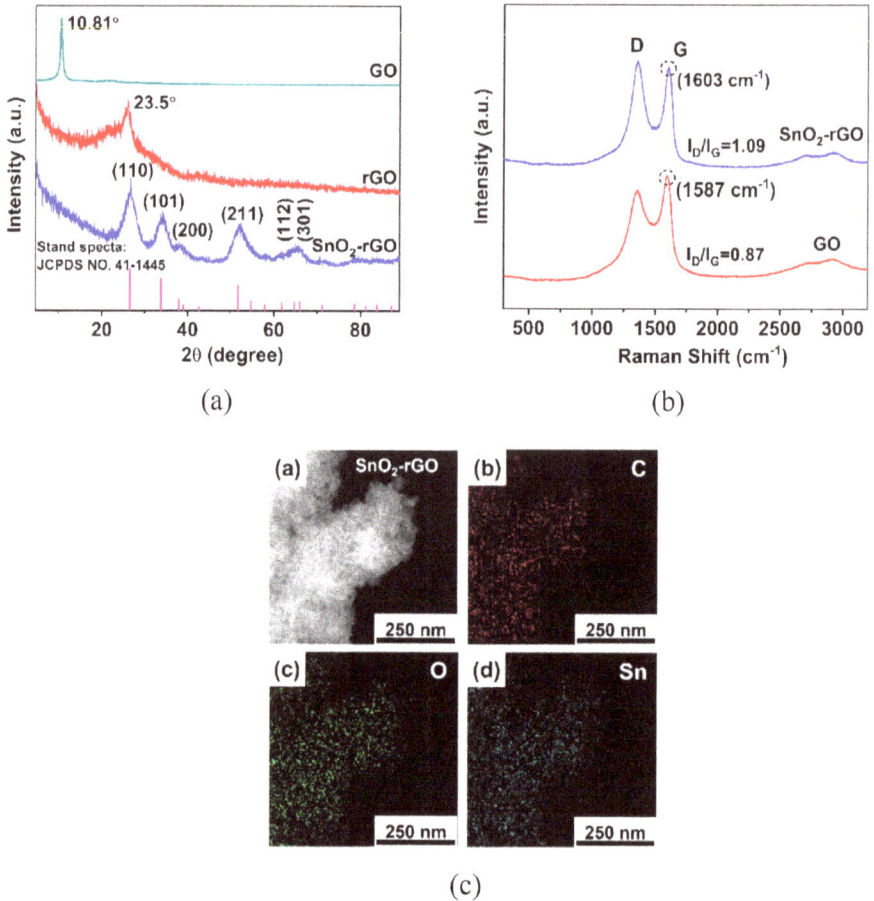

FIGURE 2.36 (a) XRD spectra of GO, rGO and SnO$_2$-rGO nanohybrid. (b) Raman spectra of GO and SnO$_2$-rGO nanohybrid. (c) EDS elemental mapping of the SnO$_2$-rGO nanohybrid. Reproduced from Ref. [4] with permission from Elsevier.

section when the frequency and out-of-plan vibration velocity of the radiation surface are 19.88 kHz and 2.4 m/s (0–p), respectively, which are identical to those used in the experiments. The computation shows that there exists sound pressure inside the tube, which is up to 1.15×10^5 Pa. This indicates that acoustic cavitation can be generated in the sonicated tube. Micro jets generated by the dynamic bubbles in the acoustic cavitation have a very high speed up to several tens m/s, which can eject the SnO$_2$-rGO nano scale materials onto the soft tube and form the SnO$_2$-rGO nanohybrid sensing coating. Figure 2.37(c) shows the sound pressure distribution along the centerline of the +y direction inside the tube, and Fig. 2.37(d) shows the sound pressure distribution along the +z direction inside the tube at different horizontal positions. The results show that the sound pressure distribution in the tube is quite uniform, which is beneficial to the fabrication of a uniform coating layer on the inner wall.

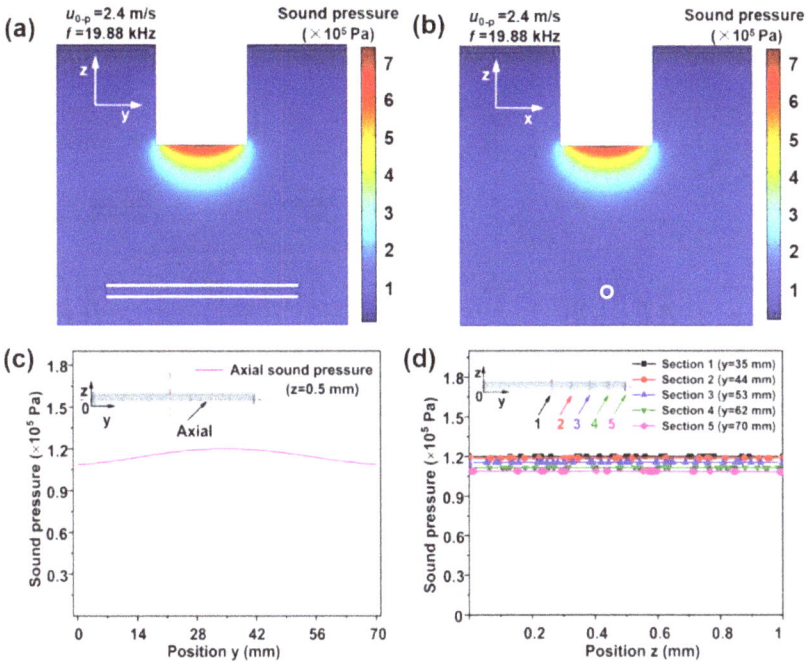

FIGURE 2.37 Computed sound pressure in the working domain of the liquid-borne ultrasound. (a,b) Computed sound pressure field of liquid-borne ultrasound under the horn. (c) Sound pressure distribution along the centerline of the tube. (d) Sound pressure distribution along +z direction in the tube at different cross-sections. In the computation, the ultrasonic frequency and out-of-plane vibration velocity of the radiation surface are 19.88 kHz and 2.4 m/s (0−p), respectively. Reproduced from Ref. [4] with permission from Elsevier.

Figure 2.38(a) exhibits the dynamic response of the gas sensor (inner diameter × length = 1 mm × 70 mm) to different NH_3 concentrations, measured at a carrier gas flow rate of 100 ml min^{-1}. It can be seen that when the sensor is exposed to NH_3, the sensor responds rapidly and its relative resistance changes significantly. When clean air replaces NH_3 gas, the sensor resistance can restore to the initial value with insignificant baseline drift, which indicates that the gas sensor has good reversibility.

Figure 2.38(b) compares the sensing performance of the tubular and planar gas sensors with the same sensing area to different NH_3 concentrations. The tubular and planar gas sensors were fabricated by the same process and with the same sensing materials, and the size of the planar gas sensor was 3.2 mm × 70 mm (length × width). It is seen that the response of the tubular sensor is much higher than that of the planar sensor for the NH_3 concentration range from 1 to 300 ppm. It is also noted that the measured LDL of the tubular gas sensor is 1 ppm and that of the planar gas sensor is only 5 ppm.

Figure 2.38(c) shows a comparison of the measured response and recovery time of the two sensors, which indicates that the tubular gas sensor has a faster response and recovery process. For 5 ppm NH_3, the sensitivity, response time and recovery time of the tubular gas sensor are 345% higher, 28.6% less and 54.4% less than the planar one,

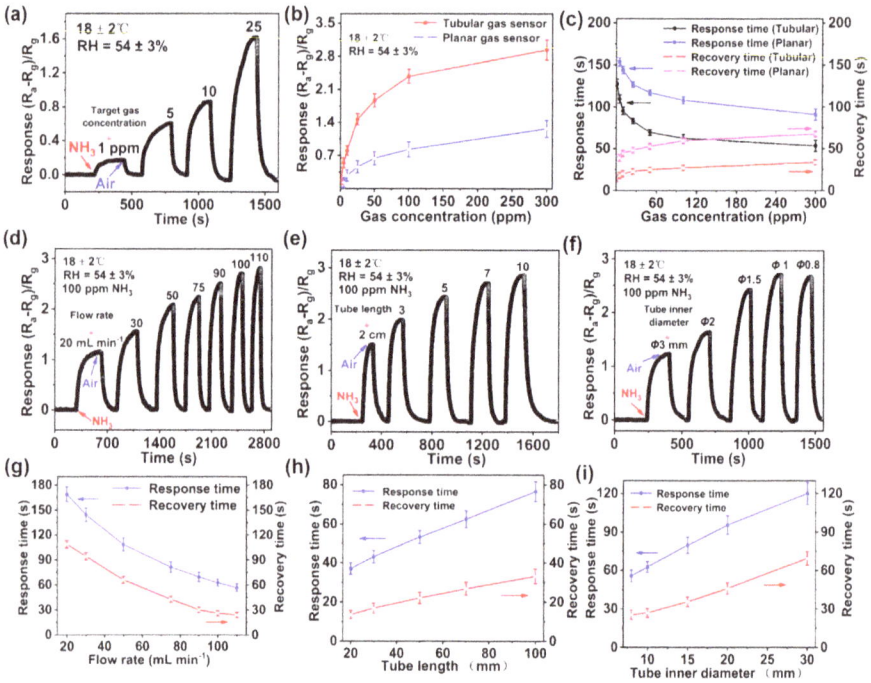

FIGURE 2.38 Sensing characteristics of the room temperature tubular gas sensor for NH$_3$. (a) Dynamic response of the tubular gas sensor to different concentrations of NH$_3$ gas. (b) Response and (c) response-recovery time of the tubular and planar gas sensors with the same sensing area for different NH$_3$ concentrations, respectively. (d) Dynamic responses to 100 ppm NH$_3$, and (g) the response-recovery time of the tubular gas sensor at different flow rates. (e) Responses to 100 ppm NH$_3$ and (h) the response-recovery time of the tubular gas sensor with different lengths. (f) Dynamic response to 100 ppm NH$_3$, and (i) the response-recovery time of the tubular gas sensor with different inner diameters gas. Reproduced from Ref. [4] with permission from Elsevier.

respectively. Thus, the sensing performance of the tubular gas sensor is significantly better than the planar one. This is because coating the sensing materials onto the inner wall of the thin tube increases the utilization rate of the sensing area and more target gas molecules move onto the sensing surface of the thin tube per unit time, which increases the sensing performance. The sensing performance of the tubular NH$_3$ sensor is compared to the recently reported NH$_3$ sensors based on the SnO$_2$-rGO sensing material, as listed in Table 2.6, which shows that the gas sensor in this work has the best LDL (1 ppm). This superiority in sensitivity is attributed to the liquid-ultrasound-based coating method, which makes it possible to coat a sensing layer onto the inner wall of a flexible tube.

Figure 2.38(d, g) show the dynamic response and recovery time of the tubular gas sensor (inner diameter × length = 1 mm × 70 mm) to 100 ppm NH$_3$ at different flow rates, respectively. The response increases with the increase of carrier gas flow, and response or recovery time decreases with the increase of carrier gas flow. The response

TABLE 2.6

Sensing Performance of Room Temperature NH$_3$ Sensors Based on SnO$_2$-rGO and Fabricated by Various Methods

Work	Materials	Synthesis/ Fabrication Method	Substrate	LDL (ppm)	Sensor Response $((R_a-R_g)/R_g)$	$T_{response}$ and $T_{recovery}$ (s)
[13]	SnO-graphene	CVD/drop-casting method	Insulation plate	5	0.21 (50 ppm)	15 and 30
[14]	SnO$_2$-rGO	Hydrothermal/ drop-casting method	Ceramic tube	20	0.3 (200 ppm)	8 and 13
[15]	SnO$_2$-rGO	Hydrothermal/ drop-coating method	Alumina ceramic plate	–	0.25 (50 ppm)	14 and >300
[16]	SnO$_2$-graphene	Hydrothermal/ drop-casting method	Alumina ceramic plate	10	0.159 (50 ppm)	<60 and <60
[17]	SnO$_2$-rGO	Mini-arc reactor	Si wafer	–	0.46 (1% in air)	30 and >300
[18]	SnO$_2$:rGO (8:10)	Hydrothermal/ drop-casting method	Alumina ceramic plate	25	0.4 (25 ppm)	210 and 150
[19]	SnO$_2$-rGO@Au	Radio-frequency magnetron sputtering	Si wafer	5	0.58 (10 ppm)	20 and 41
This work	SnO$_2$-rGO	Hydrothermal/ liquid-borne ultrasound deposition	Silicone rubber tube	1	0.16 (1 ppm) 0.79 (10 ppm)	127 and 14 95 and 19

increase and response time decrease are because of a faster replacement of target gas near the sensing surface as the flow rate of target gas increases. The recovery time decreases because as the clean air flows faster, the NH$_3$ molecules adsorbed on the sensing surface leave more rapidly.

Figure 2.38(e, h) show the dynamic responses to 100 ppm NH$_3$ gas and the recovery time of the tubular gas sensor (inner diameter = 1 mm; carrier gas flow rate: 100 ml min^{-1}) with different lengths. The results show that the response, and response and recovery time decrease as the length of the tubular gas sensor becomes shorter. As the sensor becomes shorter, some NH$_3$ gas molecules pass through the tube without diffusing to the sensing interface, which results in a smaller response. The shorter response and recovery processes are because of less reaction sites at a shorter length.

Figure 2.38(f, i) show the dynamic response to 100 ppm NH$_3$ gas, the response and recovery time of the tubular gas sensor (tube length = 70 mm; carrier gas flow rate = 100 ml min^{-1}) with different inner diameters. It is seen that a thinner sensor has a larger response and faster response/recovery process. As the inner diameter decreases, the gas flow in the tube becomes faster, which is beneficial to the gas refreshing on the sensing surface. This explains that the sensing response increases and the response/recovery process becomes faster as the inner diameter decreases.

From Fig. 2.38(f), it is also seen that the tubular sensors with inner diameters of 0.8 and 1.0 mm have close sensing response, which means that the benefit of larger sensing response achieved by reducing the inner diameter is limited when the inner

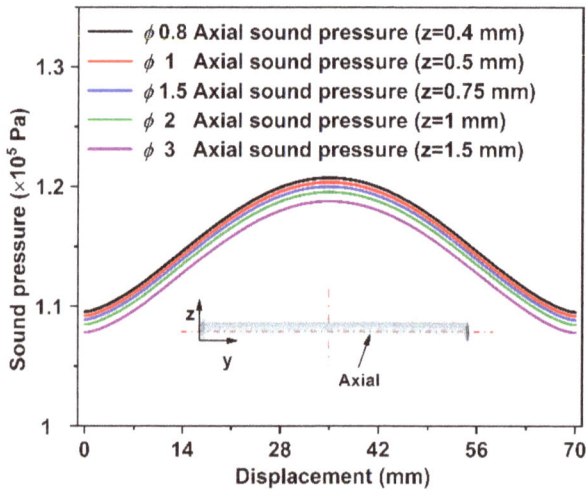

FIGURE 2.39   Sound pressure distribution along the centerline of the tubes with different inner diameters. Reproduced from Ref. [4] with permission from Elsevier.

diameter is too small. The reason for this phenomenon was investigated in this work. As a too weak sound pressure may cause a weak acoustic cavitation effect and a bad coating quality, the sound pressure distributions inside the tubes with different inner diameters were computed by the FEM, and the results are shown in Fig. 2.39. It is found that as the inner diameter decreases, the sound pressure inside the tube actually increases a bit. Thus, the change of sound pressure caused by the inner diameter change cannot be the reason.

As another possible reason, the sluggish renewal of nano solution inside the tube, which is caused by large flow resistance of a thin tube, might worsen the quality of the sensing layer. To confirm this reason, the coating fabrication equipment was improved, as shown in Fig. 2.40(a). In the improved equipment, the tube to be coated and an external nano solution reservoir are connected in series with a micro pump via flexible pipes, and the renewal of nano solution in the coated tube is enhanced by the flow driven by the micro pump. The flow rate of the nano solution is controlled by the micro pump, and the ultrasonic conditions used in the sensing coating fabrication process are the same as those for Fig. 2.38(f). Figure 2.40(b) shows the measured results of the dynamic sensing response of the tubular gas sensor fabricated by the improved equipment, to 100 ppm NH_3. It can be seen that the response value and response/recovery speed can be improved by enhanced nano solution renewal, which confirms that a sluggish renewal of nano solution inside the tube might worsen the quality of the coating. However, Fig. 2.40(b) also shows that the sensing performance increase caused by a faster nano solution flow is quite limited. This is because as the flow rate increases, some of the acoustic bubbles in the sensor tube do not have sufficient time to develop to the collapsing phase.

Figure 2.41(a) shows the response of the fabricated tubular gas sensor to various gases, including NH_3, ethanol, acetone and some typical volatile organic compounds

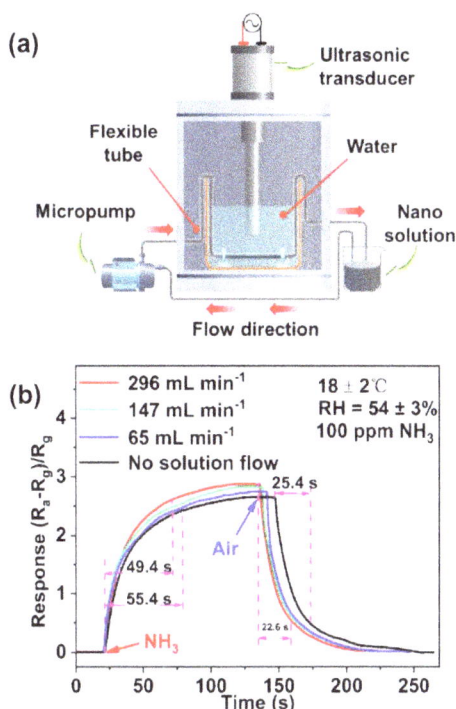

FIGURE 2.40 Improved experimental setup for the ultrasonic coating and measured dynamic responses. (a) Experimental setup with recycling nano solution. (b) Dynamic response-recovery processes with different flow rates of nano solution for 100 ppm NH_3 gas. Reproduced from Ref. [4] with permission from Elsevier.

(VOCs). It is seen that the response to NH_3 is much higher than that to other gases, which means that the sensor has a very good selectivity. This results from NH_3's low adsorption energy (25 meV) on rGO and strong H-bonding with oxygen atoms in rGO. Figure 2.41(b) shows the stability of the tubular gas sensor for NH_3 target gas of 100 ppm in 30 days. The ambient temperature and relative humidity in the measurement were $20 \pm 4°C$ and 48–86%, respectively. After each measurement, the sensor was stored in a box in the laboratory without any special protection. It shows that the sensor has very good stability, and its response remains at 2.6 ± 0.3 during the 30 days. A comparison between the sensing response fluctuation and relative humidity indicates that the small increase of sensing response is caused by the increase of the relative humidity, which results from more NH_4^+ ions generated by the ionization of $NH_3 \cdot H_2O$ at a higher relative humidity, as more NH_4^+ ions on the rGO surface may decrease the heterojunction thickness in the SnO_2-rGO nanohybrid.

Figure 2.42(a) shows a measured result of stability of the sensing layer for mechanical stretching. In the measurement, repeated loading-unloading of 40% strain was applied to the sample with a frequency of 0.5 Hz for 2000 consecutive cycles, and the sensor resistance in the process was recorded. It shows that the flexible tubular gas sensor's resistance changes little in the experiments with loading-unloading process,

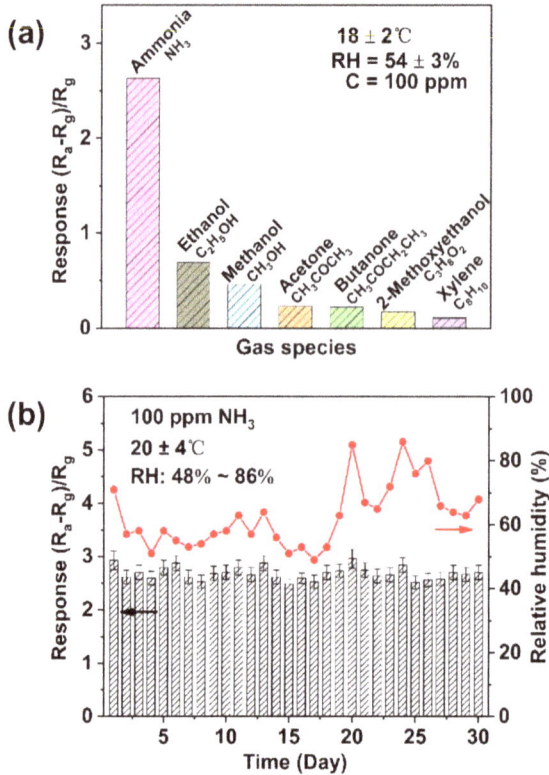

FIGURE 2.41 Sensing characteristics of the flexible tubular gas sensor at room tempera-ture. (a) Selectivity of the tubular gas sensor to different 100 ppm target gases. (b) Long-term stability of the tubular gas sensor for 100 ppm NH$_3$ gas. Reproduced from Ref. [4] with per-mission from Elsevier.

indicating that the SnO$_2$-rGO coating layer has very good mechanical stability for the tensile process. Figure 2.42(b) shows the dynamic response-recovery curves of the tubular gas sensor to 1 ppm and 100 ppm NH$_3$ after 0, 200 and 2000 loading-unloading cycles, respectively. The sensing response of the tubular gas sensor remained stable after 200 loading-unloading cycles and only decreased by 3.6% and 4.2% after 2000 loading-unloading cycles for 1 and 100 ppm NH$_3$, respectively. These results indicate that the ultrasonic coating method may be used to fabricate a sensing layer on the flex-ible substrate with very high mechanical stability.

Apart from the above-described sensing performance enhancement, the liquid-borne ultrasound-based coating method also has the following features: (i) It works at room temperature and normal pressure, which makes the operation of the coating equipment easy and energy effective; (ii) only a small amount of dispersant needs to be used for the nano solution as ultrasound itself has dispersing function; (iii) the nano materials can be deposited on a flexible substrate in one step, which makes the fabrication processes such as heating, drying, and coating transfer completely not necessary; (iv) it can be applied to the coating process on a flexible substrate with

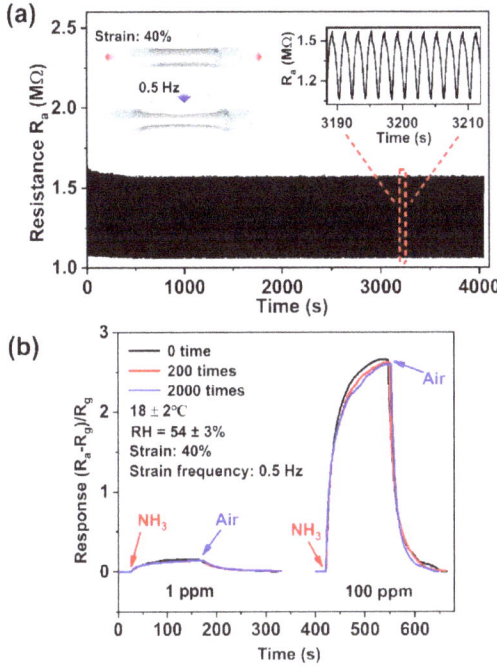

FIGURE 2.42 Mechanical stability of the flexible tubular gas sensor. (a) Resistance changes of the tubular gas sensor under repeated loading-unloading of 40% strain with a frequency of 0.5 Hz for 2000 cycles. (b) Dynamic response-recovery curves of the tubular gas sensor to 1 ppm and 100 ppm NH_3 gas after 0, 200 and 2000 loading-unloading cycles. Reproduced from Ref. [4] with permission from Elsevier.

curved surface, such as an arc surface and inner/outer surface of a cylinder, which cannot be achieved by most of the existing coating techniques; (v) the remaining nano materials in the solution can be recycled; (vi) the coating equipment has a simple structure and very low cost.

2.5.3 CONCLUSIONS

In this work, a tubular structure with a gas sensing layer on the inner wall was proposed to construct highly sensitive and stable flexible gas sensors and the liquid-borne ultrasound was utilized to coat nano sensing materials onto the inner wall. The SnO_2-rGO NH_3 sensors with the tubular structure and fabricated by liquid-borne ultrasound are highly sensitive (1 ppm LDL) and have very good stability for mechanical stretching (3.6% sensing response decrease for 1 ppm NH_3 after 2000 loading-unloading cycles at 40% strain and 0.5 Hz). Apart from those merits, the inner wall sensing structure can protect the sensing layer of a flexible gas sensor from ambient contaminants and mindless touching in practical applications. The superior sensing performance is ascribed to the tubular sensor structure with its sensing layer on the inner wall and the liquid-borne ultrasound coating method. The liquid-borne ultrasound coating method in this work has not only diversified topological structure of the flexible gas sensors that can be fabricated at room temperature and normal pressure, but also provided an environment-friendly and cost-effective way to fabricate high-performance flexible gas sensors.

2.6 REMARKS

In the examples in this chapter, the ultrasonic micro roller and knife are described, and it is demonstrated that single NWs or nanoparticles of malleable materials such as silver on a solid substrate can be rolled into a piece of nanosheet and single NWs of malleable materials on a solid substrate can be cut. They have potential applications in nano sensor fabrication, nano sample treatment, micro/nano assembly, etc., and will enhance the capability of nano fabrication. The strong enough vibration velocity and proper vibration locus at the tip of the micro fabrication tool are two key points to realize the rolling and cutting functions. Although fundamental structures of the ultrasonic micro roller and knife have been given by the two examples, the fabrication performance can be further improved by the optimization of the shape, size and vibration locus of the micro fabrication tool, working frequency of the ultrasonic transducer, and the VTN.

A green coating technology based on power ultrasound in liquid is also demonstrated in this chapter. The technology can be applied not only in the coating process of flexible nano sensors as shown, but also in that of other flexible nano electronic devices such as flexible nano electrodes. In addition to its green features, which include room temperature and normal pressure operation, recyclable nano solution, low equipment cost, no generation of nano pollutant aerosols and no need to use large amount of organic solvent, it can also provide a very strong bond between the flexible substrate and nano functional material, which makes the device stable and reliable. The strong bond is caused by the high-speed jet generated by the acoustic cavitation and may be raised further by strengthening the ultrasonic cavitation effect.

The main challenge in this green nano fabrication technology is to scale up the area of the nano material film deposited onto the substrate. This is because after scaling up the radiation surface, one faces the problem of non-uniformity acoustic field in between the flexible substrate and radiation surface, which may cause a non-uniform thickness and density of the deposited film. The possible solutions to this problem are to design the radiation head's vibration mode properly or to move the water tank or ultrasonic transducer back and forth with a constant speed.

REFERENCES

1. X. Wang and J. Hu, "A flexible ultrasonic micro tool based AgNS fabrication process," *Appl. Nanosci.*, 8(6), pp. 1579–1586, 2018.
2. X. Wang and J. Hu, "Nanowire cutting by an ultrasonically vibrating micro tool," *Precis. Eng.*, 48, pp. 152–157, 2017.
3. H. Xue and J. Hu, "A liquid power-ultrasound based green fabrication process for flexible strain sensors at room temperature and normal pressure," *Sens. Actuator A Phys.*, 329, p. 112822, 2021.

4. H. Xue and J. Hu, "A tubular flexible gas sensor fabricated by liquid-borne ultrasound for higher sensitivity and stability," *Sens. Actuators B Chem.*, 379, p. 133281, 2023.

5. W. T. Thomson, *Theory of Vibration With Applications*, *3rd ed.* (Prentice-Hall, New Jersey, 1993).

6. S. Bae, Y. Lee, B. K. Sharma, H. Lee, J. Kim and J. Ahn, "Graphene-based transparent strain sensor," *Carbon*, 51, pp. 236–242, 2013.

7. T. Yamada, Y. Hayamizu, Y. Yamamoto, Y. Yomogida, A. Izadi-Najafabadi, D. N. Futaba and K. Hata, "A stretchable carbon nanotube strain sensor for human-motion detection," *Nat. Nanotechnol.*, 6, pp. 296–301, 2011.

8. B. Liang, Z. Zhang, W. Chen, D. Lu, L. Yang, R. Yang, H. Zhu, Z. Tang and X. Gui, "Direct patterning of carbon nanotube via stamp contact printing process for stretchable and sensitive sensing devices," *Nanomicro Lett.*, 11, p. 92, 2019.

9. T. Yang, X. Li, X. Jiang, S. Lin, J. Lao, J. Shi, Z. Zhen, Z. Li and H. Zhu, "Structural engineering of gold thin films with channel cracks for highly sensitive strain sensing," *Mater. Horiz.*, 3, pp. 248–55, 2016.

10. Z. Chen, Z. Wang, X. Li, Y. Lin, N. Luo, M. Long, N. Zhao and J. Xu, "Flexible piezoelectric-induced pressure sensors for static measurements based on nanowires/graphene heterostructures," *ACS Nano*, 11, pp. 4507–4513, 2017.

11. Z. Wang, S. Wang, J. Zeng, X. Ren, A. J. Y. Chee, S. Yiu, W. C. Chung, Y. Yang, A. C. H. Yu, R. C. Roberts, A. C. O. Tsang, K. W. Chow and P. K. L. Chan, "High sensitivity, wearable, piezoresistive pressure sensors based on irregular microhump structures and its applications in body motion sensing," *Small*, 12, pp. 3827–3836, 2016.

12. S. J. Kang, C. Kocabas, T. Ozel, M. Shim, N. Pimparkar, M. A. Alam, S. V. Rotkin and J. A. Rogers, "High-performance electronics using dense, perfectly aligned arrays of single-walled carbon nanotubes," *Nat. Nanotechnol.*, 2, pp. 230–236, 2007.

13. R. Kumar, N. Kushwaha and J. Mittal, "Superior, rapid and reversible sensing activity of graphene-SnO hybrid film for low concentration of ammonia at room temperature," *Sens. Actuators B Chem.*, 244, pp. 243–251, 2017.

14. Y. Chen, W. Zhang and Q. Wu, "A highly sensitive room-temperature sensing material for NH_3: SnO_2-nanorods coupled by rGO," *Sens. Actuators B Chem.*, 242, pp. 1216–1226, 2017.

15. Q. Feng, X. Li and J. Wang, "Percolation effect of reduced graphene oxide (rGO) on ammonia sensing of rGO-SnO_2 composite based sensor," *Sens. Actuators B Chem.*, 243, pp. 1115–1126, 2017.

16. Q. Lin, Y. Li and M. Yang, "Tin oxide/graphene composite fabricated via a hydrothermal method for gas sensors working at room temperature," *Sens. Actuators B Chem.*, 173, pp. 139–147, 2012.

17. S. Mao, S. Cui, G. Lu, K. Yu, Z. Wen and J. Chen, Tuning gas-sensing properties of reduced graphene oxide using tin oxide nanocrystals. *J. Mater. Chem.*, 22(22), 2012.

18. R. Ghosh, A. K. Nayak, S. Santra, D. Pradhan and P. K. Guha, "Enhanced ammonia sensing at room temperature with reduced graphene oxide/tin oxide hybrid films," *RSC Adv.*, 5(62), pp. 50165–50173, 2015.

19. R. Peng, Y. Li, T. Liu, Q. Sun, P. Si, L. Zhang and L. Ci, "Reduced graphene oxide/SnO_2@Au heterostructure for enhanced ammonia gas sensing," *Chem. Phys. Lett.*, 737, p. 136829, 2019.

3 Ultrasonic Concentration

Controlled concentration of nano scale samples has potential applications in high-sensitivity sensing of biological substances, crystal growth, culture of artificial tissues, separation and filtering process, etc. Although the ultrasonic standing-wave field, in which micro scale objects are pushed to the nodes or anti-nodes of sound pressure by the acoustic radiation force [1, 2], provides an effective way to concentrate micro scale samples [3–6], it cannot be used to concentrate nano scale objects. To meet this challenge, the author's group once proposed and developed a method, in which ultrasonic field in a droplet at the center of an ultrasonic stage was used to concentrate nano scale samples in the droplet [7]. In many practical applications, it is desired that the nano scale samples in aqueous suspension are concentrated on a common substrate such as a glass slide or silicon substrate, which is not in vibration. This chapter describes three techniques to ultrasonically concentrate nano scale samples on such a substrate [8–11].

3.1 LINEAR CONCENTRATION OF MICRO/NANO SCALE SAMPLES

This section demonstrates a method to concentrate or enrich micro/nano scale samples in a film of aqueous suspension or nano solution on a stationary substrate (not in vibration), which employs the acoustic streaming generated by an ultrasonically vibrating needle parallel to and above the stationary substrate surface [8]. Concentrated yeast particles with a diameter of 4–6 µm may form a series of lobed zones on the stationary substrate if the position of the vibrating needle has no change during the sonication, and form a continuous linear line if the vibrating needle is moved back and forth along a linear trajectory. Smaller objects such as AgNWs with a length of 30–40 µm and diameter of 300 nm and ZnO particles with a diameter of about 1 µm can also be concentrated by the method. For the yeast and ZnO micro particles, boundaries of the concentration zones are very distinct.

3.1.1 EXPERIMENTAL SETUP, PHENOMENA AND PRINCIPLE

Figure 3.1 shows the experimental setup for concentration or enrichment of yeast particles under an ultrasonically vibrating needle. A layer of aqueous suspension film with a thickness of about 2.5 mm is dispersed on a silicon substrate, and a stainless steel needle is inserted into the suspension film horizontally, as shown in Fig. 3.1(a). The suspension is formed by deionized water with dispersed yeast particles. During the concentration process, the silicon substrate is stationary (not in vibration), and the stainless steel needle vibrates in a parallel direction to the substrate. The suspension film and lobed patterns were observed by a microscope (VHX-1000E, Keyence).

DOI: 10.1201/9781003404705-3

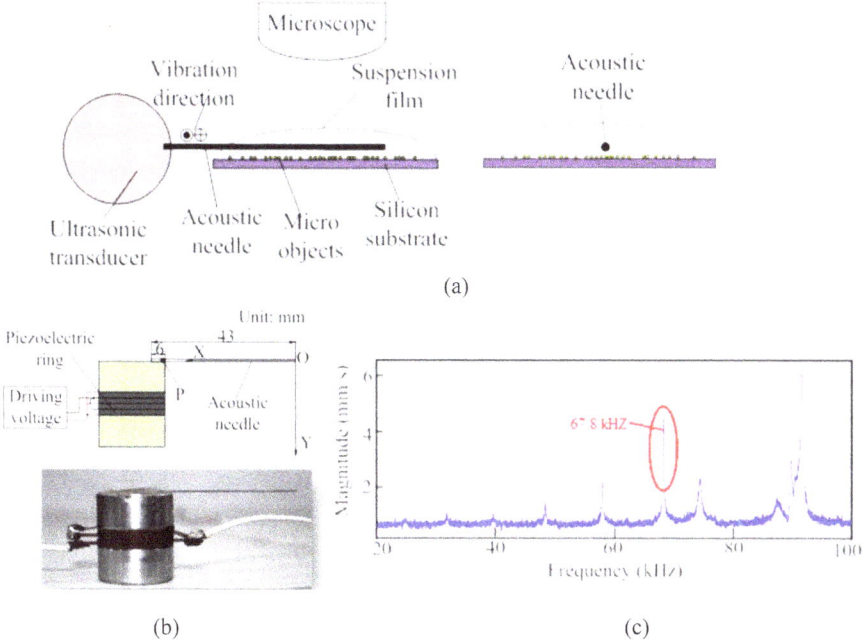

FIGURE 3.1 Experimental setup for generating the lobed concentration patterns in aqueous suspension film of yeast particles on a silicon substrate. (a) Schematic diagram. (b) Construction of the ultrasonic transducer. (c) Vibration magnitude versus operating frequency of the needle. Reproduced from Ref. [8] with permission from Elsevier.

The device used in our experiments is composed of a piezoelectric transducer, and the stainless steel needle is bonded to the radiation surface of the piezoelectric transducer with a resonance frequency of 74.5 kHz, as shown in Fig. 3.1(b). In this piezoelectric transducer, four piezoelectric rings are aligned and pressed together by two cylindrical aluminum covers with a bolting structure, with the poling directions and electrode configuration shown in Fig. 3.1(b). The outer diameter, inner diameter and thickness of each piezoelectric ring are 20 mm, 6 mm and 1 mm, respectively. The electromechanical quality factor Q_m, piezoelectric coefficient d_{33} and relative dielectric constant $\varepsilon_{33}^T/\varepsilon_0$ of the piezoelectric ring are 2000, 325×10^{-12} m/v and 1450, respectively. Each cylindrical aluminum block at the two ends of the transducer has a diameter of 20 mm and a thickness of 10 mm. The stainless steel needle has a total length of 45 mm and a uniform diameter of 0.35 mm. The length of the needle bonded onto the piezoelectric transducer is 7 mm. The driving voltage is sinusoidal, and the transducer works at resonance frequency of the needle (=67.8 kHz), as shown in Fig. 3.1(c). The piezoelectric transducer shown in Fig. 3.1(b) utilizes the vibration of piezoelectric stack to excite a flexural vibration mode in the needle in the xy plane; thus in Fig. 3.1(a), the needle vibration is perpendicular to the page and parallel to the suspension film.

Household baking yeast particles are the main samples used in the experimental aqueous suspension. Yeast particle concentration is 0.048 mg/ml, and the diameter of

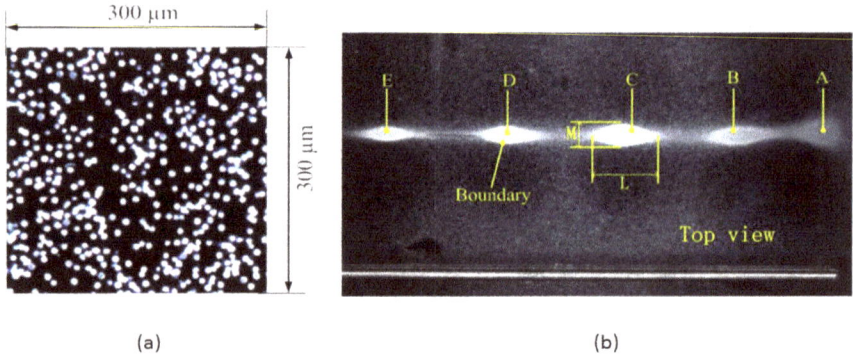

(a)   (b)

FIGURE 3.2   (a) Aqueous suspension of yeast particles used in the experiments. The real size of the image is 300 µm × 300 µm. (b) Lobed patterns in aqueous suspension film of yeast particles on a silicon substrate, generated by ultrasonic vibration of the fine stainless needle. The vibration displacement of the needle at point P (Fig. 3.1(b)) is 180 nm. The separation between the needle and substrate is 1.5 mm, and the needle length inserted into the water is 20 mm. Reproduced from Ref. [8] with permission from Elsevier.

yeast particles is 4–6 µm. Figure 3.2(a) shows the yeast particles dispersed in the suspension before sonication. The average number of yeast particles in a 300 µm × 300 µm square is about 400 with a standard deviation of 27. Figure 3.2(b) shows the observed concentration pattern in an aqueous suspension film of yeast particles on a silicon substrate under the ultrasonically vibrating needle with a 180 nm vibration amplitude at point P (see Fig. 3.1(b)) at 67.8 kHz. In the experiment, the separation between the needle and substrate is 1.5 mm. In the pattern, there are four lobed concentration zones which are termed lobes B,C,D and E, and an end zone A which is like the end of a match stick. In experiments, it takes 2–3 min to form the patterns after the sonication onset, and after this time duration, length L and width M of the patterns become stable as well as the yeast particle concentration beyond the lobed and end zones. The pattern is two-dimensional and symmetric about the axis of the needle.

Figure 3.3 gives the zoom-in images of the suspension around the boundary of zone A in the stable state at different vibration displacements. Figure 3.3(a) shows the location where the images are taken, and Fig. 3.3(b) gives images b1–b6 at the vibration displacement of 80, 100, 120, 140, 160 and 180 nm, respectively. It shows that the boundary becomes quite clear as the vibration increases. The size of the squares in Fig. 3.3(b) is 300 µm × 300 µm. Our experiments in this work also show that whether the concentration process can occur under the vibration needle depends on the distance between the needle and the substrate. When this distance is less than 0.5 mm, micro particles can hardly concentrate under the vibrating needle. Actually, in this case, a series of clear water zones, as shown in Fig. 3.4, can be observed. This acoustic pattern was reported and discussed in Ref. [12].

The principles of the above phenomena are analyzed as follows. The y-directional vibration distribution of the needle along the x-axis was calculated by ANSYS finite element method (FEM) software, and the result is shown in Fig. 3.5. The material

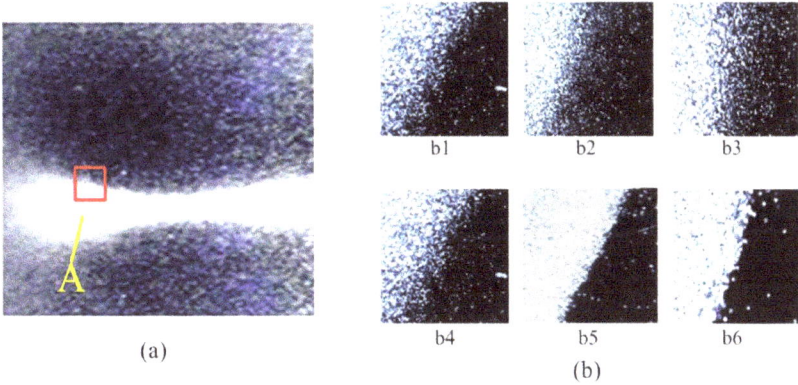

FIGURE 3.3 Images of the boundary of lobe A at different vibration amplitudes. The vibration amplitude at point P is 80, 100, 120, 140,160 and 180 nm for b1, b2, b3, b4, b5 and b6, respectively. The measurement area is the red square in (a), which has a size of 300 μm × 300 μm. Reproduced from Ref. [8] with permission from Elsevier.

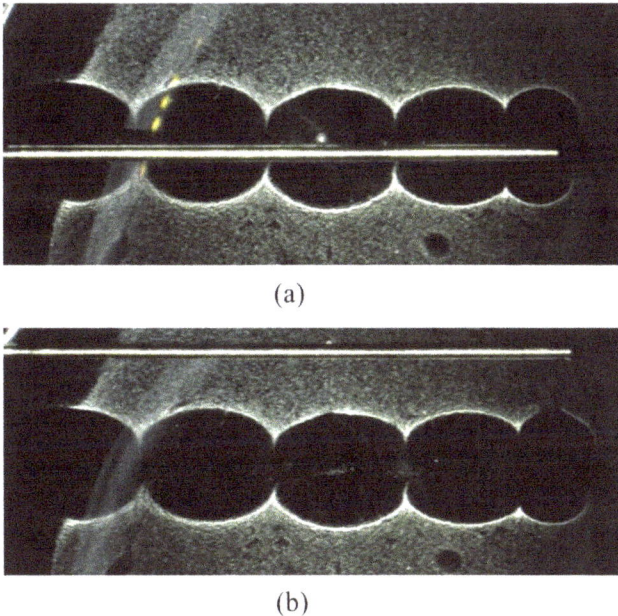

FIGURE 3.4 Five clear water zones in aqueous suspension film of yeast particles on a silicon substrate, generated by ultrasonic vibration of the needle. The vibration displacement of the needle at point P (see Fig. 3.1(b)) is 180 nm. The needle is in contact with the substrate, and the needle length inserted into the water is 20 mm. Reproduced from Ref. [8] with permission from Elsevier.

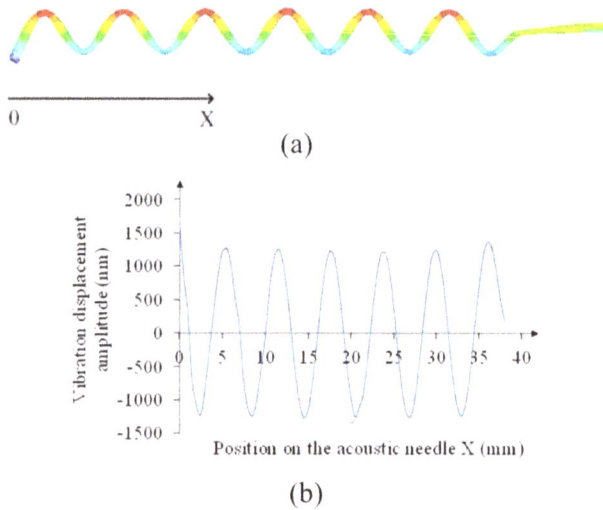

FIGURE 3.5 Distribution of vibration displacement of the acoustic needle along the x direction. (a) Mode pattern. (b) Vibration displacement amplitude versus x. Reproduced from Ref. [8] with permission from Elsevier.

constants used in the calculation are listed in Table 3.1. It shows that the needle vibrates in a flexural mode. Figure 3.6 shows a comparison between the concentration pattern and the y-directional vibration distribution of the needle measured by laser Doppler vibrometer (POLYTEC PSV-300 F). It indicates that the lobes' centers are at the locations right under the vibration anti-nodes (the location with a maximum vibration displacement) of the needle.

Acoustic streaming [13, 14] around the vibrating needle was observed by using the yeast particles and smaller objects such as silver nanowires (AgNWs) and 1 μm-diameter ZnO particles under the microscope. Based on our observation, it is known that when the distance between the vibrating needle and substrate is large enough, acoustic streaming eddies on the two sides of the needle can be generated by the vibrating needle. Also, it is observed that the streaming is the strongest under the vibration antinodes of the needle and the weakest under the vibration nodes. Based on the analyses in Section 3.4 of this chapter, it is known that the acoustic streaming is responsible for the above concentration phenomena.

TABLE 3.1

Material Constants of the Stainless Steel Needle

Material	Density	Poisson's Ratio	Modulus of Elasticity	Constant Material Damping Coefficient
0C$_r$18N$_i$9	7930 kg/m^3	0.3	1.7×10^{11} Pa	0.003

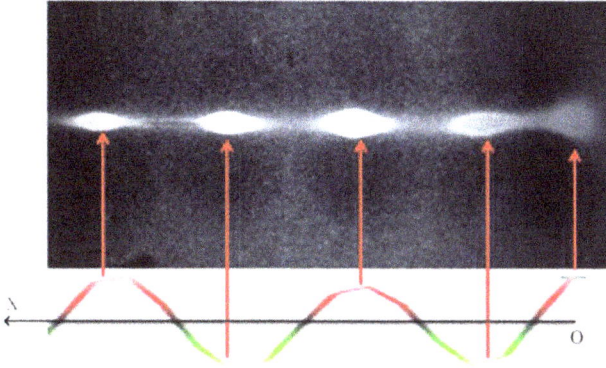

FIGURE 3.6 Comparison of the measured distributions of needle vibration and lobed patterns. Reproduced from Ref. [8] with permission from Elsevier.

3.1.2 CHARACTERISTICS AND DISCUSSION

In the following results, unless otherwise specified, the experimental conditions are as follows: The characteristics are measured at the stable state; the separation between the needle and substrate is 1.5 mm; the vibration displacement (0–p) is measured at point P, located at the root of the needle; the driving voltage is sinusoidal; a 300 μm × 300 μm square area is used to measure the concentration change of micro particles.

Length L and width M of the concentration zones (see Fig. 3.2(b)) in the steady state were measured for different vibration amplitude, and the results are shown in Fig. 3.7. It is seen that with the increase of the vibration, L and M increase first and then decreases. The increase of the dimensions is because of the increase of the acoustic streaming velocity. The decrease of the dimensions is because when the acoustic streaming velocity is too large, the driven micro particles may overlap each other.

The ratios of average particle concentration beyond zones A, B, C, D and E to that before sonication ($Cn/C0$, $n = A$, B, C, D and E) were measured for different vibration amplitude, and the result is shown in Fig. 3.8. Here, $Cn/C0$ was calculated by the ratio of average number of yeast particles in a square area of 300 μm × 300 μm after and before the sonication. The yeast particle number was counted after the pattern became stable. It is seen that the areas beyond the concentration zones become clear as the vibration increases. This indicates that more micro particles are flushed into the concentration zones as the vibration increases.

Other samples in aqueous suspension can also be concentrated by the above method. Figure 3.9(a) shows the concentration of AgNW with a length of 30–40 μm and diameter of 300 nm in a film of water suspension (50 mg/L) on the substrate surface, and Fig. 3.9(b) shows the concentration of 1 μm-diameter ZnO particles in a film of water suspension (48 mg/L). The experiments show that the micro particles can be concentrated more efficiently than the nanowires, which indicates that it is easier to drive the micro particles than the nanowires by the acoustic streaming. A possible reason for this

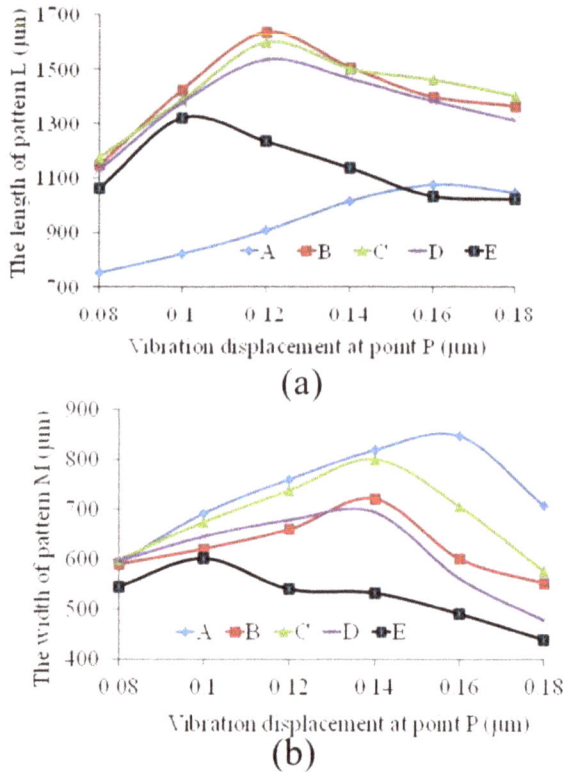

FIGURE 3.7   The effect of vibration displacement at point P of the needle on the size of the lobed patterns under the needle. (a) The length L of the lobed concentration patterns at different vibration amplitude when the needle is in resonance. (b) The width M of the lobed concentration patterns at different vibration amplitude when the needle is in resonance. Reproduced from Ref. [8] with permission from Elsevier.

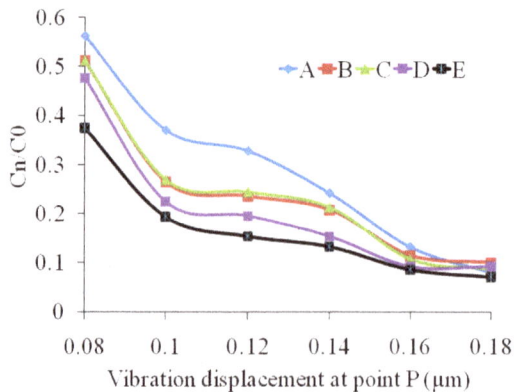

FIGURE 3.8   Ratio of particle concentration beside lobed concentration patterns A, B, C, D and E under sonication to that before sonication for different vibration amplitudes. Reproduced from Ref. [8] with permission from Elsevier.

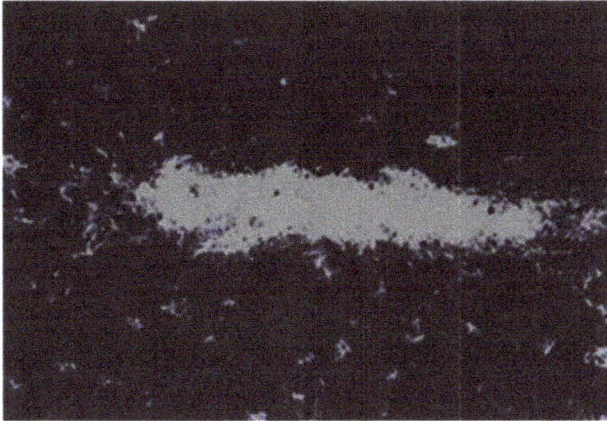

(a)

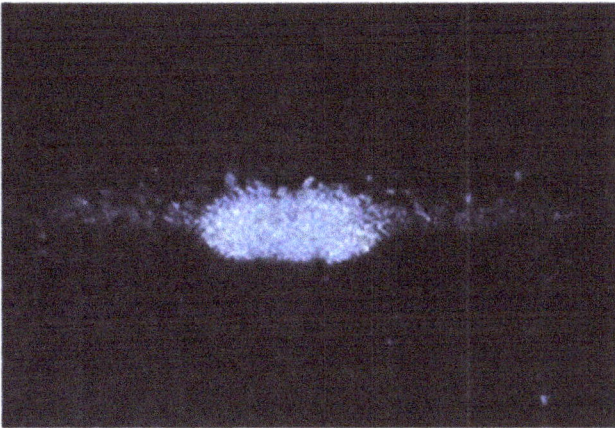

(b)

FIGURE 3.9   The image of concentration patterns for other micro/nano material. Vibration amplitude at point *P* used in the experiment is 180 nm, and the distance between the needle and substrate is 1.5 mm. (a) The image of concentration pattern formed by silver nanowires (length = 30–40 µm, diameter = 300 nm). (b) The image of concentration pattern formed by ZnO micro particles (diameter = 1 µm). Reproduced from Ref. [8] with permission from Elsevier.

phenomenon is that the friction between the nanowires and substrate is larger than that between the micro particles and substrate, which results from the bending and surface roughness of the nanowires.

In the above experiments, the concentration zones form an intermittent line under the vibrating needle on the substrate surface. To form a continuous line of the concentration particles, the vibrating needle was moved back and forth along a linear trajectory with a speed of several millimeters per second. Figure 3.10 shows the image of a continuous line of yeast particles on the substrate surface, formed by this method.

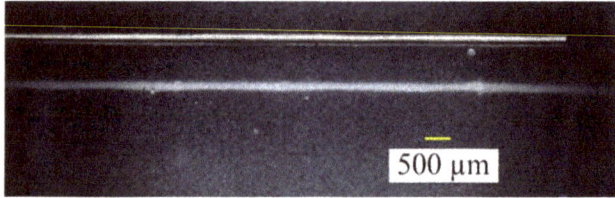

FIGURE 3.10   Yeast particle aggregation on a continuous linear line on the substrate surface. The vibration amplitude at point *P* used in the experiment is 180 nm, and the distance between the needle and substrate is 1.5 mm. Reproduced from Ref. [8] with permission from Elsevier.

Experiment also shows that the length of the concentration line can be increased by increasing the length of the vibrating needle submerged in the water film.

3.1.3 SUMMARY

In this example, a method of concentrating or enriching micro/nano scale samples in a film of aqueous suspension on a stationary substrate is demonstrated. The method employs the acoustic streaming which is generated by an ultrasonically vibrating needle parallel to and above the stationary substrate, to drive the micro/nano scale samples. Concentrated yeast particles with a diameter of 4–6 μm may form a series of lobed zones on the stationary substrate if the position of the vibrating needle has no change during the sonication and form a continuous linear line if the vibrating needle is moved back and forth along a linear trajectory. Smaller samples such as AgNWs with a length of 30–40 μm and diameter of 300 nm, and 1 μm-diameter ZnO particles, can also be concentrated by the method. Micro/nano scale particles have distinct boundaries between the concentration zones and surrounding suspension. A detailed analysis of the acoustic streaming field in the droplet around the vibrating needle is given in Section 3.4 of this chapter.

3.2   NANO CONCENTRATION BY COMPLEX SPIRAL VORTEX OF ACOUSTIC STREAMING

In this section, a strategy to concentrate or enrich nano scale materials on the boundary between a nano suspension droplet and non-vibration substrate is demonstrated and analyzed [9]. It employs the spiral vortex of acoustic streaming, generated by an ultrasonically vibrating needle parallel to and above the non-vibration substrate. The vortex drags nano scale materials to the center of itself, forming a concentration spot. For 250 nm-diameter SiO_2 particle suspension with an initial concentration of 0.09 mg/ml, the diameter of the concentration spot can be up to several hundred microns. The dependency of the spiral vortex field on the vibration distribution of the acoustic needle in the droplet is also investigated, and the concentration or enrichment conditions are obtained by analyzing the nano particle (NP) dynamics in the spiral vortex.

3.2.1 DEVICE AND EXPERIMENTAL SETUP

The ultrasonic needle-droplet-substrate system used in our experiments is composed of a stationary silicon substrate, a stainless steel needle (ultrasonic needle) excited by a piezoelectric plate and a suspension droplet dispersed on the substrate. Figure 3.11(a) shows the experimental setup for the concentration of NPs on the silicon substrate. The suspension droplet, formed by deionized water with dispersed SiO_2 NPs (Beijing DK Nano Technology Co. Ltd, China), had a maximum thickness of about 1.5 mm and an average diameter of 10 mm. The ultrasonic needle was inserted into the suspension droplet horizontally. The NP concentration was 0.09 mg/ml, and the average diameter of the NPs was 250 nm. During the concentration process, the silicon substrate did not vibrate and the ultrasonic needle vibrated in a parallel direction to the substrate surface. The experiments were conducted under an optical microscope (VHX-1000E, Keyence). The substrate diameter was about 6 cm. Compared to the stable diameter of the concentration spot (<1 mm), the droplet diameter (about 10 mm) was so large that the boundary effects at the droplet edge could be ignored.

The size parameters of the ultrasonic needle-droplet-substrate system are shown in Fig. 3.11(b). The width, length and thickness of the piezoelectric plate are 10 mm, 10 mm and 1 mm, respectively. The electromechanical quality factor Q_m, piezoelectric coefficient d_{33} and relative dielectric constant $\varepsilon_{33}^T/\varepsilon_0$ of the piezoelectric plate are 2000, 325×10^{-12} m/V and 1450, respectively. The piezoelectric plate is polarized in the thickness direction. The ultrasonic needle has a total length of 70 mm with two segments of different diameters. The thicker part has a diameter of 1 mm, used for vibration excitation. The thinner part has a diameter of 0.3 mm, used for vibration transmission. The length of the ultrasonic needle bonded onto the piezoelectric plate is 10 mm. The distance between the piezoelectric plate and the thicker tip of the ultrasonic needle is 10 mm.

The ultrasonic needle has a resonance frequency of 108 kHz, measured by a laser Doppler vibrometer (POLYTEC PSV-500). The driving voltage applied to the piezoelectric plate is sinusoidal, and the working frequency of the device is around 108 kHz. It utilizes the in-plane vibration (or k_{31} effect) of the piezoelectric plate to excite a flexural

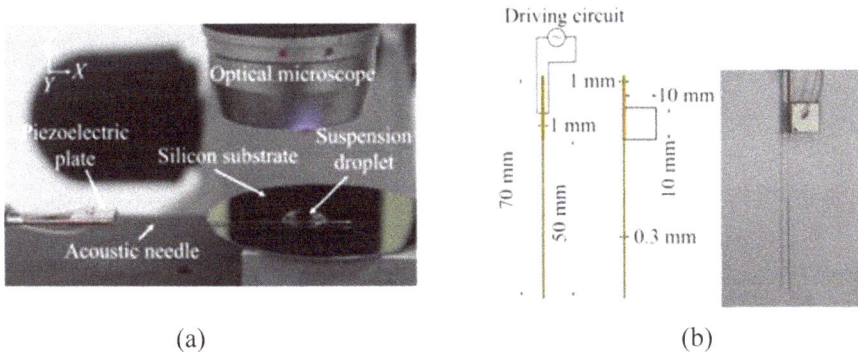

(a)   (b)

FIGURE 3.11   (a) Experimental setup for the NP concentration at the interface between a droplet and silicon substrate by the spiral vortices generated by ultrasound. (b) Structure and size. Reproduced from Ref. [9] with permission from AIP Publishing.

vibration of the ultrasonic needle in the horizontal plane, which is parallel to the substrate surface. The distance between the ultrasonic needle and the substrate is about 0.15 mm. Based on our FEM simulation and experiments, it is known that the distance between the ultrasonic needle and substrate mainly affects the vortices in the planes perpendicular to the substrate and needle, and has less influence on the main vortices parallel to the substrate. The measured and calculated wavelength of the flexural vibration in the ultrasonic needle is 4.6 mm at 108 kHz, respectively. The ultrasonic needle length in the aqueous suspension droplet is about 8 mm. Thus, the ultrasonic needle length in the droplet is one- and three-quarters wavelength in the experiments. The needle is about 3.1 mm away from the center of the aqueous suspension droplet.

3.2.2 ENRICHMENT EFFECT

It was observed that spiral vortices parallel to the substrate surface could be generated in the droplet and the NPs at the interface between the droplet and substrate could be driven to the centers of the spiral vortices and concentrated. Image (a) in Fig. 3.12(a) shows the NPs dispersed in the suspension before the sonication. Images (b) and (c) in Fig. 3.12(a) show the observed concentration spot of NPs in the droplet on the silicon

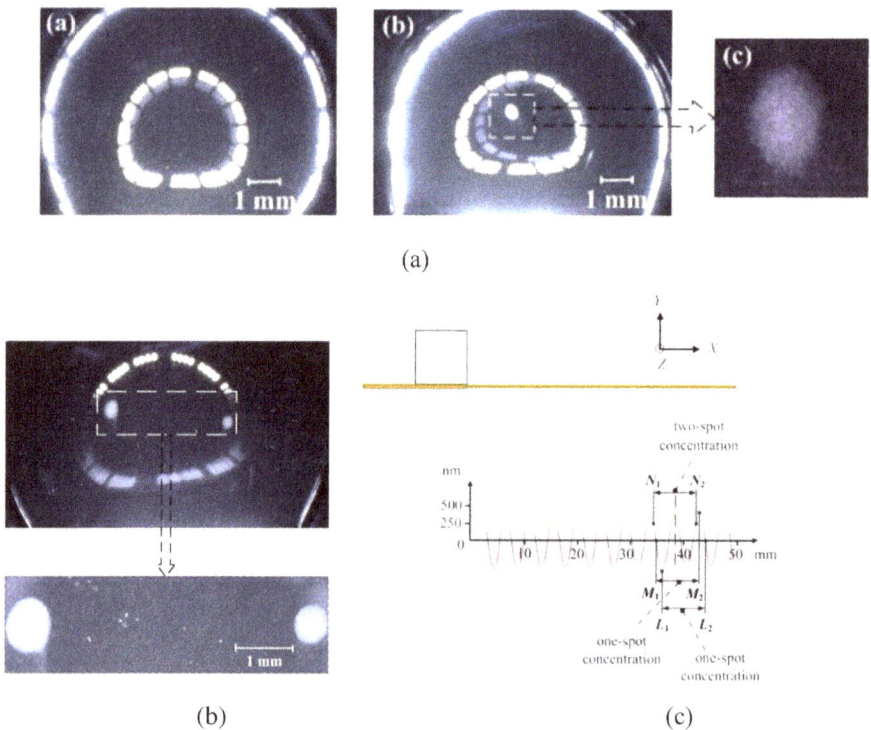

(a)

(b)   (c)

FIGURE 3.12   (a) NP concentration at the interface between the droplet and silicon substrate. (b) Two concentration spots generated by Two spiral vortices with opposite directions and similar scale. (c) Ultrasonic needle vibration distribution in parallel with the substrate. Reproduced from Ref. [9] with permission from AIP Publishing.

substrate. In the experiment, the working frequency and voltage were 108 kHz and 65 V_{p-p}, respectively; the left intersection point between the ultrasonic needle and droplet was the vibration antinode and the right one was the vibration node (see $[L_1, L_2]$ in Fig. 3.12(c)). For this way of sound field excitation, there are two spiral vortices on the side of the ultrasonic needle in opposite directions. Only the concentration caused by the major vortex is shown in image (b). The concentration spot takes on an elliptical shape with a long axis length of 414 µm and a short axis length of 366 µm approximately. Defining the center of the orthographic projection of the ultrasonic needle in the droplet onto the substrate as the origin O of a Cartesian coordinate system XYZ, as shown in Fig. 3.11(a), the concentration spot center is at $X = -0.7$ and $Y = 4$ mm, and the center of droplet is at $X = 0$ and $Y = 3.1$ mm. In the Cartesian coordinate system, the X-axis is along the longitudinal direction of the needle, the Y-axis is perpendicular to the needle and parallel to the substrate and the Z-axis is perpendicular to the substrate.

The two spiral vortices may result in similar enrichment effect when the vibration distribution of the ultrasonic needle inserted into the droplet is symmetric about the central axis (see $[N_1, N_2]$ in Fig. 3.12(c)). Figure 3.12(b) shows two concentration spots generated by the two spiral vortices in opposite directions in the droplet. The maximum diameter of the left concentration spot is about 1.3 times of the right one. Our measurement also shows that the main vortex is clockwise when the left intersection point between the ultrasonic needle and droplet is the vibration node and the right one is the vibration antinode (see $[M_1, M_2]$ in Fig. 3.12(c)). Detailed vibration distribution of the ultrasonic needle is shown in Fig. 3.12(c), which was measured by the laser Doppler vibrometer (POLYTEC PSV-500).

The acoustic streaming in the droplet was computed by COMSOL MULTIPHYSICS FEM Software, based on the method proposed by the authors' group [15, 16]. The parameters used in the computation are listed in Table 3.2.

TABLE 3.2

Parameters of the Ultrasonic Needle-Droplet-Substrate System

The Droplet's Maximum Thickness H (mm)	The Droplet's Radius R_f (mm)	The Ultrasonic Needle's Total Length L_N (mm)
1.5	5	60
The ultrasonic needle's radius R_N (mm)	The distance between the ultrasonic needle central axis and substrate d_0 (mm)	The excited segment length of the ultrasonic needle L_E (mm)
0.15	0.15	10
Density of droplet ρ (kg/m³)	Speed of sound in droplet c (m/s)	Shear viscosity of droplet η (Pa·s)
1000	1500	0.001
Volume-to-shear viscosity ratio of droplet η'/η	Nonlinear parameter of droplet at room temperature B/A	Density of ultrasonic needle ρ_N (kg/m³)
2.1	5	7850
Young modulus of ultrasonic needle E_N (Pa)	Shear modulus of ultrasonic needle G_N (Pa)	
2.05×10^{11}	2.846×10^{10}	

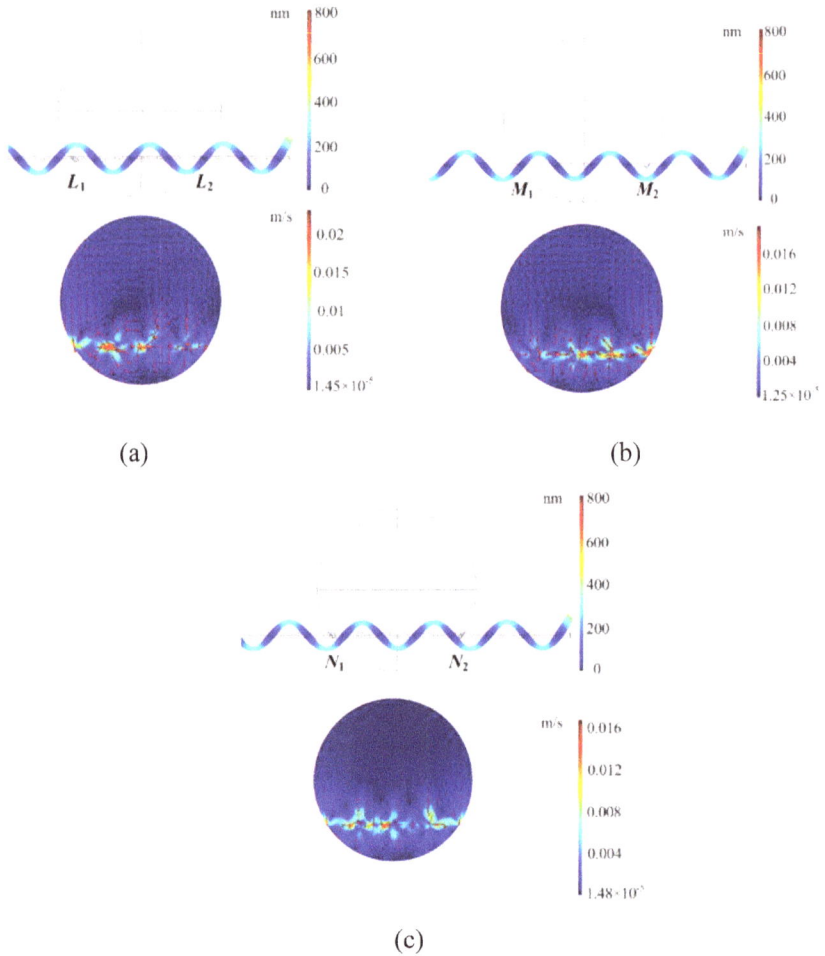

(a)

(b)

(c)

FIGURE 3.13 Computed acoustic streaming field, generated by different vibration distributions of the ultrasonic needle in the droplet. (a) Acoustic streaming field with a dominant anticlockwise spiral vortex. (b) Acoustic streaming field with a dominant clockwise spiral vortex. (c) Acoustic streaming field with two spiral vortices of similar scale and opposite directions. Reproduced from Ref. [9] with permission from AIP Publishing.

The computed acoustic streaming vortex fields, generated by different vibration modes of the ultrasonic needle in the droplet, are listed in Fig. 3.13. They agree with the experimental observation described above qualitatively. Therefore, it is confirmed that the vortex direction and distribution depend on the vibration mode of the ultrasonic needle in the droplet.

3.2.3 Dynamics of NPs in the Droplet

In the experiments, NPs on the substrate were driven to the center of the spiral vortex of acoustic streaming, along different spiral curves, as shown in Fig. 3.14(a).

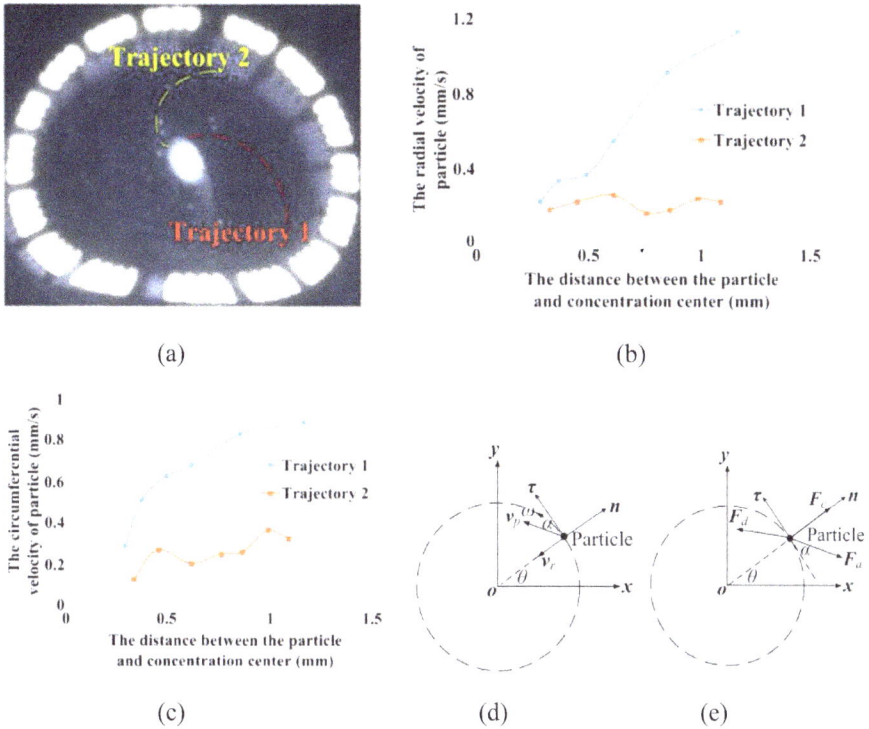

FIGURE 3.14   (a) Two typical trajectories of the NPs driven to the center of the spiral vortex of acoustic streaming on the substrate. (b) Measured radial velocity a NP versus its distance from the center of the spiral vortex. (c) Measured circumferential velocity of a NP versus its distance from the center of the spiral vortex. (d) A mathematical model proposed to analyze the spiral vortex, in which the center of the concentration spot is defined as the origin o of the Cartesian coordinate system xyz. (e) Force diagram of the NP. Reproduced from Ref. [9] with permission from AIP Publishing.

In the figure, trajectories 1 and 2 show the motion paths of an NP on the substrate, with an initial position close to the X and Y axis, respectively. Figure 3.14(b, c) is the measured radial and circumferential velocities of the NP versus the distance from the center of the spiral vortex for the trajectories. Figure 3.14(b, c) indicates that the spiral vortex on the substrate surface contains a Rankine vortex component and a complex radial flow component, respectively. Thus, a mathematical model shown in Fig. 3.14(d) was used to analyze the spiral vortex, in which the center of the concentration spot was defined as the origin o of the Cartesian coordinate system xyz. The acoustic streaming velocity is

$$v_f = r\omega\tau - v_r n, \tag{3.1}$$

where τ and n are the circumferential and radial unit vectors, respectively, r is the distance between the centers of the particle and acoustic streaming eddy, ω (= $\omega(\theta)$).

Here, θ is the polar angle from the x-axis) is the angular frequency of the Rankine vortex and v_r $(= v_r(r, \theta))$ is the radial velocity of the spiral vortex. The particle velocity is

$$v_p = v_p \cos \alpha \tau - v_p \sin \alpha n, \tag{3.2}$$

where α is the angle between the particle velocity v_p and τ. Figure 3.14(e) shows the force diagram of the NP. The drag force on the particle is estimated by the Stokes' law [17]

$$F_d = 6\pi\eta R(v_f - v_p), \tag{3.3}$$

where η and R represent the fluid shear viscosity and the particle radius, respectively. The adhesive force F_a between the particle and substrate is opposite to the particle velocity v_p, and its magnitude is measured by the method described in Ref. [9]. The particle also experiences the centrifugal force F_c, which is

$$F_c = \rho_p \frac{4}{3} \pi R^3 \frac{(v_p \cos \alpha)^2}{r} n. \tag{3.4}$$

For NPs, the acoustic radiation force on the particle can be neglected. Thus, from Eqs. (3.1)–(3.4), there are

$$6\pi\eta R(r\omega - v_p \cos \alpha) = F_a \cos \alpha, \tag{3.5}$$

and

$$6\pi\eta R(v_r - v_p \sin \alpha) = F_a \sin \alpha + \rho_p \frac{4}{3} \pi R^3 \frac{(v_p \cos \alpha)^2}{r}, \tag{3.6}$$

where ρ_p is the density of the particle. From Eq. (3.5), there is

$$\cos \alpha = \frac{6\pi\eta Rr\omega}{6\pi\eta Rv_p + F_a}. \tag{3.7}$$

And from Eqs. (3.5) and (3.6), there is

$$\frac{R\rho_p F_a^2}{27\pi\eta^2 r} \cos^2 \alpha - \frac{4R^2 \rho_p \omega F_a}{9\eta} \cos \alpha + 6\pi\eta Rr\omega \tan \alpha = 6\pi\eta Rv_r - \frac{4\pi R^3 \rho_p r\omega^2}{3}. \tag{3.8}$$

Based on the parameters in Table 3.2, the order of magnitude of $\dfrac{R\rho_p F_a^2}{27\pi\eta^2 r}$ and $\dfrac{4R^2 \rho_p \omega F_a}{9\eta}$ is 10^{-21} (N) and 10^{-19} (N), respectively, which are much less than the order of magnitude of $6\pi\eta Rr\omega$ ($\approx 10^{-11}$ (N)). Thus Eq. (3.8) can be simplified as

$$6\pi\eta Rr\omega \tan \alpha = 6\pi\eta Rv_r - \frac{4\pi R^3 \rho_p r\omega^2}{3}. \tag{3.9}$$

The necessary and sufficient condition to achieve the NP concentration is $0 < \alpha \le \pi/2$. As $\cos \alpha \ge 0$ (see Eq. (3.7)), $\tan \alpha > 0$ is the necessary and sufficient condition to achieve the NP concentration. Based on Fig. 3.14(b), it is assumed that $v_r = kr$. Here k (= $k(\theta)$) is a parameter related to the 3D flow pattern of the acoustic streaming, and the stronger the height (Z-) directional flow, the larger k is. Therefore, the necessary and sufficient condition to concentrate the NPs is

$$\frac{k}{\omega^2} > \frac{2R^2\rho_p}{9\eta}.$$

(3.10)

Our experimental observation shows that it is more difficult to concentrate the particles that have a diameter 50%–100% larger than the average one (250 nm). This is because the spiral vortex field needs larger k/ω^2 to concentrate larger particles. Thus Eq. (3.10) indicates that it is theoretically possible to separate NPs in a droplet on a substrate by the spiral vortex, based on their size or density.

3.2.4 CHARACTERISTICS

Figure 3.15(a) shows the diameter change of the concentration spot versus sonication time under different driving voltages (50, 65 and 80 $V_{p\text{-}p}$) and at resonance point (around 108 kHz). The vibration magnitude of the ultrasonic needle at 50, 65 and 80 $V_{p\text{-}p}$ was 424, 432 and 435 nm, respectively. The sonication started at $t = 0$ and stopped at $t = 10$ min. In the experiments, the average length of the long and short axes was used as the spot diameter. It is seen that the spot diameter decreased as the sonication time increased in the beginning (during the first 4 min), and then it became stable. After the first 4 min, the boundary of the spot was quite clear. After the sonication was switched off, the diameter of the concentration spot increased a bit first due to the diffusion effect, and then the boundary became stable.

The images in Fig. 3.15(b), taken at the sonication time of 1, 4, 7 and 10 min, respectively, show the concentration process. Those in Fig. 3.15(c), taken at 1, 4, 7, and 10 min after the sonication, show the diffusion process after the sonication is switched. Both experiments were conducted at 65 $V_{p\text{-}p}$ and resonance. The thickness of the spot after drying is about 50 μm, which was measured by optical microscope VHX-1000E (Keyence). Possible applications of the concentration process include nano sensor fabrication, high-sensitivity biological sensors, analyses of ultra-low concentration biological samples, and nano separation.

3.2.5 SUMMARY

This work demonstrates that the spiral vortex induced by ultrasound is capable of concentrating NPs at the interface between a droplet and substrate. NPs in the spiral vortex on the substrate can be dragged to the vortex center, forming round concentration spots. The diameter of the concentration spots can be up to several hundred microns. The distribution of the spiral vortices in the droplet depends on the vibration mode of the vibrating needle in the droplet. The sufficient and necessary condition to achieve the concentration has also been clarified by analyzing the dynamics of the NPs. Potential applications of this method include nano material fabrication, high-sensitivity sensing, nano separation, etc. Compared with methods based on other physical principles, the

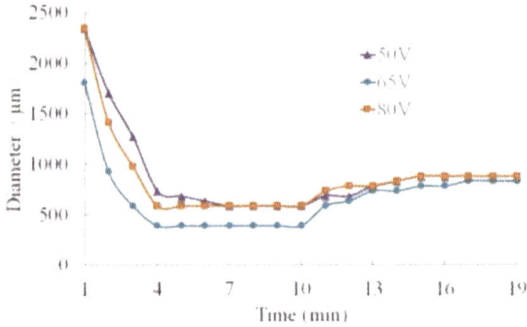

(a)

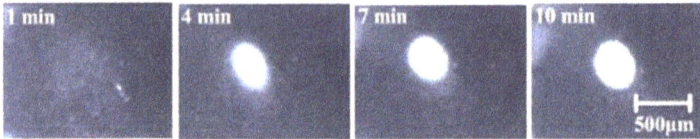

(b)

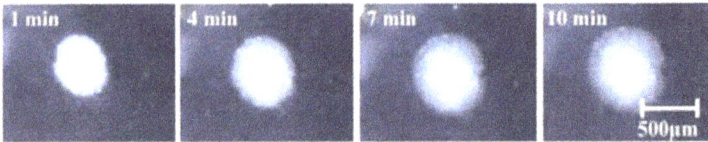

(c)

FIGURE 3.15 (a) Diameter change of the concentration spot versus sonication time under different driving voltages (50, 65 and 80 V_{p-p}) and at resonance point (around 108 kHz). (b) Concentration process shown by a series of Concentration spots at the sonication time of 1 min, 4 min, 7 min, and 10 min, respectively. (c) Diffusion process shown by a series of concentration spots 1 min, 4 min, 7 min, and 10 min after the sonication is switched off. Reproduced from Ref. [9] with permission from AIP Publishing.

method has the merits of being harmless to the manipulated samples, biocompatible to living samples and not selective to the material properties of samples.

3.3 A DUAL FUNCTIONAL DEVICE FOR CONCENTRATION AND TRANSPORTATION OF NPs

Controlled concentration of nano scale materials on the surface of a smooth substrate without vibration excitation mechanism and micro channels (termed as plain substrate), and transportation of the concentrated nano material on the surface, have large potential applications in the fabrication of nano sensors and electrodes, decoration and assembly of nano materials, etc. However, implementation of these two nano manipulation functions by one single device has been a big challenge. In this section, a strategy to concentrate NPs at an arbitrary location at the interface between a plain substrate and water droplet and to transportation the concentrated nano material freely at the interface is described [10]. It employs the acoustic streaming, which is generated by a micro manipulating probe (MMP) vibrating linearly above the substrate. In total, 500 nm-diameter silicon nano particles (SiNPs) can be concentrated under the MMP at a desired location, forming a round spot of nano materials with a diameter of up to 230 μm. The concentrated nano material can be transported through an arbitrary path at the interface by shifting the device and has little change in size and shape during the transportation. The dependency of the acoustic streaming field around the MMP on device parameters is clarified by numerical computation and verified by experiments.

3.3.1   Materials and Methods

3.3.1.1   Preparation of NP Suspension

500 nm-diameter SiNPs (Beijing DK Nano Technology Co., Ltd, China) were dispersed in deionized water by 20 min sonication (Shanghai Ouhor Mechanical Equipment Co., Ltd, China) to form the experimental NP suspension with a concentration of 0.06, 0.08, 0.1, 0.21 and 0.38 mg/ml, respectively.

3.3.1.2   Device and Experimental Setup

Figure 3.16 shows the device structure and experimental setup for concentration and transportation of NPs at the interface between an NP suspension droplet and substrate surface. Figure 3.16(a) is a schematic diagram of the device structure and experimental setup. The device consists of three main components, that is, piezoelectric plate (Haiying Enterprise Group Co., Ltd, China), vibration transmission needle (VTN, Shanghai Dongfeng Co., Ltd, China) and MMP (Nanjing Fiberglass Research & Design Institute Co., Ltd, China). The VTN, made of stainless steel, is bonded onto the edge of the longer side of the piezoelectric plate by modified Acrylate adhesive (Geliahao New Material Co., Ltd, China), and the VTN's root is fixed in a supporting jig. The MMP is bonded to the VTN's tip (point P) by glue 502 (Guangdong Aibida Adhesive Co., Ltd, China). The piezoelectric plate, perpendicular to the MMP, is used for vibration excitation, and the excited vibration passes through the VTN to the MMP. The adhesive is applied at room temperature for 3 min after mixing its components A and B with a weight ratio of 1:1. The bonding becomes stable after 24 h.

The VTN has a diameter of 1 mm and is 26 mm long out of the piezoelectric plate. The MMP has a uniform radius of 10 μm, a length of 3 mm and makes an angle θ (see Fig. 3.16) with the VTN. In this paper, unless otherwise specified, θ is 90°. The length, width and thickness of the piezoelectric plate are 20 mm, 10 mm and 0.78 mm, respectively. The piezoelectric constant d_{33}, electromechanical coupling factor k_{33}, mechanical quality factor Q_m, dielectric dissipation factor $tan\delta$ and density are 200×10^{-12} C/N, 0.6, 800, 0.5% and 7.45×10^3 kg/m³, respectively.

Figure 3.16(b) is a photograph of the experimental setup. The concentration and transportation processes were observed and recorded by an optical microscope system (VHX-1000E, Keyence, Japan). Detailed parameters of the fabricated devices and experimental system are listed in Table 3.3.

(a)

(b)

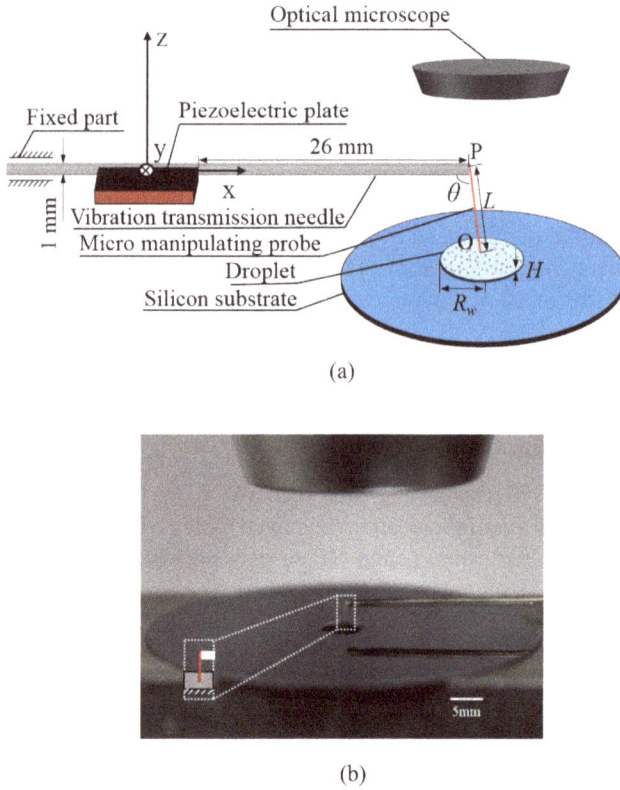

FIGURE 3.16   Device structure and experimental setup for concentration and transportation of NPs at the interface between a NP suspension droplet and substrate surface. (a) schematic diagram of the experimental setup and device. (b) Photograph of the experimental setup. Detailed values of the changeable device dimensions and experimental system parameters are listed in Table 3.3. Reproduced from Ref. [10] with permission from Elsevier.

TABLE 3.3
Related Parameters of the Fabricated Devices and Experimental System

Device Dimensions and Experimental Setup Parameters		Material Constants	
MMP length L (mm)	2–3.5	Water density ρ (kg/m³)	1000
MMP radius R (μm)	6–21.5	Sound velocity in water c (m/s)	1500
Water film thickness H (mm)	0.15–1.5	Shear viscosity of water η (Pa·s)	0.001
Water film radius R_w (mm)	1–3.5	MMP density (kg/m³)	2200
Distance d between the MMP's tip and substrate (μm)	10–120	Poisson's ratio of the MMP	0.3
Angle θ between the MMP and the VTN (°)	85–100	Young's modulus of the MMP (Pa)	7.4×10^{10}

3.3.1.3   Vibration Measurement

The vibration velocity of the VTN's tip (point *P*) was measured by a laser Doppler vibrometer (POLYTEC PSV-300F, Germany), which was used to represent the vibration strength of the device. Vibration trajectory of the measured point was obtained by measuring the orthogonal vibration components of the measurement point.

3.3.1.4   Experiments of Nano Concentration and Transportation

Firstly, a silicon substrate (Zhejiang Liji Photoelectric Technology Co., Ltd, China) with a diameter of 50 mm and thickness of 500 μm was put on the optical microscope's platform. Then a droplet of SiNP suspension was dispensed onto the substrate by a pipette. After 5 min, most of the NPs were evenly deposited on the substrate surface. The height and diameter of the SiNP suspension film were about 0.15 mm and 7 mm (with a volume of 5.8 mm^3), respectively, and the distance between the MMP and substrate surface was about 40 μm. The VTN's root was clamped to a manually manipulated three-dimensional movement platform (LD125-LM-2, Shengling Precise Machinery Co., Ltd, China), which had a precision of 0.01 mm. In the operation, a signal generator (Tektronix AFG3022B, USA) and power amplifier (Nanjing Foneng Technology Industrial Co., Ltd, China) were used to apply the driving voltage to the piezoelectric plate, and an oscilloscope (Tektronix, TDS2014C, USA) was used to monitor the electric signal. When the driving frequency and voltage were properly chosen, the S_iNPs on the substrate surface could move from the periphery to the location just below the MMP's tip (point *O*), to form a round spot of concentrated nano material. At a proper working frequency, the MMP can generate a desired pattern of acoustic streaming field. At a proper driving voltage, the acoustic streaming field can drive the NPs to the region under the MMP's tip while keeping the concentrated nano material stable.

3.3.1.5   Finite Element Analyses

In order to understand the concentration mechanism and characteristics, the MMP vibration, ultrasonic field and acoustic streaming field in the water film (SiNP suspension film) was computed by the finite element method (FEM). In the computation, COMSOL Multiphysics software (version 5.2a) was employed. The FEM model only consisted of the MMP and water film. Measured amplitude and phase of the orthogonal vibration components at the MMP's root were used as the excitation for the MMP vibration. To control the computation error while reducing the computation time, meshes near the MMP were much smaller than those in the outside region.

Boundary conditions for the ultrasonic field and flow (acoustic streaming) field in the droplet were as follows: The interface between the droplet and substrate was treated as a hard acoustic boundary for the ultrasonic field and as a slip boundary for the flow field. The interface between the droplet and air was treated as a soft acoustic boundary for the ultrasonic field and as a slip boundary for the flow field. The interface between the MMP and droplet was treated as a sound/solid boundary for the ultrasonic field and as a slip boundary for the flow field.

The FEM computation had three steps [15]. In the first step, the MMP vibration and ultrasonic field in the droplet were computed by the sound/solid coupling module of the software. In the second step, the time average of spatial gradients of the Reynolds stress and second-order sound pressure were computed by the post-processing functions of the software, which was used as the driving force of the acoustic streaming. In the third step, steady-state Navier-Stokes equations for the acoustic streaming were solved by the laminar flow module.

In the FEM computation, unless otherwise specified, the working frequency was 75.46 kHz, orthogonal vibration velocity components at the MMP's root were $V_x = 5.2\angle-13.8°$ mm/s (0–p), $V_y = 61.6\angle-12.5°$ mm/s (0–p), $V_z = 2.2\angle172.5°$ mm/s (0–p), and parameters of the water film and MMP shown in Table 3.3 were used.

3.3.2 RESULTS AND DISCUSSION

3.3.2.1 Concentration and Transportation of NPs

Images a1–a8 in Fig. 3.17(a) show a concentration process for 500 nm-diameter SiNPs at a driving frequency of 75.46 kHz and driving voltage of 25 V_{p-p}. The concentration occurs under the vibrating MMP. Sonication time for the images is 20, 60,100, 140, 180, 220, 260 and 300 s, respectively.

It is seen that the diameter of concentrated nano material becomes stable after a sonication of 5 min. The darker area at the center of each concentration spot is because the central area is relatively thick. Figure 3.17(b) is an image of the stable spot of concentrated nano material, with a diameter of 105 µm. Images c1–c6 in Fig. 3.17(c) show a transportation process of concentrated nano material when the driving frequency is 75.46 kHz and the driving voltage is 25 V_{p-p}. The average speed of device movement is 0.01 mm/s, and the movement trajectory of the MMP's tip is indicated by the red arrow in the figure.

Sonication time for the images in Fig. 3.17(c) is 120, 220, 280, 380, 440 and 500 s, respectively. From c1 to c5, concentrated nano material is transported along the four sides of the dash rectangle. From c5 to c6, concentrated nano material is transported to the center of the rectangle. Thus, concentrated nano material can be trapped under the vibrating MMP, and then simply transported by moving the device. In the transportation process, small amounts of NPs were sucked onto the concentration spot while the concentration spot lost some NPs during the movement. The final diameter of the transported nano material spot in image c6 is 130 µm.

3.3.2.2 Mechanism of the NP Concentration

In the experiments, the orthogonal vibration velocities at the MMP's root V_x, V_y and V_z were 5.2$\angle-13.8°$, 61.6$\angle-12.5°$ and 2.21$\angle-72.5°$ mm/s, respectively. As the magnitude of the y-directional vibration velocity is much larger than that of the x- and z-directional vibration velocities, vibration trajectory at the MMP's root is approximately linear. Under the conditions that the MMP has a radius of 10 µm and length of 3 mm, and vibrates at 75.46 kHz, the computed ratio of vibration velocity (or amplitude) at the MMP's tip to that at the MMP's root is 0.85, 0.84 and 0.95 for the x, y and z components, respectively.

To understand the concentration mechanism, the acoustic streaming field in the droplet was investigated by the FEM with a 3D mesh model shown in Fig. 3.18(a). Figure 3.18(b, c) shows the acoustic streaming field near the MMP. Computed acoustic

(a)

(b)

(c)

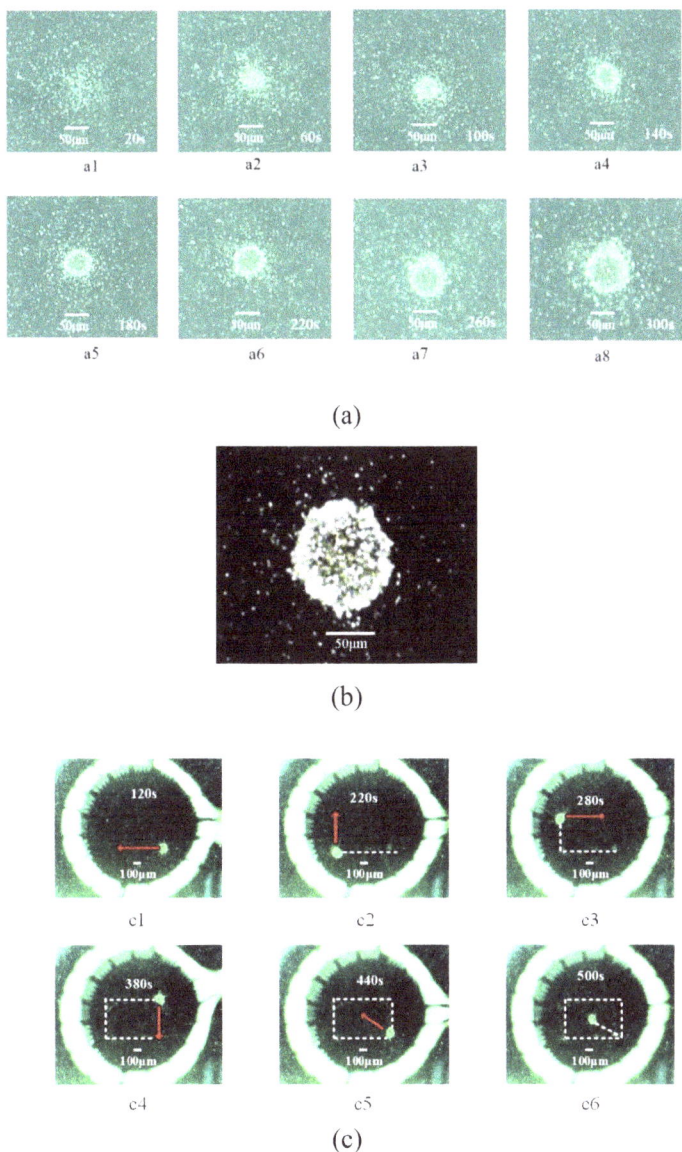

FIGURE 3.17 (a) a concentration process for 500 nm-diameter SiNPs. (b) Image of a stable state of the concentrated nano material. (c) A transportation process of the concentrated nano material. Sonication time for the images in Fig. 3.2(a) is 20, 60, 100, 140, 180, 220, 260 and 300 s, respectively, and sonication time for the images in Fig. 3.2(c) is 120, 220, 280, 380, 440 and 500 s, respectively. The initial concentration density of the SiNPs in Fig. 3.2(a, c) is 0.06 mg/ml and 0.1 mg/ml, respectively. The MMP radius in the experiment is 10 μm. The aureole in (c) is caused by the light reflection on the droplet surface. Reproduced from Ref. [10] with permission from Elsevier.

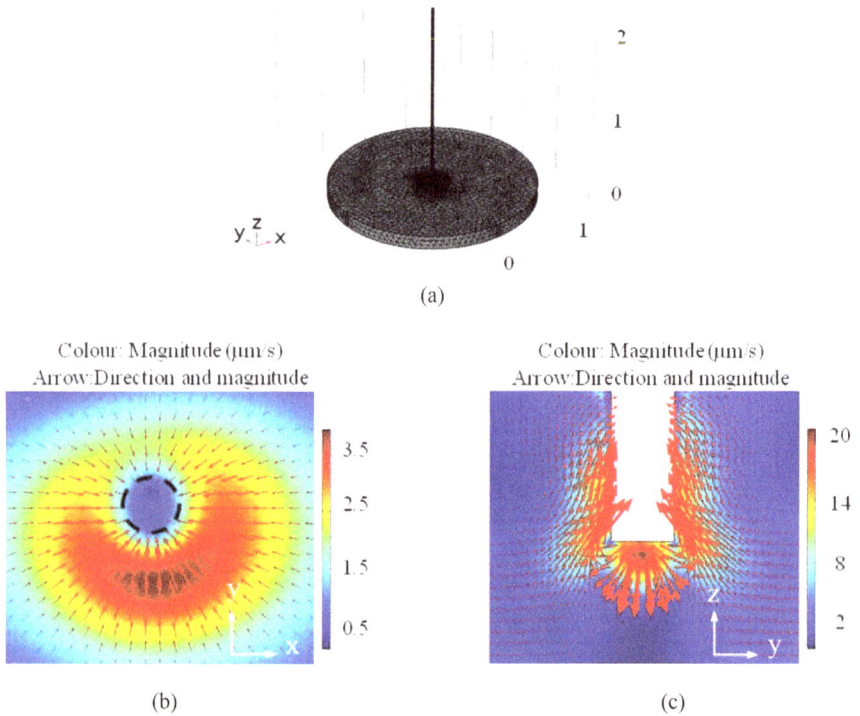

FIGURE 3.18   (a) Meshed 3D FEM computation model. (b) Acoustic streaming field on the substrate surface. (c) Acoustic streaming field in the *yz* plane, which is perpendicular to the substrate and passing the center axes of the MMP and VTN. In Fig. 3.18(b), the region enclosed by the dotted circle represents the projection of the MMP's tip on the substrate. Reproduced from Ref. [10] with permission from Elsevier.

streaming field on the substrate surface under the MMP is shown in Fig. 3.18(b), and that in the *yz* plane which is perpendicular to the substrate surface and passing through the center axes of the MMP and VTN, is shown in Fig. 3.18(c). In the figures, the length and direction of the arrows represent the largeness and direction of acoustic streaming velocity, respectively.

Figure 3.18(b) indicates that on the substrate surface, the acoustic streaming can flush NPs to a round region under the MMP, and concentrate them in the region. Figure 3.18(c) indicates that although there is acoustic streaming flowing from the MMP's tip toward the substrate, it cannot reach the substrate and does not affect the concentration process. Due to the trapping capability of the acoustic streaming field and the fact that the acoustic streaming field moves together with the MMP, the transportation of concentrated nano material, as shown in Fig. 3.17(c), can be implemented.

3.3.2.3   Effects of NP Suspension Density and Vibration Velocity

Figure 3.19(a) shows the measured diameter of concentrated nano material versus sonication time at five different NP suspension densities when the *y*-directional vibration velocity at the MMP's root is 61.6 mm/s. It shows that it takes several minutes for the

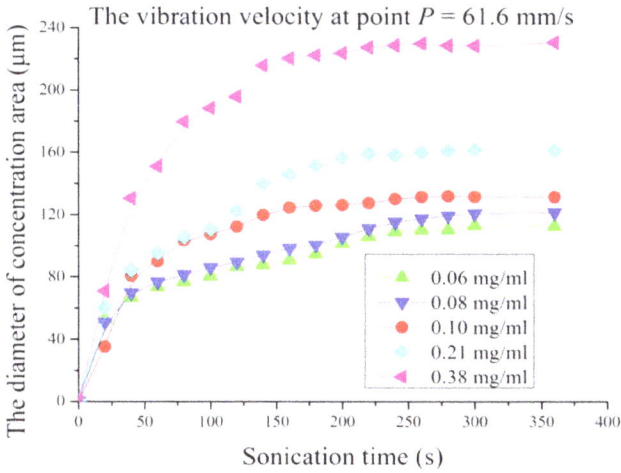

(a)

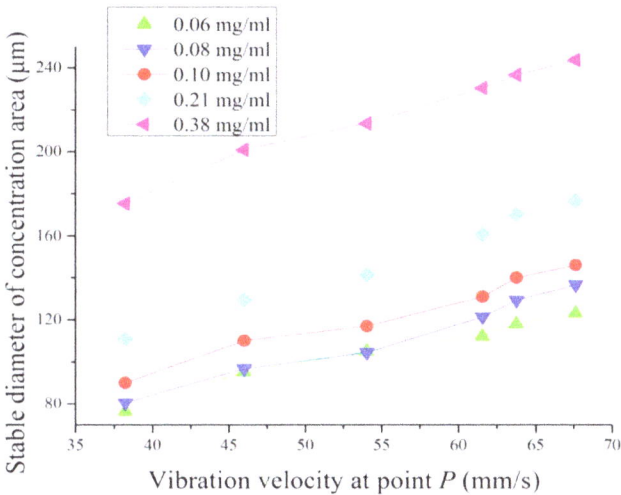

(b)

FIGURE 3.19 (a) Measured diameter of the concentrated nano material versus sonication time under different suspension densities. (b) Measured stable diameter of the concentrated material versus vibration velocity at the MMP's root under different suspension densities. Reproduced from Ref. [10] with permission from Elsevier.

diameter of concentrated nano material to become stable (which means the diameter of concentrated nano material does not increase anymore.), and the stable diameter can be more than 200 µm. It also shows that the stable diameter is affected by the initial NP suspension density, which is due to the total amount of NPs on the substrate surface. In addition, it was observed in the experiments that the NPs flushed into the concentration region in the earlier stage could block the motion of those in the later stage.

Figure 3.19(b) shows the stable diameter versus the y-directional vibration velocity at the MMP's root at five different suspension densities. It is seen that the stronger the ultrasonic vibration, the larger the stable diameter. This is because acoustic streaming velocity on the substrate surface becomes higher as the ultrasonic vibration increases, and a higher acoustic streaming velocity can flush the further NPs to the concentration region.

3.3.2.4 Concentration and Transportation Capabilities

To clarify the dependency of concentration and transportation capabilities on the device and working parameters, which is very important in the practical design of the device, the dependency of acoustic streaming field on different parameters was investigated by the FEM. The parameters investigated in this work include the MMP radius R, distance d between the MMP's tip and substrate surface, angle θ between the VTN and MMP, and MMP length L. When the effect of one parameter was being investigated, the others were kept constant.

The effect of the MMP radius R on the concentration and transportation capabilities was analyzed by the FEM, and the results are summarized in Fig. 3.20. Figure 3.20(a) shows different working regions (for the concentration and transportation capabilities) caused by a change of the MMP radius R, as well as the maximum acoustic streaming velocity V_{max} (on the substrate surface). To verify the computational results, the dependency of the manipulation capabilities and maximum acoustic streaming velocity V_{max} on the MMP radius R was measured, and the results are listed in Table 3.4 and Fig. 3.20(a), respectively.

Table 3.4 shows that the computed concentration and transportation capabilities are consistent with the measured ones. The acoustic streaming velocity was measured by recording the motion of SiNP markers and analyzing the video clips. It is seen that the computed maximum acoustic streaming velocities agree quite well with the measured ones.

Figure 3.20(b) lists several typical acoustic streaming fields on the substrate surface, corresponding to the working regions shown in Fig. 3.20(a). For 7.5 μm < R < 20 μm, the acoustic streaming field has a pattern shown in b3, in which the water flows into the round location under the MMP's tip. Thus, the device has normal concentration and transportation capabilities in this region. The peak of the maximum acoustic streaming velocity indicates that the MMP works in resonance at $R = 13.5$ μm. As the radius increases, the acoustic streaming field changes into the pattern shown in b4. The water flows out of the round region under the MMP's tip, which makes the concentration impossible. For 6 μm < R < 7.5 μm, the acoustic streaming field changes into the pattern shown in b2, in which part of the round region under the MMP's tip has weak or very weak acoustic streaming. In this case, the device still has the concentration capability, but its transportation capability weakens. As the radius decreases further, the acoustic streaming field becomes non-convergent under the MMP's tip, causing the concentration impossible.

Figure 3.21 lists photos of some experimental MMPs with different radii. The effect of the distance d between the MMP's tip and substrate surface on the concentration and transportation capabilities was analyzed by the FEM, and the results are summarized in Fig. 3.22.

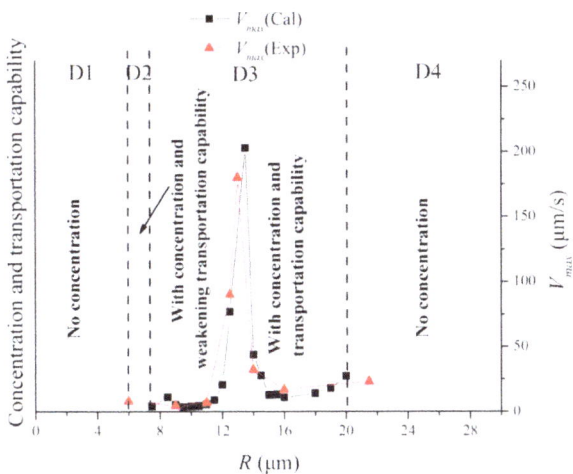

(a)

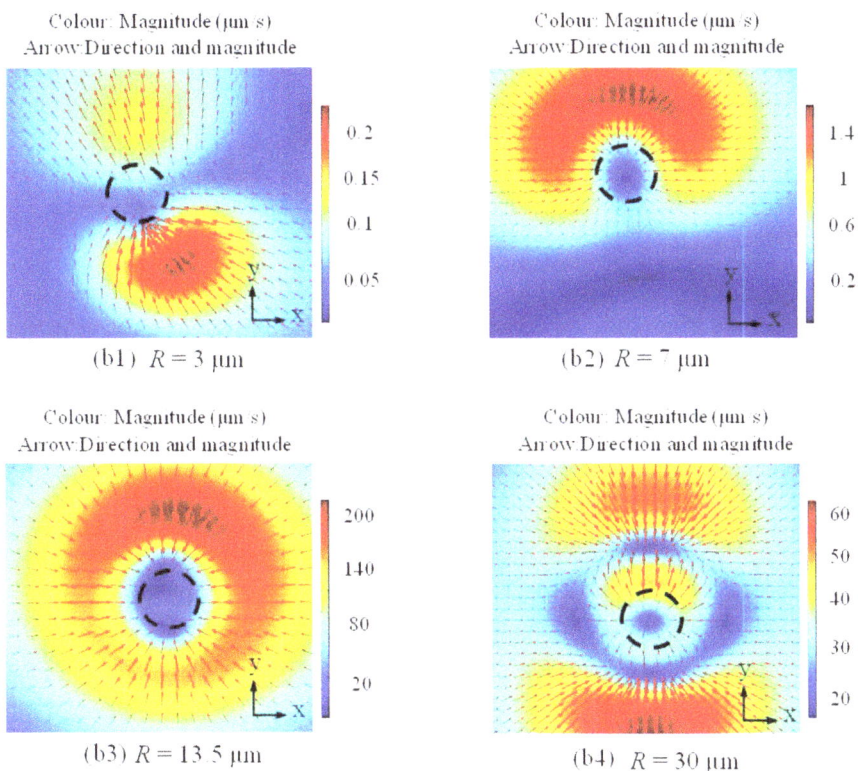

(b)

FIGURE 3.20 (a) computed dependency of the manipulation capabilities and maximum acoustic streaming velocity vmax (on the substrate surface) on the MMP radius r. Regions D1 and D4: No concentration; Region D2: With concentration and weakening transportation capabilities; Region D3: With concentration and transportation capabilities. (b) computed acoustic streaming fields at different MMP radius r at $V_x = 5.2\angle{-13.8°}$ mm/s (0–p), $V_y = 61.6\angle{-12.5°}$ mm/s (0–p), $V_z = 2.2\angle{172.5°}$ mm/s (0–p). Measured dependency of the maximum velocity vmax on the MMP radius r is also listed in (a). Reproduced from Ref. [10] with permission from Elsevier.

TABLE 3.4

Measured Dependency of the Concentration and Transportation Capabilities on the MMP Radius R

R (µm)	6	7.5	9, 11, 12.5, 13, 14, 16	21.5
Concentration capability	No	Yes	Yes	No
Transportation capability	No	Weakening	Yes	No

Figure 3.22(a) shows different working regions for the manipulation capabilities, caused by a change of the distance d, as well as the maximum acoustic streaming velocity V_{max}. To verify the computational results, the dependency of the manipulation capabilities and maximum acoustic streaming velocity V_{max} on the distance d was measured, and the results are listed in Fig. 3.22(a) and Table 3.5, respectively. They indicate that the computed concentration and transportation capabilities are consistent with measured ones. Moreover, Fig. 3.22(a) indicates that the computed maximum acoustic streaming velocities agree quite well with the measured ones.

Figure 3.22(b) lists several typical acoustic streaming fields on the substrate surface, corresponding to the working regions shown in Fig. 3.22(a). For 10 µm < d < 90 µm, the acoustic streaming field is convergent as shown in b2, in which the water flows into the round region under the MMP's tip. Thus, the device has normal concentration and transportation capabilities. As the distance increases, the acoustic streaming field changes into the ones shown in b3 and b4, and the concentration and transportation capabilities weaken and eventually disappear for the same reason given in the discussion for Fig. 3.20. For 0 µm < d < 10 µm, the acoustic streaming

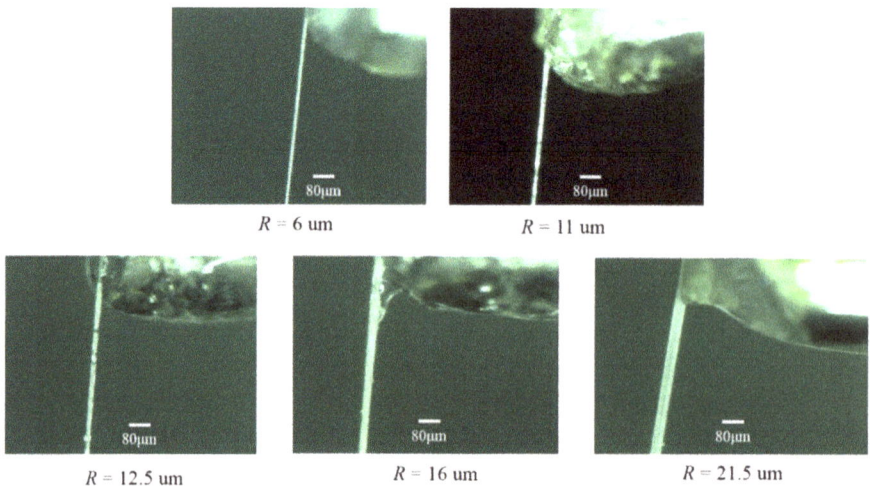

R = 6 um   R = 11 um

R = 12.5 um   R = 16 um   R = 21.5 um

FIGURE 3.21 Photos of some experimental MMPs with different radii. Reproduced from Ref. [10] with permission from Elsevier.

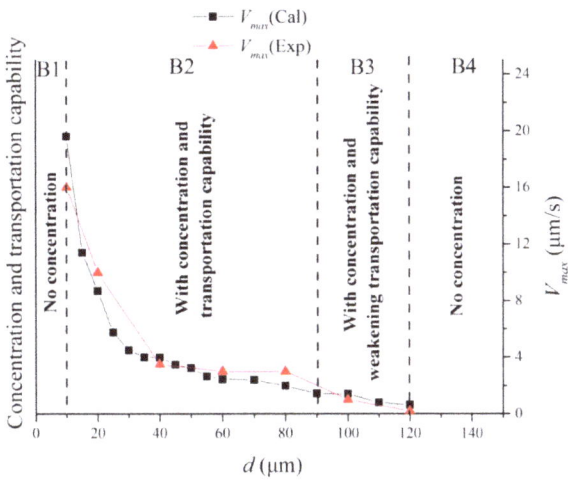

(a)

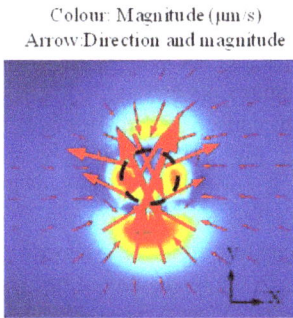

(b1) $d = 3$ μm

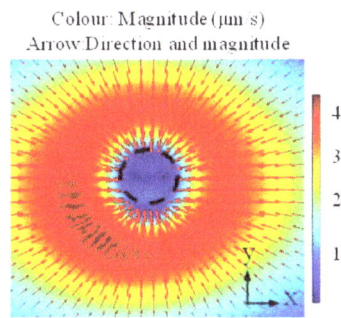

(b2) $d = 30$ μm

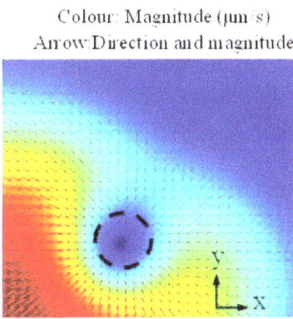

(b3) $d = 100$ μm

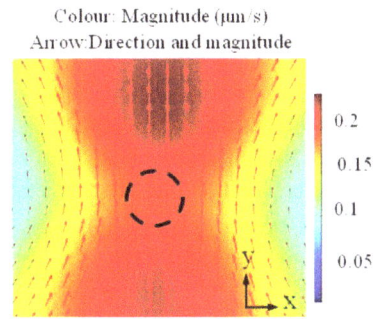

(b4) $d = 130$ μm

(b)

FIGURE 3.22 (a) Computed dependency of the manipulation capabilities and maximum acoustic streaming velocity V_{max} (on the substrate surface) on the distance d between the MMP's tip and substrate surface. Regions B1 and B4: No concentration; Regions B3: With concentration and weakening transportation capabilities; Region B2: With concentration and transportation capabilities. (b) Computed acoustic streaming fields at different distance d between the MMP's tip and the substrate surface at $V_x = 5.2\angle-13.8°$ mm/s (0–p), $V_y = 61.6\angle-12.5°$ mm/s (0–p), $V_z = 2.2\angle172.5°$ mm/s (0–p). Measured dependency of the maximum velocity V_{max} on the distance d is also listed in (a). Reproduced from Ref. [10] with permission from Elsevier.

TABLE 3.5

Measured Dependency of the Concentration and Transportation Capabilities on Distance d between the MMP's Tip and Substrate Surface

d (μm)	10	20, 40, 60, 80	100	120
Concentration capability	No	Yes	Yes	No
Transportation capability	No	Yes	Weakening	No

has a pattern shown in b1. In this case, there is quite strong outward acoustic streaming under the MMP's tip, which flushes NPs out of the round region. When the distance between the MMP's tip and substrate is small, the ultrasonic field between them becomes strong rapidly as the distance d decreases. This causes a strong downward flow from the MMP's tip to the substrate, which generates the outward flow on the substrate as shown in b1.

The effect of angle θ between the VTN and MMP on the concentration and transportation capabilities was analyzed by the FEM, and the results are summarized in Fig. 3.23. Figure 3.23(a) shows different working regions caused by a change of θ, as well as the maximum acoustic streaming velocity V_{max}. Figure 3.23(b) lists several typical acoustic streaming fields on the substrate surface, corresponding to working regions shown in Fig. 3.23(a).

For $70° < \theta < 110°$, the acoustic streaming field has a pattern shown in b3, in which the water flows into the round region under the MMP's tip. Thus, the device has normal concentration and transportation capabilities. As the angle increases or decreases, the acoustic streaming field changes into the pattern shown in b2 or b4, in which part of the round region under the MMP's tip has weak or very weak acoustic streaming. In this case, the NPs on the substrate still can be flushed into the round region, but cannot be well transported across the substrate surface with the movement of the MMP. As the angle increases or decreases further, the acoustic streaming field becomes non-convergent under the MMP's tip, as shown in b1 and b5, and the concentration becomes impossible.

The effect of the MMP length L on the concentration and transportation capabilities was analyzed by the FEM, and the results are summarized in Fig. 3.24. Figure 3.24(a) shows different working regions caused by a change of L, as well as the maximum acoustic streaming velocity V_{max}. Figure 3.24(b) lists several typical acoustic streaming fields on the substrate surface, corresponding to the working regions shown in Fig. 3.24(a). For 0 mm $< L <$ 10 mm, the acoustic streaming field has a pattern shown in b1, in which the water flows into the round region under the MMP's tip. Thus, the device has normal concentration and transportation capabilities. As the length increases, the acoustic streaming field changes into the pattern shown in b2, in which part of the round region under the MMP's tip has weak or very weak acoustic streaming. In this case, the concentration and transportation capabilities weaken. To clarify the reason for the weakening of transportation capability at large MMP length, the ratio of vibration velocities in the z and y directions at the MMP' tip $V_{z,O}/V_{y,O}$ versus

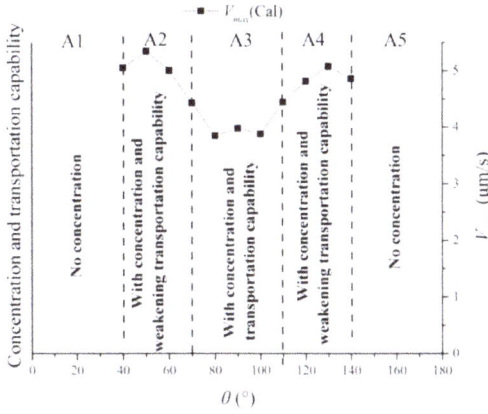

(a)

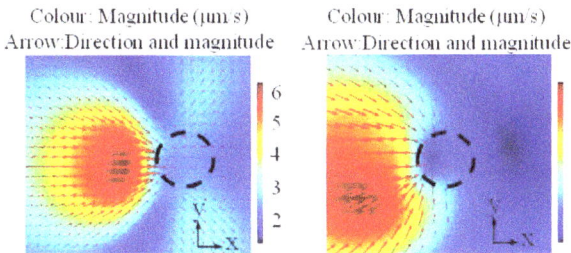

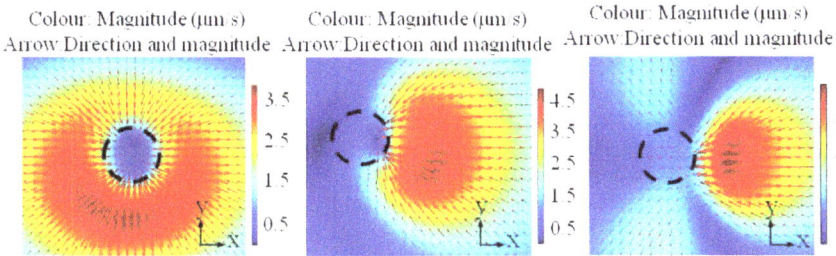

(b)

FIGURE 3.23 (a) Computed dependency of the manipulation capabilities and maximum acoustic streaming velocity V_{max} (on the substrate surface) on angle θ between the MMP and VTN. Regions A1 and A5: No concentration; Regions A2 and A4: With concentration and weakening transportation capabilities; Region A3: With concentration and transportation capabilities. (b) Computed acoustic streaming fields at different angle θ between the MMP and VTN at $V_x = 5.2\angle-13.8°$ mm/s (0–p), $V_y = 61.6\angle-12.5°$ mm/s (0–p), $V_z = 2.2\angle172.5°$ mm/s (0–p). Reproduced from Ref. [10] with permission from Elsevier.

(a)

(c)

(b1) $L = 8$ mm

(b2) $L = 12$ mm

(b)

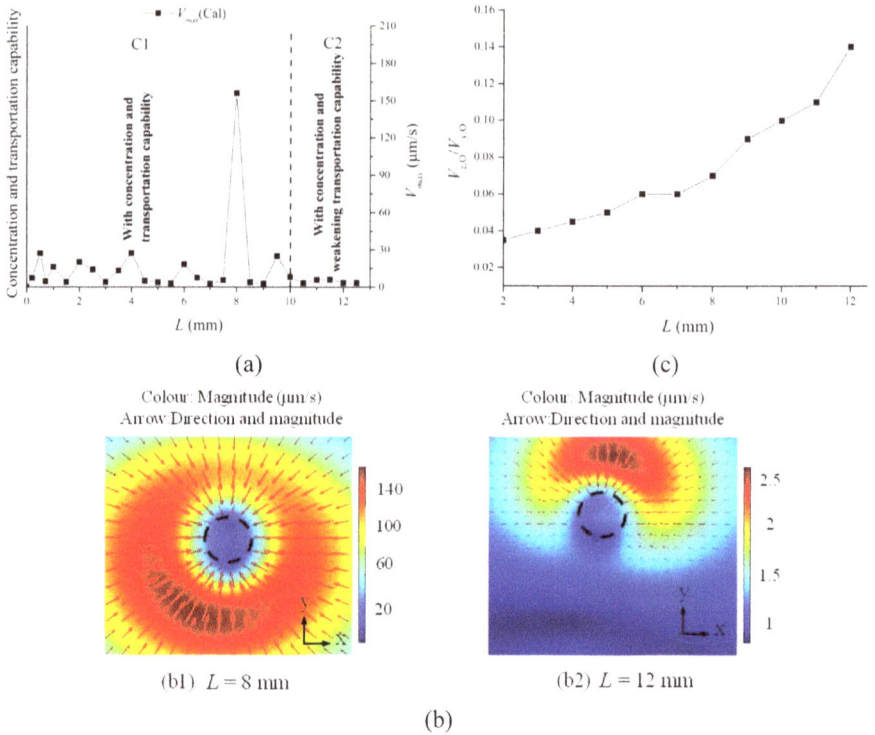

FIGURE 3.24 (a) Computed dependency of the manipulation capabilities and maximum acoustic streaming velocity V_{max} (on the substrate surface) on the MMP length L. Region C2: With concentration and weakening transportation capabilities; Region C1: With concentration and transportation capabilities. (b) Computed acoustic streaming fields when the MMP length L is 8 mm and 12 mm at $V_x = 5.2\angle-13.8°$ mm/s (0–p), $V_y = 61.6\angle-12.5°$ mm/s (0–p), $V_z = 2.2\angle172.5°$ mm/s (0–p). (c) Computed ratio of vibration velocities in the z and y. directions at the MMP' tip versus the MMP length. Reproduced from Ref. [10] with permission from Elsevier.

the MMP length is computed, and the result is shown in Fig. 3.24(c). It is seen that as the MMP length increases, the relative effect of the z-directional vibration velocity at the MMP's tip increases. In P. Liu's experimental work reported in Ref. [16], it has been clarified that the z-directional vibration velocity at the MMP's tip generates acoustic streaming field flowing outwards. Thus, as the MMP length increases, undesired acoustic streaming pattern shown in image b2 of Fig. 3.24(b) happens.

The peak of the maximum acoustic streaming velocity V_{max} at $L = 8$ mm indicates that the MMP works in resonance at this point. As the vibration velocity of the MMP increases, sound field around the MMP becomes strong, and so is the acoustic streaming. Thus, increasing the MMP vibration can increase the Stokes force applied to the NPs, which is generated by the acoustic streaming. The Stokes force applied on the NPs is the driving force during the concentration process and the holding force during the transportation process. Therefore, the resonance phenomenon may be employed to increase the concentration and transportation capabilities. It can be implemented by properly designing the MMP length and/or tuning the working frequency. However, one needs to shift the working point a bit away from the resonance point to avoid the instability of the acoustic streaming field and concentration process. This is because when the working point is exactly at the resonance point, the MMP vibration amplitude will be very sensitive to the variation of the MMP resonance frequency, which may be caused by a change of the MMP length submerged in the droplet.

From Figs. 3.20 to 3.24, it is also known that the maximum acoustic streaming velocity varies with the parameters when the device works within the region with normal concentration and transportation capabilities, no matter whether there is resonance or not. This provides many possible methods to enhance the concentration and transportation capabilities. This is because a stronger acoustic streaming field can enhance the trapping capability by increasing the Stokes force applied onto the NPs, and the enhancement of trapping capability makes the transportation of concentrated nano material easier.

3.3.3 CONCLUSIONS

A strategy that is capable of concentrating or enriching NPs at the interface between a droplet and plain substrate (a smooth substrate without vibration excitation mechanism and micro channels) and of transporting the concentrated nano material across the interface has been proposed and investigated. It offers a reproducible and facile method for the concentration and transportation of NPs, without the employment of complex equipment. The concentration can be implemented at any desired location at the interface, and the transportation can be implemented through an arbitrary or selected path across the interface. The method has potential applications in the fabrication of nano devices such as nano sensors, nano electrodes, nano transistors and super capacitors, and in the nano decoration and various nano assembly processes. It is especially competitive in the on-site manipulations of nano materials on the substrate of a device.

3.4 A COMPUTATIONAL EXAMPLE OF ACOUSTOFLUIDIC FIELD FOR CONCENTRATION

A quantitative analysis is necessary to understand the working mechanism of ultrasonic concentration process of micro/nano scale material and to design the ultrasonic device properly. This section gives an example of analyses of acoustofluidic field for micro/nano scale material concentration [11]. The FEM is used in this example to compute the acoustic field and the acoustic streaming field which is generated by the acoustic field. The computation of the acoustofluidic field is implemented by the FEM software COMSOL Multiphysics (version 4.3).

3.4.1 COMPUTATIONAL MODEL AND METHOD

A math-physical model for the ultrasonic needle-liquid-substrate system is shown in Fig. 3.25, and its meshed FEM model is shown in Fig. 3.26. This math-physical

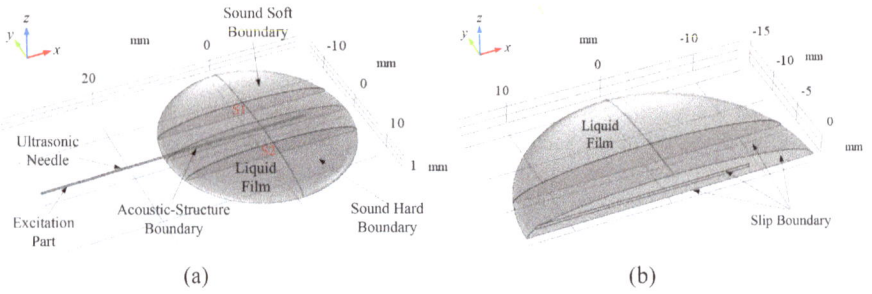

FIGURE 3.25 A 3D acoustofluidic field model for the ultrasonic needle-liquid-substrate system. (a) Boundary conditions for the sound field. (b) Boundary conditions for the acoustic streaming field. Reproduced from Ref. [11] with permission from Springer Nature.

model is based on the ultrasonic device reported in Ref. 11. The mesh size of the acoustofluidic field near the ultrasonic needle is smaller than that in the rest region, in order to decrease the computational error of the acoustofluidic field near the needle, which is more important to the analyses and discussion in this work. In Fig. 3.25(a), S1 and S2 are two cross-sections of the water film, which are symmetric about and

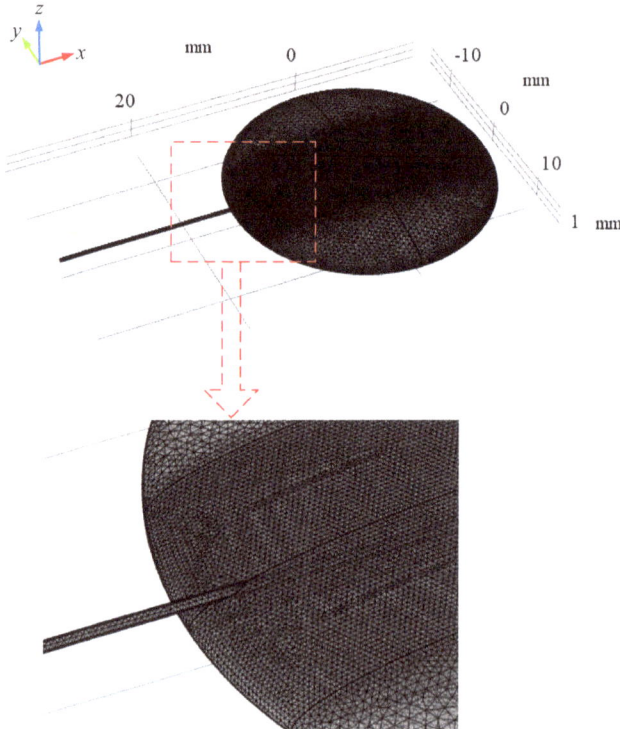

FIGURE 3.26 Meshed model for the acoustofluidic field. Reproduced from Ref. [11] with permission from Springer Nature.

parallel to the needle and have a separation of one-third of the water film diameter. Unless otherwise specified, the detailed mesh sizes of different regions are as follows. Within the region of the liquid film between S1 and S2, the maximum element size is 0.18 mm (about 0.81% of the wavelength of the sound field in water at 67.8 kHz). In the rest region of the liquid film, the maximum element size is 0.34 mm (about 1.54% of the wavelength of the sound field in water at 67.8 kHz). Also, the maximum element size of the needle is 0.075 mm (about 21.4% of the needle's diameter). It was confirmed that the numerical results were mesh-independent and convergent around these conditions.

The detailed boundary conditions of the ultrasonic needle-liquid-substrate system for the sound field and acoustic streaming calculation in the liquid film are shown in Fig. 3.25. Only the steady-state acoustofluidic field is computed. The computational process consists of the following three steps.

In the first step, the sound field is solved by the sound-structure coupling module with the boundary conditions shown in Fig. 3.25(a). Boundary conditions of the sound field are as follows: The normal acceleration is continuous at the interfaces between the ultrasonic needle and liquid film ($a_n^n = a_n^l$, where a_n^n represents the normal acceleration of the ultrasonic needle at the interfaces, and a_n^l represents the normal acceleration of the liquid film at the interfaces), which can be defined as acoustic-structure boundary. The interfaces between the liquid film and air are sound soft (sound pressure $p = 0$) for the reason that ultrasound attenuates quickly in the air; The interface between the liquid film and substrate is sound hard ($\frac{\partial p}{\partial \mathbf{n}} = \mathbf{0}$, where $\mathbf{n}$ denotes the unit normal vector of the boundary). The following wave equation is used to solve the sound field:

$$\rho_f \frac{\partial^2 p}{\partial t^2} = \rho_f c_f^2 \nabla^2 p + b \nabla^2 \frac{\partial p}{\partial t}, \tag{3.11}$$

where p is the sound pressure, ρ_f is the fluid density without sound field and c_f is the sound speed, and the acoustic dissipation factor b is computed by

$$b = \frac{4}{3}\eta + \eta', \tag{3.12}$$

where η and η' are the shear and bulk viscosity coefficients of the acoustic medium, respectively. The vibration velocity u_i (where subscript i represents x, y or z) of the sound field can be calculated according to

$$u_i = i \frac{1}{\rho_f \omega} \frac{\partial p}{\partial x_i}, \tag{3.13}$$

where $i = \sqrt{-1}$ is the imaginary unit, and ω is the angular frequency.

In the second step, computed vibration velocity and sound pressure of the sound field are used to calculate the spatial gradients of the Reynolds stress and mean pressure (the second order pressure in the sound field), by the post-processing functions of the FEM software. The spatial gradients of the Reynolds stress and mean pressure result in the force driving the acoustic streaming. The spatial gradient of the Reynolds stress F_j is computed by

$$F_j = -\partial \langle \rho_f u_i u_j \rangle / \partial x_i, \tag{3.14}$$

where u_i and u_j are the vibration velocities of the sound field, repeated suffixes i and j represent x, y and z in the three-dimensional model, and $< >$ represents the time average over one time period. The mean pressure $\bar{p}_2$ is computed by

$$\bar{p}_2 = \frac{1}{2\rho_f c_f^2} \frac{B}{A} \langle p^2 \rangle \tag{3.15}$$

where $\dfrac{B}{A}$ is the nonlinear parameter of the medium.

In the last step, the steady acoustic streaming is solved by the fluidic dynamics module of the FEM software. Due to the symmetrical characteristic of the acoustic streaming field produced by ultrasonic vibration in the y direction in our simulation model, only half of the liquid film is used to save the workstation's memory used for the calculation, as shown in Fig. 3.25(b.) The steady acoustic streaming satisfies the following equation:

$$\rho_f (\bar{u}_i \partial \bar{u}_j / \partial x_i) = F_j - \partial \bar{p}_2 / \partial x_j + \eta \nabla^2 \bar{u}_j, \tag{3.16}$$

where $\bar{u}_i$ is acoustic streaming velocity. The acoustic streaming also satisfies the continuity equation

$$\partial \bar{u}_i / \partial x_i = 0. \tag{3.17}$$

Slip boundary condition ($\bar{u}_t \neq 0$ and $\bar{u}_n = 0$, which means that the tangential flow exists while the normal flow velocity is zero) is used in the FEM computation of the acoustic streaming, as shown in Fig. 3.25(b). This is because our experiment shows that tangential acoustic streaming velocity can exist at the interfaces between the liquid film and substrate, the liquid film and needle and the liquid film and air. In the central plane (the zx plane), the flow velocity in the y direction is zero due to the symmetry, and thereby only the slip boundary condition is used.

3.4.2 EXPERIMENTAL VERIFICATION

In order to experimentally verify the FEM simulation results, an ultrasonic needle-liquid-substrate system for the concentration of micro particles in a water film on a silicon substrate is constructed. The experimental micro scale particles (yeast cells) have an average diameter of 4–6 μm, and particle concentration in aqueous suspension is 0.048 mg/ml. Figure 3.27(a) shows a photo of the experimental setup. Figure 3.27(b) shows the detailed size and structure of the ultrasonic needle-liquid-substrate system with a piezoelectric transducer to excite the vibration of the ultrasonic needle. The ultrasonic needle which is mechanically excited by the piezoelectric transducer is inserted into the aqueous suspension film on the

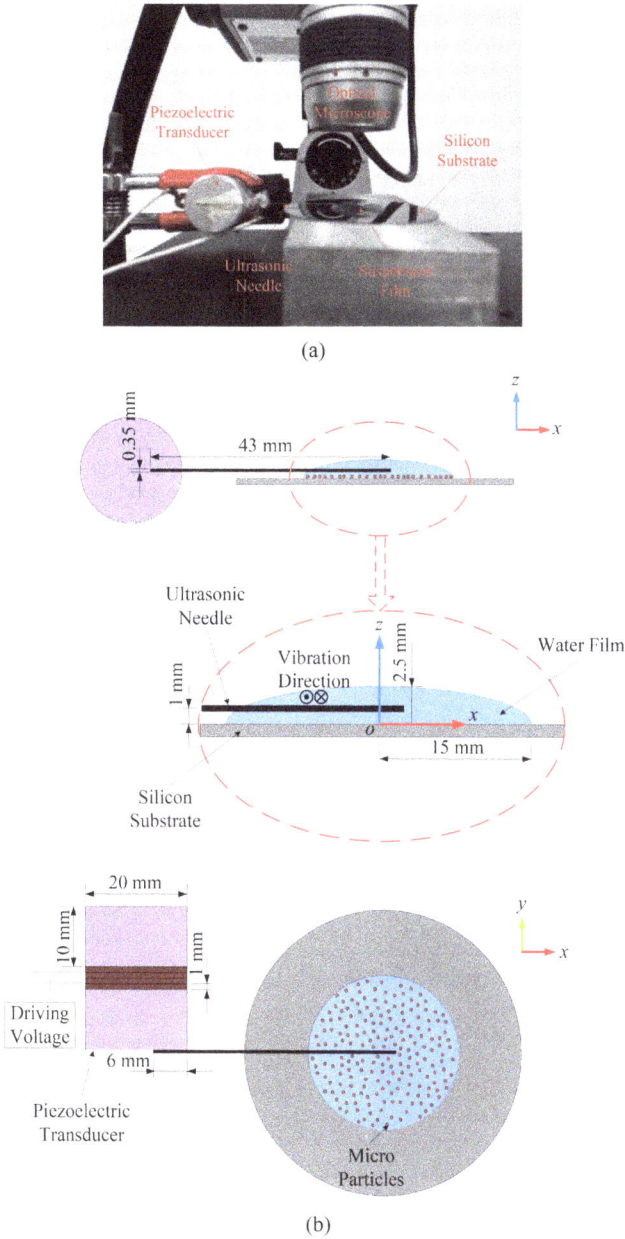

FIGURE 3.27 Experimental setup and the device to verify calculation results. (a) Photo of the experimental setup. (b) Device structure and size. Reproduced from Ref. [11] with permission from Springer Nature.

substrate. In the piezoelectric transducer, four piezoelectric rings are aligned and pressed together by two cylindrical aluminum covers via a bolting structure.

Its vibration direction and electrode configuration are also shown in Fig. 3.27(b). The outer and inner diameters and the thickness of each piezoelectric ring in the transducer are 20 mm, 6 mm and 1 mm, respectively. The electromechanical quality factor Q_m, piezoelectric coefficient d_{33} and relative dielectric constant $\varepsilon_T^{33}/\varepsilon_0$ of the piezoelectric ring are 2000, 325×10^{-12} m/V and 1450, respectively. Each cylindrical aluminum cover at the two ends of transducer has a diameter of 20 mm and thickness of 10 mm. The steel needle is 43 mm long and 0.35 mm in diameter. The length of the needle bonded onto the piezoelectric transducer is 6 mm. The length of the needle inserted in the water film is about 22 mm. The distance between the ultrasonic needle center and substrate surface is controlled by an *XYZ* stage and is 1 mm in Fig. 3.27b. The maximum height and radius of the water film are 2.5 and 15 mm, respectively. The driving voltage is sinusoidal, and the transducer woks at resonance frequency of the needle (67.8 kHz). In Fig. 3.27(b), the needle is excited by the transducer in the *y* direction. Thus, its vibration is parallel to the substrate. In the computation, the center of the interface between the water film and substrate is defined as the origin *o* of the *xyz* coordinate system.

3.4.3 RESULTS AND DISCUSSION

To simplify the computation, the piezoelectric transducer used to excite the needle's vibration is not included in the FEM model. Also, unless otherwise specified, property parameters of the needle, liquid film (water) and micro particles (yeast cells) shown in Table 3.6 are used.

To explain the experimental phenomena of micro particle concentration, the acoustic streaming field around the ultrasonic needle is computed, and the flow pattern on the substrate is shown in Fig. 3.28(a). In Fig. 3.28(a), the color denotes the

---

TABLE 3.6

Parameters of the Ultrasonic Needle-Liquid-Substrate System

Water Film's Height H_w (mm)	Water Film's Radius R_w (mm)	Needle's Total Length L_{nt} (mm)
2.5	15	43
Needle's radius R_n (mm)	Distance between needle center and substrate surface D_{ns} (mm)	Excitation part length of needle L_{ne} (mm)
0.175	1	6
Density of water ρ_f (kg/m³)	Sound speed in water c_f (m/s)	Shear viscosity of water η (Pa·s)
998	1479	0.001
Volume-to-shear viscosity ratio in water η'/η	Nonlinear parameter of water at room temperature B/A	Density of needle ρ_n (kg/m³)
2.1	5	7850
Young modulus of needle E_n (Pa)	Poisson ratio of needle γ_n	Density of yeast cell ρ_p (kg/m³)
2.05×10^{11}	0.28	1114
Sound speed in yeast cell c_p (m/s)	Average radius of yeast cell R_p (μm)	Excitation amplitude of needle A_n (nm)
1606	3	150

magnitude of the acoustic streaming velocity, and the arrow denotes the direction and magnitude of the acoustic streaming velocity. It is seen that there are some small regions near the needle (regions A–F), in which the flow velocity is much larger than that in the region farther away from needle. From Fig. 3.28(a), it is seen that flow near the needle can drive the particles on the substrate to the location under the needle,

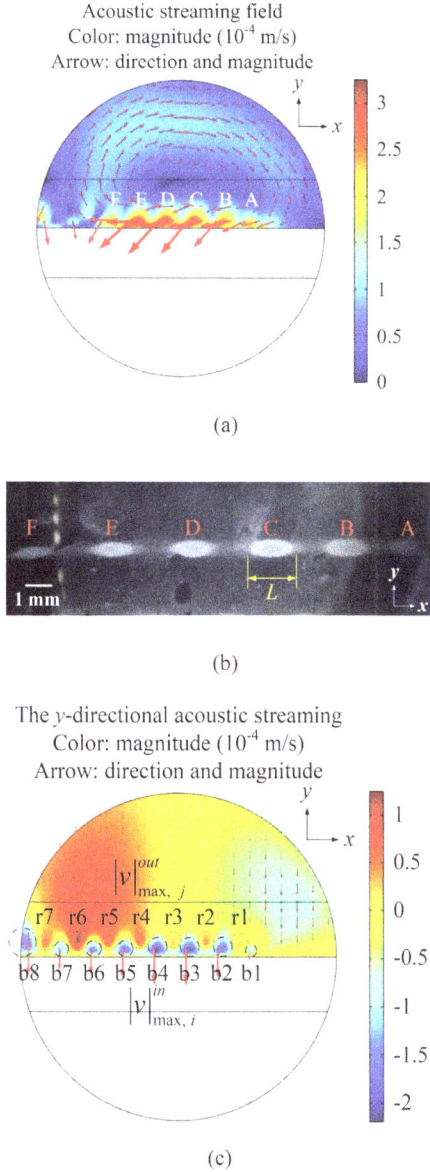

(a)

(b)

(c)

FIGURE 3.28 Acoustic streaming field on the substrate surface at 67.8 kHz and a photo of the concentration pattern. (a) Calculated acoustic streaming field on the substrate. (b) Observed concentration pattern. (c) Distribution of the y-directional acoustic streaming velocity. Reproduced from Ref. [11] with permission from Springer Nature.

and the symmetry of the flow on the two sides of the needle makes the particle concentration under the needle feasible.

Figure 3.28(b) is the measured concentration pattern that was obtained with the use of yeast cells. Comparing Fig. 3.28(a, b), it is seen that regions $A–F$ in Fig. 3.28(a) correspond to the lobed concentration spots $A–F$ in Fig. 3.28(b), respectively, and the distance between the neighboring locations with maximum inward flow velocity in Fig. 3.28(a) (= 3 mm) is very close to that between the neighboring locations with maximum width of the lobed concentration in Fig. 3.28(b) (= 2.9 mm). The average length L of concentration spots in the experiment is about 1.5 mm, while the average length of red regions in Fig. 3.28(a), in which there is inward flow, is about 1.8 mm. The difference is because it is difficult to drive the particles on the substrate at the locations where the flow velocity is small.

Figure 3.28(c) shows the distribution of the y-directional acoustic streaming velocity. It is seen that there are inward and outward flows (for the needle) on the substrate. Dashed circles b1–b8 and r1–r7 are inserted into Fig. 3.28(c) to indicate the regions of inward and outward flows, respectively. A ratio of the average of maximum inward flow velocities ($v|^{in}_{\max,i}$ in the blue regions b1–b8) to the average of maximum outflow velocities ($v|^{out}_{\max,j}$ in the red regions r1–r7) is defined to quantify the concentration capability. The concentration ability γ is

$$\gamma = \frac{\dfrac{1}{M}\displaystyle\sum_{i=1}^{M}|v|^{in}_{\max,i}}{\dfrac{1}{N}\displaystyle\sum_{j=1}^{N}|v|^{out}_{\max,j}}, \tag{3.18}$$

where M and N represent the total numbers of inward flow and outward flow regions, respectively.

For micro manipulation, there is always a question about which force is dominant during the manipulation process, the drag force induced by the acoustic streaming or acoustic radiation force applied on individual manipulated micro objects. We calculated the y-directional starting drag force on a 6-μm-diameter yeast cell at the interface between the water and substrate, which is caused by the y-directional acoustic streaming, and the result is shown in Fig. 3.29(a). For comparison, we also calculated the y-directional acoustic radiation force on a 6-μm-diameter yeast cell at the interface between the water and substrate, and the result is shown in Fig. 3.29(b).

The y-directional starting drag force F_y^{drag} generated by the acoustic streaming is calculated by

$$F_y^{drag} = 6\pi\eta R_p\left(\bar{u}_y - v_{py}\right), \tag{3.19}$$

where R_p is the average radius of manipulated particles (yeast cells), and v_{py} is the particle velocity in the y direction and is set to be zero in the calculation of a starting drag force. The y-directional acoustic radiation force is calculated by

$$F_y^{rad} = -\frac{4}{3}\pi R_p^3\frac{\partial}{\partial y}\left[\frac{1-\beta}{2\rho_f c_f^2}\langle p^2\rangle - \frac{D}{2}\rho_f\langle u_x^2 + u_y^2\rangle\right], \tag{3.20}$$

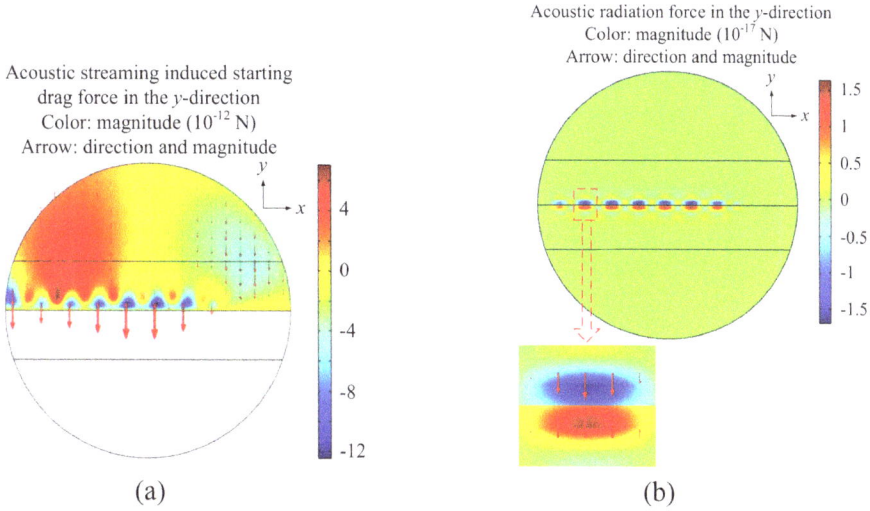

(a)   (b)

FIGURE 3.29   (a) Distribution of the y-directional acoustic streaming induced starting drag force on the yeast cell on the substrate. (b) Distribution of the y-directional acoustic radiation force on the yeast cell on the substrate. Reproduced from Ref. [11] with permission from Springer Nature.

where parameters β and D can be expressed as

$$\beta = \frac{\rho_f c_f^2}{\rho_p c_p^2}, \tag{3.21}$$

and

$$D = \frac{3(\rho_p - \rho_f)}{2\rho_p + \rho_f}, \tag{3.22}$$

where ρ_p and c_p are the density and sound speed of the micro particles (yeast cells), respectively, and ρ_f and c_f are the density and sound speed of the fluid, respectively. The detailed parameter values of the yeast cells and fluid (water) are listed in Table 3.6. According to the calculation results, it is known that the starting drag force generated by the acoustic streaming is larger than the acoustic radiation force by 10^5 times. That is the reason why the acoustic streaming is dominant in the concentration process implemented by the ultrasonic needle-liquid-substrate system.

The effects of the distance between the needle center and substrate surface D_{ns} on the concentration capability γ for different needle radii are computed at 67.8 kHz, and the result is shown in Fig. 3.30. In the calculation, the water film's radius and height are kept constant (15 and 2.5 mm, respectively). The concentration capability increases first and then decreases as the distance between the needle center and substrate surface increases. Our previous experiments show there exists a range of the distance between the needle center and substrate surface (from 0.5 to 2 mm), beyond

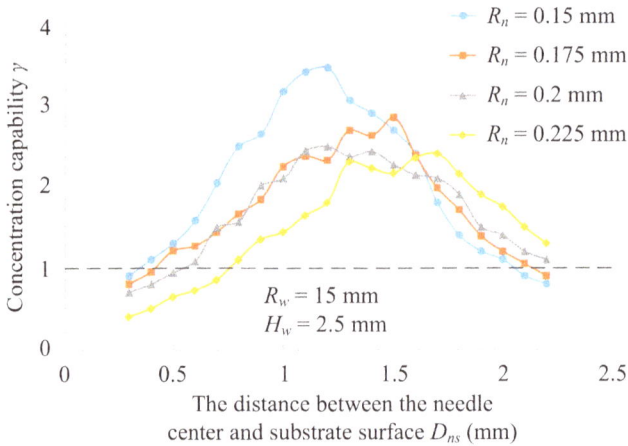

FIGURE 3.30 Effect of the distance between the needle center and substrate surface on the concentration capability for different needle radii at 67.8 kHz. Reproduced from Ref. [11] with permission from Springer Nature.

which no particle concentration can be observed. The results in Fig. 3.30 clearly indicate that the concentration capability γ is less than 1 when the distance is beyond this range, which means that the averaged inward flow velocity is smaller than the averaged outward flow velocity in this case.

The measured and calculated minimum distances for the generation of the concentration effect at different needle radii in the experiments and calculated results are shown in Table 3.7.

It is seen that there is good agreement between the computed and measured results, and the larger the needle radius is, the larger the minimum distance is. For a given distance between the needle center and substrate, the distance between the needle and substrate decreases as the needle radius increases. This causes an increase of inward flow resistance and a decrease of outward flow resistance, which lowers the concentration capability. To maintain the concentration capability, one needs to raise the needle.

The dependency of the concentration capability γ on the water film's height and radius was computed at 68.7 kHz, and the results are shown in Fig. 3.31. Figure 3.31(a) shows that the concentration capability γ keeps almost constant for different water film heights (from 1 to 4.5 mm). Figure 3.31(b) shows that the concentration capability γ keeps almost constant for different water film radius (from 10 to 20 mm). Thus, the

TABLE 3.7

Measured and Calculated Minimum Distances for the Concentration at Different Needle Radii

Needle's radius R_n (mm)	0.15	0.175	0.2	0.225
Measured minimum distance (mm)	0.3	0.5	0.7	0.8
Calculated minimum distance (mm)	0.4	0.5	0.6	0.8

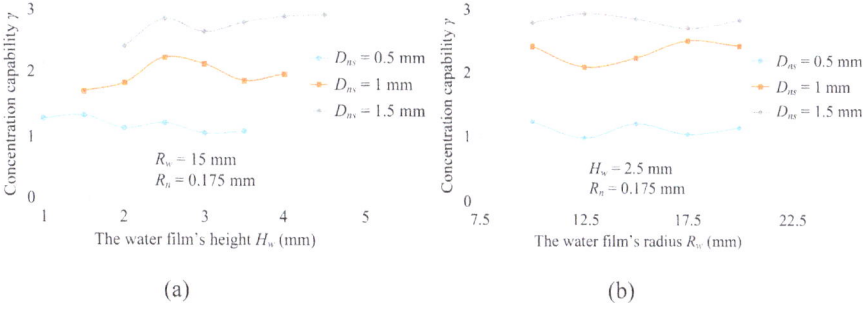

FIGURE 3.31 (a) Concentration capability versus the water film's height. (b) Concentration capability versus the water film's radius. Reproduced from Ref. [11] with permission from Springer Nature.

water film's height and radius have little effect on the concentration capability γ, which are also in agreement with the experimental phenomena. The water film's size only affects the acoustic streaming field farther away from the needle and has little effect on the acoustic field and acoustic streaming near the needle. The non-sensibility of the concentration capability to the water film's size is beneficial for lowering the requirement on the dispenser's performance in repeatability.

The effects of the needle's cross-section shape and size on the concentration capability were also calculated by the FEM. Elliptical, rectangular and rhombic cross-sections of the needle were investigated. The calculation conditions are as follows: The working frequency and excitation amplitude of the needle are 67.8 kHz and 150 nm, respectively; the water film's radius and height are 15 and 2.5 mm, respectively; the distance between the needle center and substrate surface is 1 mm. The maximum element size of the needle in the computation is kept to be about 20% of the needle's minimum structural dimension when the needle's cross-section shape and size have a change.

Figure 3.32 shows the calculated concentration capability versus half of the axis length (a_n) parallel to the substrate when the cross-section is elliptical. In the computation, the ellipse area is kept constant ($S_n = \pi \times 0.175^2$ mm$^2 \approx 0.0962$ mm^2) and a_n changes from 0.075 to 0.275 mm. Figure 3.32(a) shows the FEM model viewed from the x direction. Figure 3.32(b) shows that as a_n increases, the concentration capability increases first and then changes little. Figure 3.32(b) also includes the calculation result for the needle with a circular cross-section. It is known that the concentration capability may be improved if the commonly used needle with a circular cross-section is vertically flattened along the z direction. For the elliptical cross-section with a constant area, as a_n increases, the curvature of the needle's surface facing the substrate decreases, which lowers the needle resistance to the inward flow. When a_n is large enough, the needle's shape is close to a thin plate parallel to the substrate. In this case, the thickness change of the needle little affects the flow resistance.

Figure 3.33 shows the calculated concentration capability versus the side length L_n parallel to the substrate when the cross-section is rectangular. In the computation, the rectangle area is kept constant ($S_n = 0.35^2$ mm$^2 = 0.1225$ mm^2) and L_n changes

(a)

(b)

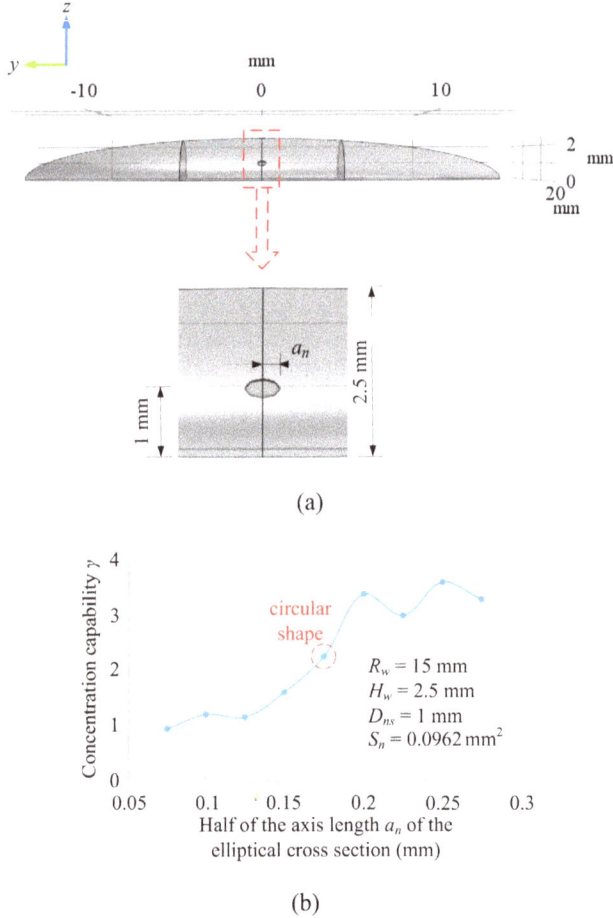

FIGURE 3.32 Calculated concentration capability versus half of the length of the axis a_n parallel with the substrate when the cross-section is elliptical. (a) The FEM model viewed from the x direction. (b) Concentration capability at 67.8 kHz. Reproduced from Ref. [11] with permission from Springer Nature.

from 0.15 to 0.55 mm. Figure 3.33(a) shows the FEM model viewed from the x direction. It is seen that as L_n increases, the concentration capability decreases first and then changes little. Figure 3.33(b) also includes the calculation result for the needle with a square cross-section. It is known that the concentration capability may be improved if the needle with a rectangular cross-section is horizontally flattened along the y direction.

Figure 3.34 shows the calculated concentration capability versus the diagonal length L_{nd} parallel to the substrate when the cross-section is rhombic. In the computation, the rhombus area is kept constant ($S_n = 0.1225$ mm^2) and L_{nd} changes from 0.21 to 0.77 mm. Figure 3.34(a) shows the FEM model viewed from the x direction, and the rhombic angle facing the substrate is defined as θ. Figure 3.34(b) shows that

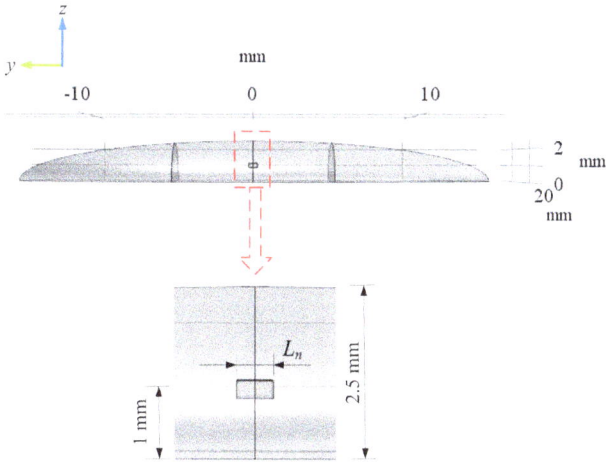

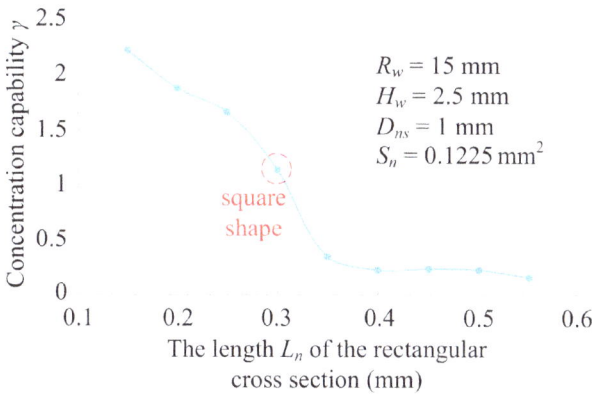

FIGURE 3.33 Calculated concentration capability versus the side length L_n parallel with the substrate when the cross-section is rectangular. (a) The FEM model viewed from the x direction. (b) Concentration capability at 67.8 kHz. Reproduced from Ref. [11] with permission from Springer Nature.

as L_{nd} increases, the concentration capability increases first and then decreases. It is known that the maximum concentration capability may be achieved when the angle θ is about $36°$ (the diagonal length of the rhombus is 0.28 mm). The phenomena shown in Figs. 3.33 and 3.34 are explained as follows.

The sharp edge on a vibrating needle can generate strong local eddies due to very large spatial gradient of the Reynolds stress nearby. When the needle vibrates in the direction parallel to the substrate, the local eddies usually flow outwards to the right or left side of the needle and flow back along the substrate surface, and generate the inward flow on the substrate, which positively contributes to the concentration capability.

(a)

(b)

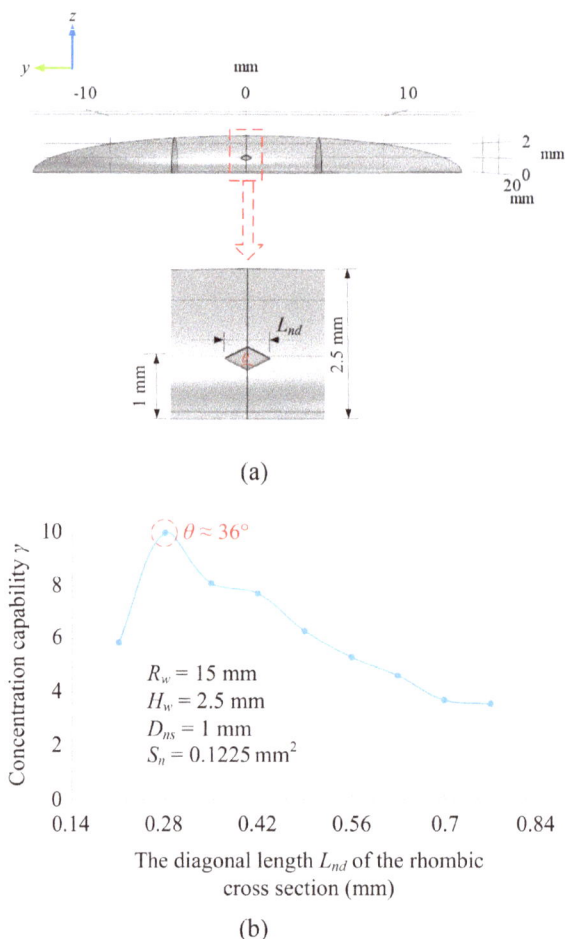

FIGURE 3.34 Calculated concentration capability versus the diagonal length L_{nd} parallel with the substrate when the cross-section is rhombic. (a) The FEM model viewed from the x direction. (b) Concentration capability at 67.8 kHz. Reproduced from Ref. [11] with permission from Springer Nature.

The rectangular cross-section has right-angle edges, which generate the local eddies and positively contribute to the concentration capability. For the cross-section with a constant area, as L_n increases, the distance between the needle's lower surface (facing the substrate) and the substrate increases, which decreases the edge-caused inward flow on the substrate and lowers the concentration capability. When L_n is large enough, the distance between the needle's lower surface (facing the substrate) and the substrate is so large that the edge-caused inward flow on the substrate affects little on the concentration capability. In this case, the concentration capability changes little with L_n. When the cross-section of the needle is rhombic, the two side vertices can also generate the local eddies, which positively contribute to the concentration capability. For the cross-section with a constant area, as L_{nd}

increases, the distance between the lower vortex and substrate surface increases. This lowers the inward flow on the substrate and causes the decrease of the concentration capability.

By a comparison of Figs. 3.32–3.34, it is known that the concentration capability of the needle with a rhombic cross-section is the highest and that of the needle with an elliptical one is the weakest when the lateral dimensions ($2a_n$, L_n and L_{nd}) are smaller. This phenomenon can also be well explained by the local eddies which are generated by the edges of the needle in vibration. The needle with a rhombic cross-section has two sharp edges on its two sides. Also, compared with the rectangular cross-section, it has smaller flow resistance to the inward flow due to its declining lower surface. Thus, it has the strongest concentration capability at the smaller lateral dimension. The needle with an elliptical cross-section has no sharp edge on its surface. Thus, it has the weakest concentration capability at the smaller lateral dimension.

By a comparison of Figs. 3.32–3.34, it is also known that the concentration capability of the needle with a rectangular cross-section is much smaller than those of the needle with the elliptical and rhombic ones when the lateral dimensions ($2a_n$, L_n and L_{nd}) are sufficiently large. This phenomenon is explained as follows. For the needle with a rectangular cross-section, the thickness of the ultrasonic field between the needle and substrate is uniform, which makes its ultrasonic field on the substrate relatively uniform compared to the ones generated by the needles with the elliptical and rhombic cross-sections. A uniform ultrasonic field on the substrate makes the spatial gradient of the Reynolds stress along the substrate very weak, which results in a very weak acoustic streaming along the substrate and very weak concentration stability. For the needle with an elliptical or rhombic cross-section, the ultrasonic field between the needle and substrate has a larger thickness at the field's edge than at the center, which causes the acoustic streaming on the substrate to flow inward and relatively strong concentration stability.

Further FEM calculation for a needle with the rhombic cross-section indicates that the optimum apex angle θ is affected by the distance between the needle center and substrate surface D_{ns} and the cross-section area of the needle S_n, as shown in Fig. 3.35. It shows that for a needle with the rhombic cross-section of given size and shape, one may change the distance between the needle and substrate to achieve the optimal concentration effect.

From Figs. 3.32 to 3.35, it is seen that the concentration capability is affected by the shape and size of the cross-section of the needle. The effects of the distance between the needle center and substrate surface, the needle's radius, the water film's height and radius and the shape of the needle's cross-section, on the concentration capability, are summarized in Table 3.8.

In addition, a comparison of the streaming fields in the yz plane of an anti-nodal position along the needles with the elliptical ($a_n = 0.075$ mm, $S_n = 0.0962$ mm^2), rectangular ($L_n = 0.15$ mm, $S_n = 0.1225$ mm^2) and rhombic ($L_{nd} = 0.28$ mm, $S_n = 0.1225$ mm^2) cross-sections is listed in Fig. 3.36. It is confirmed that the sharp edges on the needles with the rectangular and rhombic cross-sections can generate remarkable local eddies, while the needle with the elliptical cross-section generates no local eddy as the needle surface is smooth.

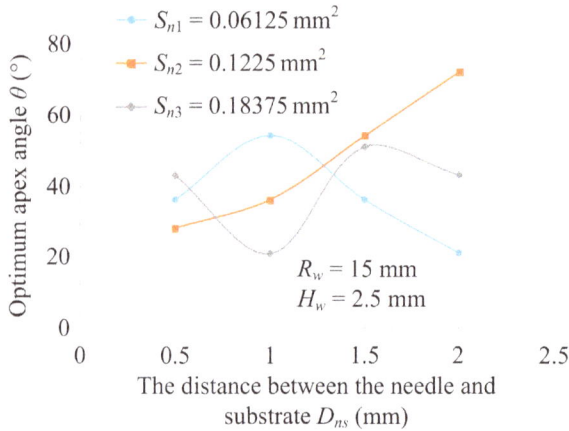

FIGURE 3.35 The dependence of the optimum apex angle θ on the cross-section area of the needle and the distance between the needle center and substrate. Reproduced from Ref. [11] with permission from Springer Nature.

TABLE 3.8

Effects of the Dimensions on Concentration Capability at 67.8 Khz

	Distance between the Needle Center and Substrate Surface D_{ns}	Needle's Radius R_n	Water Film's Height H_w	Water Film's Radius R_w	Shape of the Needle's Cross-Section
Concentration capability γ	Sensitive	Sensitive	Insensitive	Insensitive	Sensitive

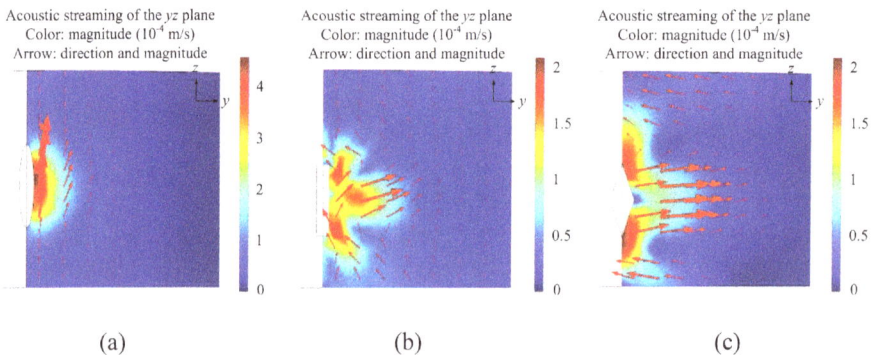

FIGURE 3.36 A comparison of the acoustic streaming fields in the yz plane for the needles with different cross-sections. (a) Elliptical ($a_n = 0.075$ mm, $S_n = 0.0962$ mm²). (b) Rectangular ($L_n = 0.15$ mm, $S_n = 0.1225$ mm²). (c) Rhombic ($L_{nd} = 0.28$ mm, $S_n = 0.1225$ mm²). Reproduced from Ref. [11] with permission from Springer Nature.

3.4.4 Summary

In this example, the acoustofluidc field in the ultrasonic needle-liquid-substrate system is numerically calculated and analyzed. The FEM calculation results of the acoustic streaming field on the substrate can well explain the concentration pattern in the experiments. With the FEM model, useful guidelines to design the device have been achieved.

3.5 REMARKS

The ultrasonic concentration methods for nano/micro scale objects, demonstrated in this chapter, give a special way to enrich nano/micro scale materials at the interface between a liquid and solid substrate, which have potential applications in nano device fabrication, high-sensitivity biological sensing systems, nano composite functional material syntheses, etc. In nano device fabrication, the enriched nano functional materials may be further transferred onto a substrate to form the devices such as nano gas sensors. In high-sensitivity biological sensing systems, the ultrasonic nano concentration method may be used to increase the local density of the biological analyte, which is an effective way to increase the output signal of a biological sensor system, such as fluorescence strength.

A basic feature of these methods is that they employ the acoustic streaming vortices to implement the concentration. This results in a quite low temperature rise at the concentration spot. Actually, in the author's experiments of ultrasonic nano concentration, no temperature-rise higher than 2 °C was observed at the concentration spot, which is invaluable in the enrichment process of biological samples. This feature also means that the enrichment process will not be affected by the electromagnetic and optical property of the micro/nano scale samples.

The key technology in the ultrasonic nano/micro concentration methods includes the device design and acoustic streaming field control. Based on the existing topological structure, the design of the ultrasonic device can be further optimized by the FEM and prototype testing. The purpose of design optimization of the ultrasonic device is to make the vibration excitation system capable of generating a desired and sufficiently strong vibration mode at the manipulation probe. The control of acoustic streaming field is to control the shape, size and relative location (to the manipulation probe) of the aggregates. The acoustic streaming pattern is mainly affected by the vibration mode of the manipulation probe. However, the distance between the manipulation probe and substrate, ultrasonic wavelength and manipulation probe's orientation also have quite large effect on the acoustic streaming pattern.

A big challenge facing the ultrasonic nano/micro concentration technology is how to greatly increase the lateral size of an aggregate. Solving this problem is beneficial to the effective fabrication of nano functional materials for various kinds of electronic devices. What acoustic streaming patterns may result in large-size aggregates and how to excite the required acoustic streaming patterns still need to be explored extensively.

REFERENCES

1. N. V. Baker, "Segregation and sedimentation of red blood cells in ultrasonic standing waves," *Nature*, 239(5372), pp. 398–399, 1972.
2. T. Hasegawa, T. Kido, T. Iizuka and C. Matsuoka, "A general theory of Rayleigh and Langevin radiation pressures," *J. Acoust. Soc. Jpn. E*, 21(3), pp. 145–152, 2000.
3. W. T. Coakley, D. W. Bardsley, M. A. Grundy, F. Zamani and D. J. Clarke, "Cell manipulation in ultrasonic standing wave fields," *J. Chem. Tech. Biotechnol.*, 44(1), pp. 43–62, 1989.
4. H. Bohm, L. G. Briarty, K. C. Lowe, J. B. Power, E. Benes and M. Davey, "Quantification of a novel h-shaped ultrasonic resonator for separation of biomaterials under terrestrial gravity and microgravity conditions," *Biotechnol. Bioeng.*, 82(1), pp. 74–85, 2003.
5. J. Hu and A. Santoso, "A π-shaped ultrasonic tweezers concept for manipulation of small particles," *IEEE Trans. Ultrason. Ferroelectr. Freq. Control*, 51(11), pp. 1499–1507, 2004.
6. J. Friend and L. Yeo, "Microscale acoustofluidics: microfluidics driven via acoustics and ultrasonics," *Rev. Mod. Phys.*, 83(2), pp. 647–704, 2011.
7. Y. Zhou, J. Hu and S. Bhuyan, "Manipulations of silver nanowires in a droplet on low-frequency ultrasonic stage," *IEEE Trans. Ultrason. Ferroelectr. Freq. Control*, 60(3), pp. 622–629, 2013.
8. B. Yang and J. Hu, "Linear concentration of microscale samples under an ultrasonically vibrating needle in water on a substrate surface," *Sens. Actuators B Chem.*, 193, pp. 472–477, 2014.
9. Q. Tang, X. Wang and J. Hu, "Nano concentration by acoustically generated complex spiral vortex field," *Appl. Phys. Lett.*, 110, p. 104105, 2017.
10. X. Qi, Q. Tang, P. Liu, I. V. Minin, O. V. Minin and J. Hu, "Controlled concentration and transportation of nanoparticles at the interface between a plain substrate and droplet," *Sens. Actuators B Chem.*, 274, pp. 381–392, 2018.
11. Q. Tang, P. Liu and J. Hu, "Analyses of acoustofluidic field in ultrasonic needle-liquid-substrate system for micro/nano scale material concentration," *Microfluid. Nanofluidics*, 22, p. 46, 2018.
12. J. Hu, H. Zhu, N. Li and C. Zhao, "Sound induced lobed pattern in aqueous suspension film of micro particles," *Sens. Actuators A Phys.*, 167, pp. 77–83, 2011.
13. O. V. Abramov, *High-Intensity Ultrasonics*, (Gordon and Breach Science Publishers, Singapore, 1998).
14. J. Lighthill, *Waves in Fluids*, (Cambridge University Press, Cambridge, 1978), p. 329 and 344–350.
15. Q. Tang and J. Hu, "Diversity of acoustic streaming in a rectangular acoustofluidic field," *Ultrasonics*, 58, pp. 27–34, 2015.
16. P. Liu and J. Hu, "Controlled removal of micro/nanoscale particles in submillimeter-diameter area on a substrate," *Rev. Sci. Instrum.*, 88, p. 105003, 2017.
17. G. K. Batchelor, *An Introduction to Fluid Dynamics*, (Cambridge University Press, 1967).

4 Ultrasonic Nano Tweezers

Ultrasonic nano tweezers are a type of ultrasonic device with manipulation functions including but not limited to selective capture of nanoscale objects and controlled move and release of the captured objects. Compared with other physical tweezers, such as optical tweezers, ultrasonic nano tweezers have the merits such as low working temperature at the manipulation point, multiple manipulation function, etc. Thus, they have potential applications in biological sample manipulation, nano sensor fabrication, assembly of nanoscale parts, etc. This chapter gives the fundamental, working principle, structure and characteristics of the probe-type ultrasonic nano tweezers.

4.1 THE FUNDAMENTAL OF ULTRASONIC NANO TWEEZERS

4.1.1 BASIC STRUCTURE AND WORKING PRINCIPLE

An ultrasonic nano tweezer is usually constructed by three parts, that is, the micro probe (micro manipulating probe or MMP), vibration transmission needle (VTN) and ultrasonic transducer [1–13], as shown in Fig. 4.1. The micro probe is bonded onto the VTN's tip, and the VTN's root is bonded onto the excitation location of the ultrasonic transducer. During the operation, vibration generated by the ultrasonic transducer is transmitted to the micro probe through the VTN, and an acoustofluidic field is excited around the micro probe to capture a nanoscale object or a bunch of nanoscale objects. The capture is carried out at the interface between the nano solution/suspension droplet and substrate under an optical microscope.

The microprobe can be made of fiberglass or a tungsten needle with a diameter from several microns to several ten microns. A too-thin micro probe cannot generate a strong enough acoustofluidic field, and a too-thick microprobe may cause difficulty in the observation of the captured object and increase the positioning error of the captured object. Thus, one has to trade-off the capture force against positioning error in the design. The VTN is made of stainless steel or other metal material, with a diameter of 1 mm or so. The ultrasonic transducer can be a piezoelectric component or sandwich-type structure formed by piezoelectric components (2 or 4 pieces) and two metal end plates. The VTN's root may be bonded onto the edge of the piezoelectric component or the corner of the end plate of the ultrasonic transducer to obtain a relatively large vibration in the VTN. In addition, the "nano" in the name of ultrasonic nano tweezers refers to the scale of the manipulated samples rather than the size of the tweezers. Thus, the apparent size of ultrasonic nano tweezers is not necessarily in the nanometer range.

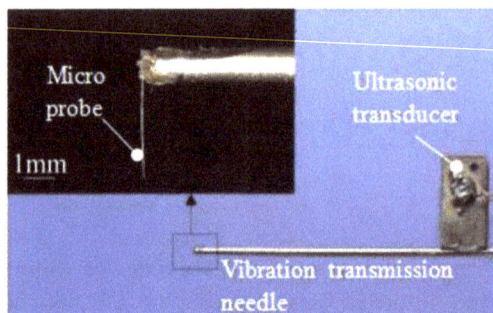

FIGURE 4.1   Typical structure of an ultrasonic nano tweezer.

When the ultrasonic tweezers are in operation, acoustic streaming eddies are generated in the acoustofluidic field around the vibrating micro probe and utilized to capture a nanoscale object or a bunch of nanoscale objects. The acoustic streaming eddies are caused by the 2nd order effect of the ultrasonic field. In ultrasonic nano tweezers, they are mainly caused by the spatial gradient of Reynolds stress [13, 14]. It is the key technology of ultrasonic nano tweezers to design and control the pattern of acoustic streaming eddies around the micro probe, which will be explained by the examples in the rest sections of this chapter.

4.1.2   CLASSIFICATION AND RELATED PROBLEMS

Based on whether the captured nanoscale objects are in contact with the micro probe, the working mode of ultrasonic nano tweezers is classified into the contact-type and noncontact-type. The noncontact-type working mode decreases the chance of cross-contamination among the manipulated nano samples greatly. However, as the transportation of a captured nanoscale sample in this working mode is implemented by dragging the sucked sample(s) on a substrate surface, and micro/nano objects in the transportation path may hinder the transportation process [2, 4]. Although there have been some efforts to solve this problem, it is still a big challenge to freely transport a captured single nanoscale sample in the noncontact-type working mode.

The ultrasonic nano tweezer which can capture a bunch of nanoscale particles at the interface of a droplet and substrate in a noncontact-type working mode, has been proposed and developed [7]. Although the micro concentrated spot of the nanoscale particles can be dragged to any location at the interface by shifting the tweezer, it is very tough to drag a single nanoscale particle to a desired location at the interface as there are other nanoscale particles and their aggregation in the path. In Refs. [9, 10], the ultrasonic nano tweezer, which can capture a bunch of nanoscale particles near the interface of a droplet and substrate in a noncontact working mode, has also been proposed and developed. It can concentrate the captured particles at a location between the micro probe's end and substrate by the acoustic radiation force. With this method, one can transport the concentrated materials onto a desired location within the droplet, and the transportation is not hindered by the objects at the interface. However, this method cannot be applied to transport a single nanoscale object captured, as the acoustic radiation force cannot balance the nanoscale object.

By tuning the working frequency, an ultrasonic nano tweezer may operate in the contact and noncontact mode, respectively. An example of this will be given in Section 3 of this chapter.

4.1.3 VIBRATION CONTROL & EXCITATION MECHANISM

The microprobe of ultrasonic nano tweezers usually vibrates elliptically or linearly [1, 4, 6–10]. So far, the elliptical and linear vibration have been generated by properly designing the shape of the VTN, choosing the transducer form (a sandwich-type transducer or single piezoelectric component), the VTN orientation relative to the vibration direction of the transducer, tuning the working frequency, etc. To obtain a linear and strong vibration at the micro probes tip, the double-parabolic-reflector wave-guided high-power ultrasonic transducer may be employed to excite the VTN's root [8–10]. Of course, one may use two piezoelectric components with vibration velocities spatially normal and 90°temporal phase difference to excite an elliptical motion in the micro probe, like the vibration excitation mechanism in ultrasonic motors [15–17].

4.1.4 BASIC FEATURES AND TECHNICAL CHALLENGES

One of the merits of ultrasonic nano tweezers is that the manipulation point usually has a very low-temperature rise. One reason for this is that they employ the acoustic streaming eddies near the microprobe to manipulate, and the heat flux from the microprobe to the surrounding fluid can be immediately taken away by the acoustic streaming eddies. Another reason is that they employ the vibration of the micro probe to excite the acoustofluidic field, and the heat flux from the microprobe is very small. For the droplet made of distilled water and nanoscale objects, the temperature rise at the manipulation point may be less than $0.1\ °C$ [1, 4, 6–10].

Another merit of ultrasonic nano tweezers is that the electromagnetic and optical properties of manipulated samples do not have a crucial effect on whether the capture is feasible or not. Theoretically, as long as the material of the manipulated nanoscale object is different from the fluid around the microprobe, the manipulation principle is feasible, provided that the acoustic streaming field is strong enough. However, mechanical properties of manipulated nanoscale objects, such as the surface smoothness, adhesion force with the substrate, and shape, may affect the characteristics of a given ultrasonic nano tweezer.

Up to now, in the operation of ultrasonic nano tweezers, the capture location has mainly been limited to the interface between an aqueous droplet and substrate and it is nearby. It is still a big challenge to develop an ultrasonic nano tweezer with the capability of manipulating the nanoscale sample(s) at the air-substrate interface. This is because the acoustic eddies in air are usually much weaker than those in liquid due to a much larger acoustic attenuation in air.

The existing ultrasonic nano tweezers can also be utilized to manipulate nanoscale samples in alcohol solution droplets as acoustic streaming eddies can be generated in alcohol solution, but has not been used to manipulate nano samples in the droplet made of sticky fluid such as oil and dense polymer solution, as it is very difficult to generate acoustic streaming in the sticky fluids.

Apart from capture, shift and release, ultrasonic manipulation tweezers have other manipulation functions. In Ref. [3, 11], it is reported that the ultrasonic twee-zer, which dynamically captures a 100 nm-thick silver nanowire on the interface between a droplet and solid substrate, can simultaneously drive the nanowire to rotate with an angular speed up to 3 rad/s, using the center or one end of the nanow-ire as the rotation center. In Ref. [12], it is reported that an ultrasonic tweezer is capable of nano concentration, nano extraction, nano decoration and nano clean-ing. It can not only capture nanowires and particles in a droplet on a solid substrate, but also extract the captured samples from the droplet into the air to form decorated nanowires, and clean the nano objects on the surface of the solid substrate as wish. These additional manipulation functions make the ultrasonic nano tweezers more helpful in the fabrication of nano sensors, handling of biological samples, assembly of micro/nano machines, etc. How to add more additional manipulation functions to ultrasonic nano tweezers is definitely an important and interesting research area in nano manipulations.

4.2 PRINCIPLE ANALYSIS OF ULTRASONIC NANO TWEEZERS

Ultrasonic nano tweezers are essentially a probe-liquid-substrate system, in which a MMP is inserted into a liquid film of nano suspension on a substrate [1]. With the MMP's vibration, a three-dimensional (3D) acoustic streaming field around the MMP is generated, which has been used to trap, transfer and rotate nano entities in the liq-uid film. In this section, the 3D acoustic streaming field in a probe-liquid-substrate system for the contact-type trapping of individual nanowires is numerically investi-gated by the finite element method (FEM) [18]. The computational results show that the MMP root elliptical vibration can generate an acoustic streaming field capable of trapping a single nanowire in the contact mode. This conclusion can well explain the experimental phenomena that the MMP can trap a single nanowire at some frequen-cies and cannot at other frequencies. The computational results clarify the effect of the distance between the MMP and substrate, the MMP's radius and length, and the water film's thickness on the acoustic streaming field.

4.2.1 COMPUTATIONAL MODEL AND METHOD

A math-physical model for the probe-liquid-substrate system is shown in Fig. 4.2(a), and its meshed FEM model is shown in Fig. 4.2(b). To take a balance between the computational error and time, the mesh size of the acoustofluidic field near the MMP is smaller than that in the rest region. Within the region enclosed by a cylindrical sur-face with a radius of $2R_P \sim 3R_P$, in which R_P is the MMP's radius, the maximum ele-ment size is 1.5 μm (30% of the MMP's radius and about 0.0135% of the wavelength of the sound field at 135 kHz). In the rest region of the acoustofluidic, the maximum element size is 0.55 mm (4.95% of the wavelength of the sound field at 135 kHz). Also, the maximum element size in the MMP is 0.2 mm.

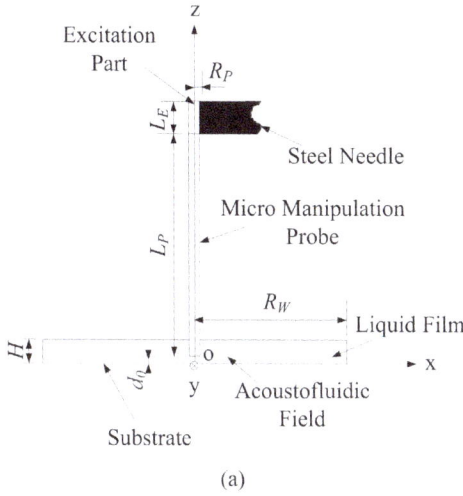

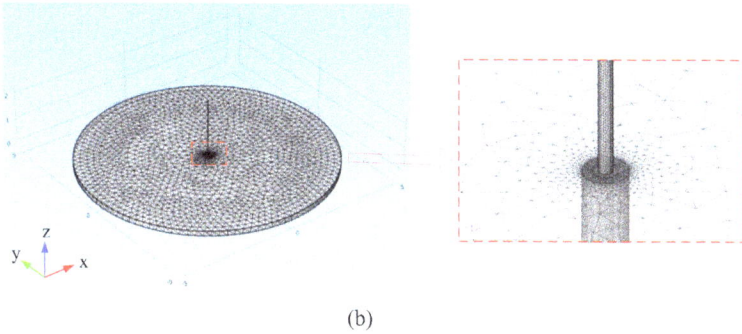

FIGURE 4.2 3D model and meshing for the probe-liquid-substrate system. (a) Math-physical model for the acoustofluidic field. (b) Meshed model for the acoustofluidic field. Reproduced from Ref. [18] with permission from Springer Nature.

The boundary conditions of the probe-liquid-substrate system for the sound field and acoustic streaming in the liquid film are shown in Fig. 4.3. The computation of the acoustic streaming field is implemented by the FEM software COMSOL Multiphysics (version 4.3a). Only the steady-state acoustic streaming field is computed. The computational process consists of the following three steps.

In the first step, the sound field is solved by the sound-structure coupling module with the boundary conditions shown in Fig. 4.3a. Boundary conditions of the sound field are as follows: The acceleration is continuous at the interfaces between the MMP and liquid film; The interfaces between the water film and air are sound soft (sound pressure $p = 0$) for the reason that ultrasound attenuates quickly in the air; The interface between the water film and substrate is sound hard. The following wave equation is used to solve the sound field:

$$\rho \frac{\partial^2 p}{\partial t^2} = \rho c^2 \nabla p + b \nabla \frac{\partial p}{\partial t}, \tag{4.1}$$

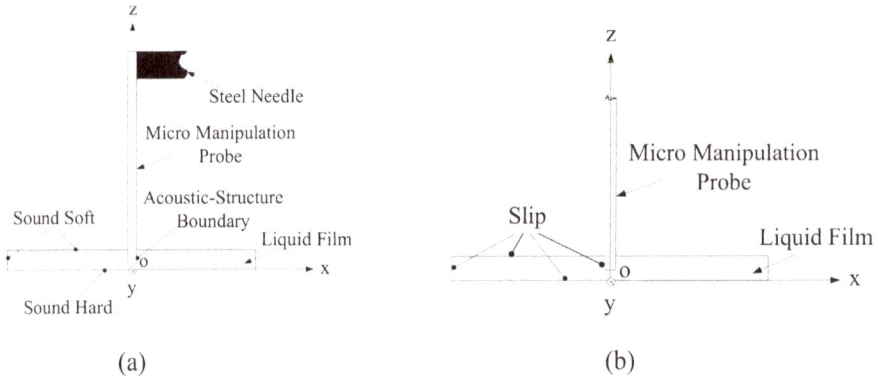

FIGURE 4.3 Boundary conditions of the acoustofluidic field. (a) Boundary conditions for the sound field. (b) Boundary conditions for the acoustic streaming field.

where p is the sound pressure, ρ is the fluid density, c is the sound speed in the acoustic medium, and the acoustic dissipation factor b is computed by

$$b = \frac{4}{3}\eta + \eta', \tag{4.2}$$

where η and η' are the shear and bulk viscosity coefficient of the acoustic medium, respectively.

In the second step, computed vibration velocity and sound pressure are used to calculate the spatial gradients of the Reynolds stress and mean pressure (the second-order pressure in the sound field) by the post processing functions of the FEM software, which generate the driving forces of acoustic streaming. The spatial gradient of the Reynolds stress is computed by [13, 19]

$$F_j = -\partial(\overline{\rho_0 u_i u_j})/\partial x_i, \tag{4.3}$$

where ρ_0 is the medium density in the undisturbed state, u_i is the vibration velocities of the sound field, repeated suffix i and j represent x, y and z in the 3D model, and the symbol "$\overline{}$" signifies the mean value over one time period. The mean pressure is computed by [13]

$$\bar{p}_2 = \frac{1}{2\rho_0 c_0^2} \frac{B}{A} \langle p_1^2 \rangle, \tag{4.4}$$

where p represents the (first-order) sound pressure, $<>$ represents the time average over one time period, c_0 is the medium sound speed in the undisturbed state, and $\frac{B}{A}$ is the nonlinear parameter of the medium [20, 21].

In the last step, the steady acoustic streaming is solved by the fluidic dynamics module of the FEM software. The steady acoustic streaming satisfies the following equation [19]:

$$\rho_0(\bar{u}_i \, \partial \bar{u}_j/\partial x_i) = F_j - \partial \bar{p}_2/\partial x_j + \eta \nabla^2 \bar{u}_j, \tag{4.5}$$

where $\bar{u}_i$ is acoustic streaming velocity. The acoustic streaming also satisfies the continuity equation

$$\rho_0 \, \partial \bar{u}_i / \partial x_i = 0. \tag{4.6}$$

Slip boundary condition is used in the FEM computation of the acoustic streaming, as shown in Fig. 4.3b. This is because our experiment shows that tangential acoustic streaming can exist at the interface between the water film and substrate [22].

4.2.2 EXPERIMENTAL VERIFICATION

To experimentally verify the FEM simulation results, a probe-liquid-substrate system to trap a single silver nanowire in a water film on a silicon substrate is constructed. The experimental silver nanowires have a diameter of 100 nm and a length of several microns up to several tens of microns. Figure 4.4(a) shows a photo of the experimental setup. Figure 4.4(b) shows the detailed size and structure of the probe-liquid-substrate system with a transducer to excite the vibration of the system.

An MMP, which is mechanically excited by a steel needle, is immersed into the nanowire suspension film on the substrate. The MMP is bonded on and excited by the

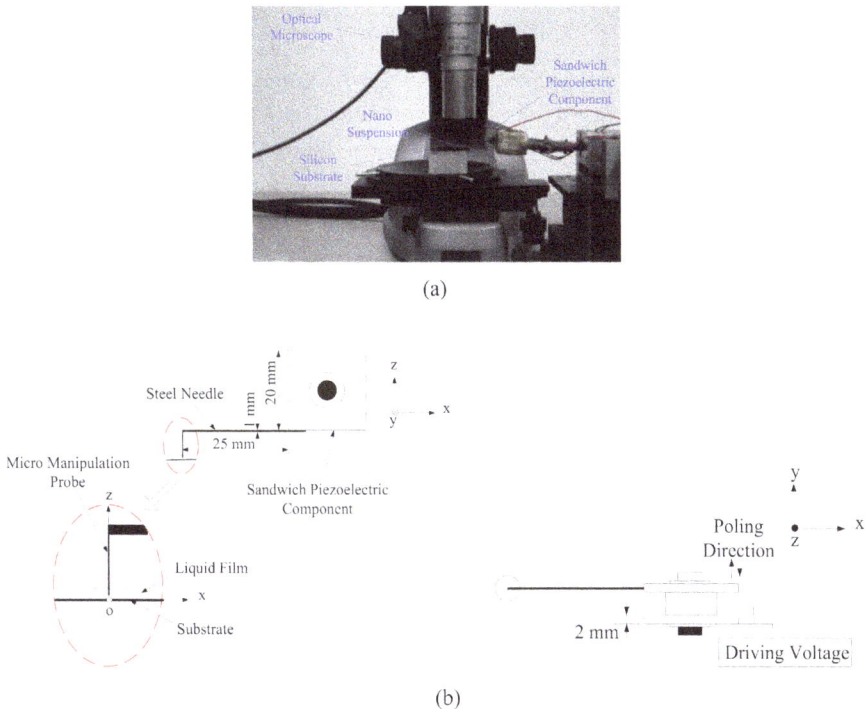

(a)

(b)

FIGURE 4.4 Experimental setup and the device to verify computational results. (a) Photo of the experimental setup. (b) Device structure and size. Reproduced from Ref. [18] with permission from Springer Nature.

tip of the steel needle, which is mechanically driven by a sandwich-type piezoelectric transducer shown in Fig. 4.4(b). The steel needle is 25 mm long and 1 mm thick. The outer and inner diameters and the thickness of each piezoelectric ring in the transducer are 12mm, 6 mm, and 1.2 mm, respectively. The piezoelectric constant d_{33} is 250×10^{-12} C/N, electromechanical coupling factor k_{33} is 0.63, mechanical quality factor Q_m is 500, dielectric dissipation factor $tan\delta$ is 0.6%, and density is 7450 kg/m³. The two stainless plates at the two ends of the transducer are square with 20 mm length and 2 mm thickness. The tightening torque applied to the transducer is 6 Nm. The resonance frequency of the sandwich transducer is 93 kHz, and the manipulation system with a working frequency around 135 kHz is not in resonance.

The tip of the steel needle is used to excite the vibration of the MMP in the probe-liquid-substrate system. The amplitudes and initial phases of the three orthogonal vibration components at the tip of the steel needle are measured by a laser vibrometer (POLYTEC PSV-300 F). The distance between the MMP tip and substrate is controlled by an XYZ stage, and the water film thickness is controlled by properly spreading out the water film and measuring its approximate height. Considering the evaporation during the experiments, the initial water film is thicker than 200 μm. It is experimentally found that the trapping performance is not sensitive to the water film thickness as long as it is larger than 50 μm.

4.2.3 RESULTS AND DISCUSSION

To simplify the computation, the sandwich transducer and the steel needle used to excite the MMP's vibration are not included in the FEM model of the probe-liquid-substrate system. Also, unless otherwise specified, property parameters of the MMP and the liquid film (water) shown in Table 4.1 are used.

Three orthogonal vibration components (the x-, y- and z-components in Fig. 4.4) at the MMP root were measured by the laser Doppler vibrometer (POLYTEC PSV-300 F)

TABLE 4.1
Parameters of the Probe-Liquid-Substrate System

Water Film's Thickness H (mm)	Water Film's Radius R_W (mm)	MMP's Length L_P (mm)
0.2	5	1.5
The MMP's radius R_P (μm)	The distance between MMP tip and substrate d_0 (μm)	The excited part length L_E (mm)
5	5	1
Density of water ρ (kg/m3)	Speed of sound in water c (m/s)	Shear viscosity of water η (Pa·s)
1000	1500	0.001
Volume-to-shear viscosity ratio in water η′/η	Nonlinear parameter of water at room temperature $\dfrac{B}{A}$	Density of MMP ρ_P (kg/m³)
2.1	5	2200
Young modulus of MMP E_P (Pa)	Shear modulus of MMP G_P (Pa)	
7.4×10^{10}	2.846×10^{10}	

TABLE 4.2

Measured Vibration Distribution and Workability at Some Frequencies

f (kHz)	125	130	135	140
A_x (nm)	17.45	9.44	5.77	13.72
A_y (nm)	11.44	35.78	18.20	14.10
A_z (nm)	6.11	18.54	15.68	9.69
Φ_x (°)	−100.0	−131.1	−134.5	100.0
Φ_y (°)	13.9	120.3	165	−23.3
Φ_z (°)	−144.9	−99.1	67.4	59.2
$\Phi_{zy}=\Phi_z\text{-}\Phi_y$ (°)	−158.8	−219.4	−97.6	82.5
$\Phi_{xy}=\Phi_x\text{-}\Phi_y$ (°)	−113.9	−251.4	−299.5	123.3
Can the device trap a single NW?	No	No	Yes	No

at the frequencies 125, 130, 135 and 140 kHz, respectively, and the results are shown in Table 4.2. The measured vibration information includes the vibration amplitude and initial phase. Also, trapping performance at these frequencies was observed, and the results are also listed in Table 4.2. It shows that only at 135 kHz can the acoustic streaming field be used to trap a single nanowire. To explain the experimental phenomena, the acoustic streaming field around the MMP is computed, and flow patterns on the silicon substrate and in the yz plane (the plane that is perpendicular to the substrate and vibration transmission needle, and passes the center axis of the MMP) are shown in Figs. 4.5 and 4.6. Figure 4.5 shows the result for 135 kHz, and Fig. 4.6 the results for 125, 130, and 140 kHz. In the computation, parameters listed in Table 4.2 were used, and the distance between the MMP tip and substrate was 5 μm.

The trapping performance listed in Table 4.2 is explained by Figs. 4.5 and 4.6 as follows. From Fig. 4.5(a), it is seen that a nanowire lying on the substrate surface within the effect range can be driven toward the point directly under the MMP tip from p_1 to p_3 through p_2, by the inward flow along the y-direction. Under the MMP tip, the sucked nanowire rotates to the x-direction due to the $\pm x$-directional flow while being lifted up by the upward flow (the z-direction flow), as shown in Fig. 4.5(b). The lifted nanowire is pushed onto the MMP's side by the y-directional flow and aligned in the x-direction by the $\pm x$-directional flow. Due to the pressing force on the trapped nanowire, there is a frictional force between the trapped nanowire and MMP, which contributes to the balance of the trapped nanowire.

The acoustic streaming shown in Fig. 4.6 can well explain why a single nanowire cannot be sucked to the MMP tip and trapped on the MMP at 125 kHz, 130 kHz and 140 kHz. Images $a1$, $b1$ and $c1$ show that a single nanowire cannot reach a force balance under the MMP tip due to the severe asymmetry of the outward acoustic streaming, and it would be flushed away by the acoustic streaming on the substrate.

From the above discussion, it is known that the acoustic streaming shown in Figs. 4.5 and 4.6 can well explain why a single nanowire can be sucked to the MMP tip and

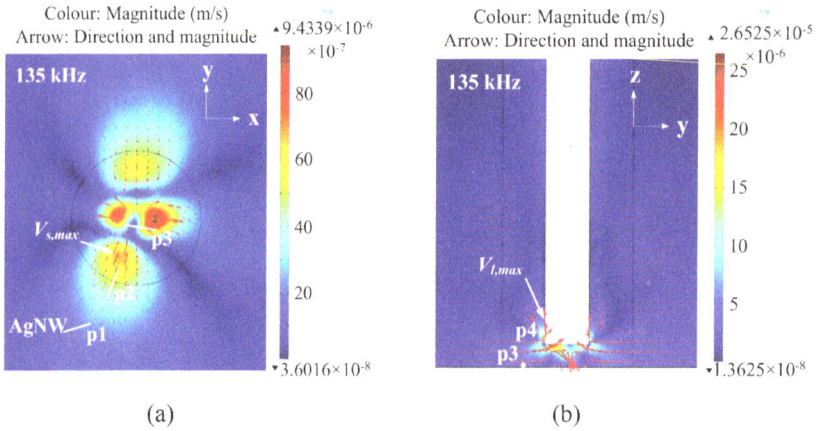

(a)   (b)

FIGURE 4.5   Acoustic streaming field on the substrate surface and in the *yz* plane (the plane that is perpendicular to the substrate and vibration transmission needle, and passes the center axis of the MMP) at 135 kHz. (a) Acoustic streaming field on the substrate. (b) Acoustic streaming field in the *yz* plane. The maximum acoustic streaming velocities on the substrate $V_{s,max}$ and in the *yz* plane $V_{l,max}$ are shown in (a) and (b), respectively. The black circle in image a, representing the boundary between the fine and coarse element regions, has a radius of 15 μm. And the white zone in image b representing the cross section of MMP has a width of 10 μm. The separation between the MMP tip and substrate is 5 μm. Reproduced from Ref. [18] with permission from Springer Nature.

trapped on the MMP at 135 kHz and why it cannot be trapped at other working frequencies. This indicates that the computed acoustic streaming field has the features which can well explain the trapping process, even if the nanowire is not included in the FEM model. This is because the diameter of the experimental nanowires is only 1/50 of the distance between the MMP tip and substrate, and ignoring the nanowires does not affect basic features of the acoustic streaming field.

The maximum acoustic streaming velocity on the silicon substrate surface is defined as $V_{s,max}$ (see Fig. 4.5(a)), and the maximum *z*-directional velocity in the *yz* plane (see Fig. 4.5(b)), which is usually along the MMP wall, is defined as $V_{l,max}$. These two velocities are used to describe the strength of the acoustic streaming field in the following discussion.

To investigate the dependency of the acoustic streaming field on the phase difference among the orthogonal vibration components at the MMP root, the acoustic streaming field is computed with different phase difference values between the *y*- and *z*-directional vibration displacements Φ_{zy} at working frequency of 130 kHz, and the computed results are shown in Fig. 4.7. In the computation, the amplitudes of the three orthogonal vibration displacement components and the phase difference values other than Φ_{zy}, as listed in Table 4.3, are used, and all the computational parameters except Φ_{zy} are kept constant to exclude the effects of these parameters. The phase difference Φ_{zy} is set to be 0°, 45° and 90°.

From images *a1*, *b1* and *c1* in Fig. 4.6, it is found that as the phase difference Φ_{zy} increases, the acoustic streaming pattern on the substrate becomes more symmetric about the *x* axis, but the outward flow on the substrate remains asymmetric about

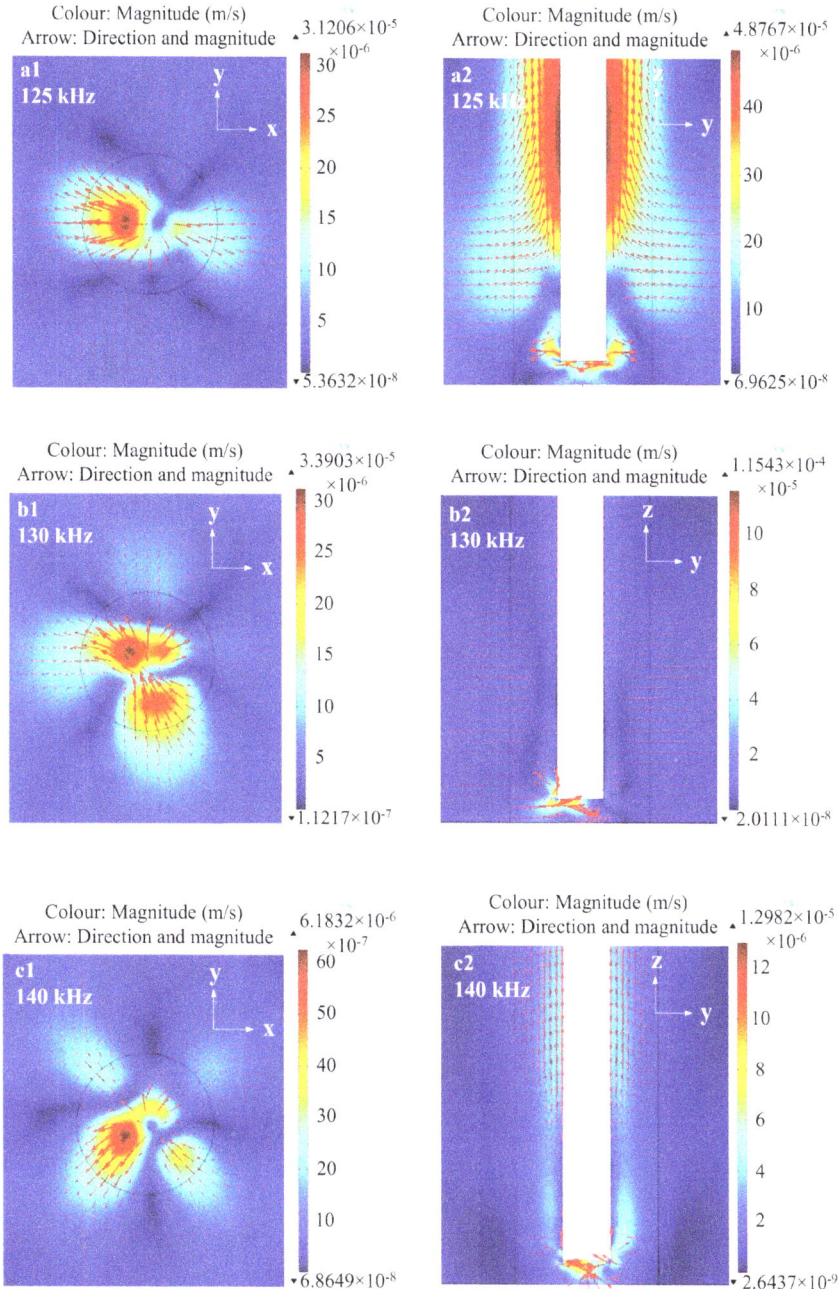

FIGURE 4.6 Acoustic streaming fields on the substrate surface and in the yz plane (the plane that is perpendicular to the substrate and vibration transmission needle, and passes the center axis of the MMP) at 125, 130 and 140 kHz. The black circles in images *a1*, *b1* and *c1*, representing the boundary between the fine and coarse element regions, have a radius of 15 μm. And the white zones in images *a2*, *b2* and *c2*, representing the cross section of MMP, have a width of 10 μm. Reproduced from Ref. [18] with permission from Springer Nature.

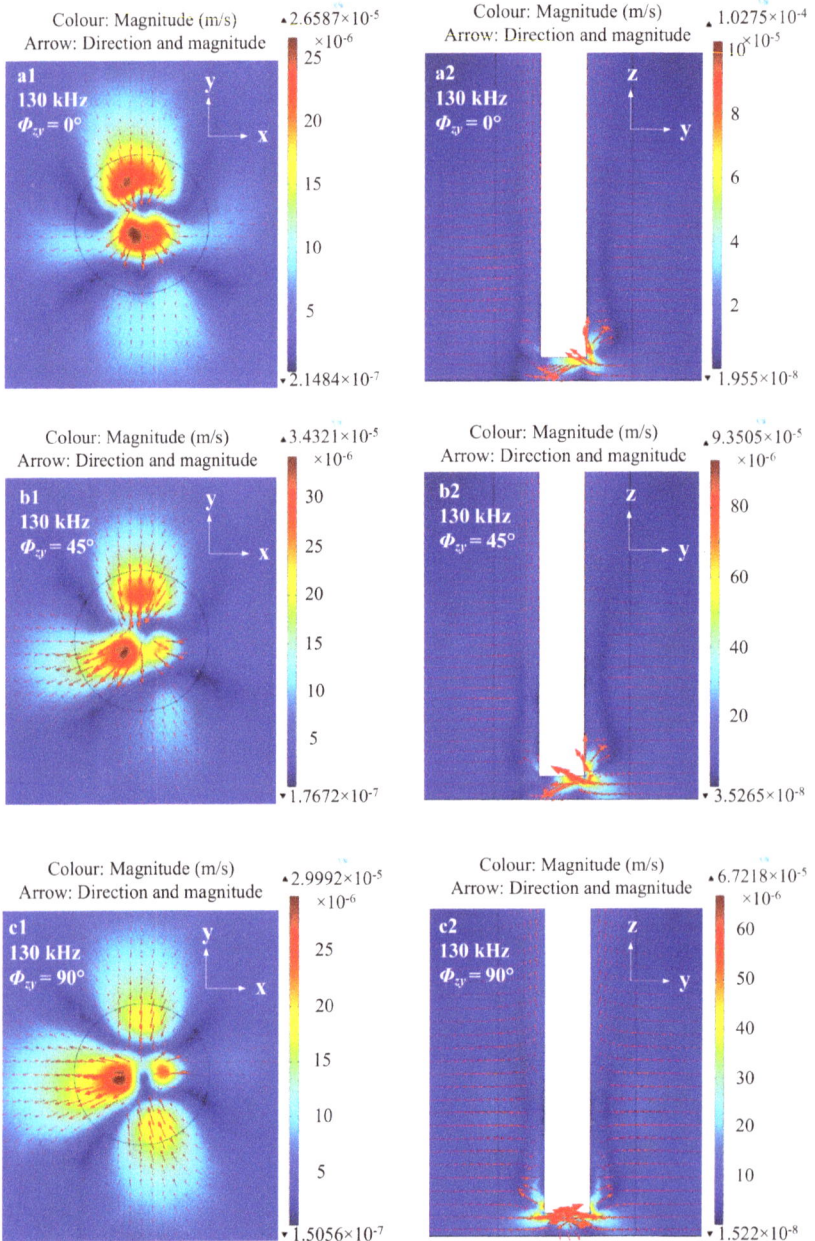

FIGURE 4.7 Acoustic streaming fields on the substrate surface and in the yz plane (the plane that is perpendicular to the substrate and vibration transmission needle, and passes the center axis of the MMP) at 130 kHz for different phase difference values between the y- and z-directional vibration displacement components Φ_{zy} (=0°, 45° and 90°). The black circles in images $a1$, $b1$ and $c1$, representing the boundary between the fine and coarse element regions, have a radius of 15 μm. And the white zones in images $a2$, $b2$, and $c2$, representing the cross section of MMP, have a width of 10 μm. Reproduced from Ref. [18] with permission from Springer Nature.

TABLE 4.3

Effects of Some Dimensional Parameters on the Acoustic Streaming Field at 135 kHz

	Distance between the MMP Tip and Substrate d_0	MMP's Radius R_P	MMP's Length L_P	Water Film's Thickness H
Acoustic streaming pattern	Affected	Affected	Insensitive	Insensitive
Acoustic streaming velocities	Affected	Sensitive near the MMP resonance point	Sensitive near the MMP resonance point	Insensitive to large H ($>50\mu m$)

the y axis. For the acoustic streaming field shown in image $c1$, it is still difficult for a sucked nanowire under the MMP tip to reach a force balance in the x direction.

In order to achieve a more symmetric acoustic streaming field that can be used to suck and trap a single nanowire, the acoustic streaming field at 130 kHz is computed for different amplitude ratios A_x/A_y at $\Phi_{zy} = 90°$, and the results are shown in Fig. 4.8. In the computation, A_x decreases from 9.44 nm to 0, and the other parameters are kept constant (see Table 4.2). The results shown in Fig. 4.8 indicate that as A_x/A_y decreases, the outward acoustic streaming becomes more symmetric about the y-axis, and a sucked and 90° rotated nanowire under the MMP tip can keep a force balance at $A_x/A_y = 0$.

Based on the results shown in Figs. 4.7 and 4.8, it comes to the conclusion that the MMP root's elliptical vibration in the yz plane can generate the acoustic streaming field capable of trapping a single nanowire in the contact mode. For this reason, the acoustic streaming field shown in Fig. 4.8(b1) and (b2) is defined as the ideal flow field for the contact type trapping of a single nanowire.

The effects of the distance between the MMP tip and substrate d_0 on the acoustic streaming pattern and the maximum acoustic streaming velocities on the substrate and in the z-direction are computed at 135 kHz, and the results are shown in Fig. 4.9. Figure 4.9(a) shows the acoustic streaming patterns at $d_0 = 2$ and 8 μm, and Fig. 4.9(b) shows the maximum acoustic streaming velocities on the substrate $V_{s,max}$ and in the z-direction $V_{l,max}$ versus d_0.

The computation shows that as the distance between the MMP tip and substrate increases, the acoustic streaming pattern on the substrate has a change, and the maximum acoustic streaming velocities on the substrate $V_{s,max}$ and in the z direction $V_{l,max}$ decrease. The decrease of $V_{s,max}$ is because of the decrease of the tangential vibration velocity on the substrate as d_0 increases. The decrease of $V_{l,max}$ is because the fluid circulates between the MMP tip and substrate as d_0 increases. Therefore, a large distance between the MMP and substrate weakens the trapping capability.

The effect of the MMP's radius on the acoustic streaming field around the MMP was computed, and the results are shown in Fig. 4.10. In the computation, the working frequency was 135 kHz, and the vibration excitation conditions listed in Table 4.2

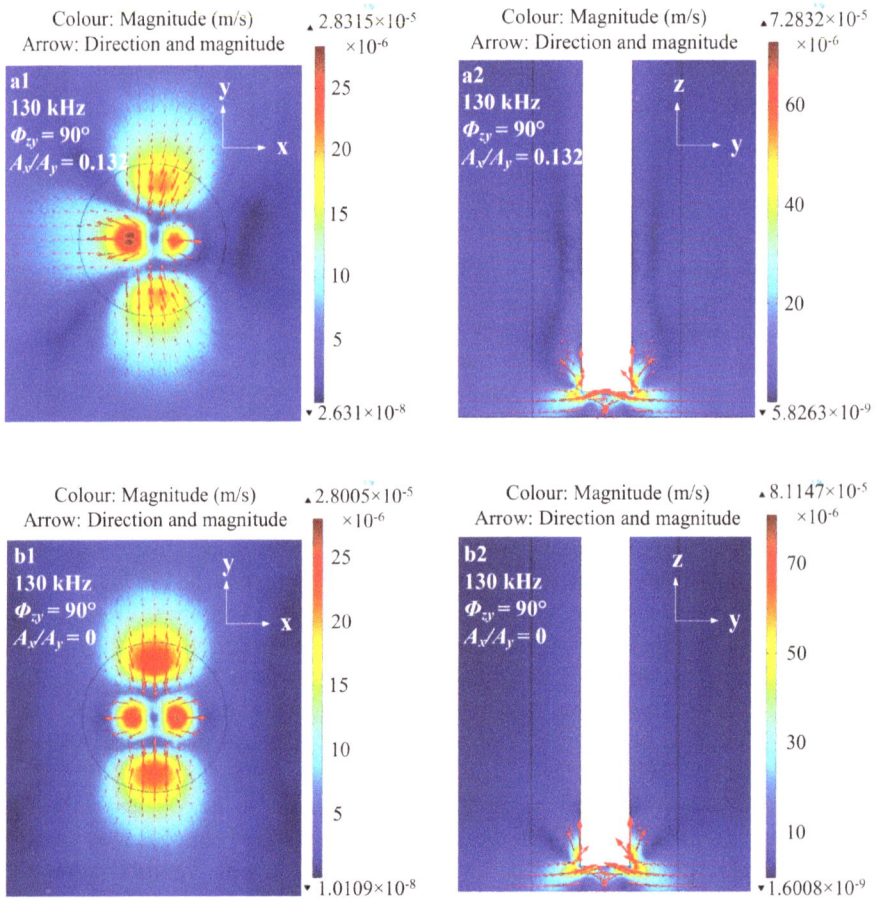

FIGURE 4.8 Acoustic streaming fields on the substrate surface and in the yz plane (the plane that is perpendicular to the substrate and vibration transmission needle, and passes the center axis of the MMP) at 130 kHz for different amplitude ratios between the x- and y-directional vibration displacement components A_x/A_y (=0.132 and 0) when the phase difference between the y- and z-directional vibration displacement components is 90°. The black circles in images $a1$ and $b1$, representing the boundary between the fine and coarse element regions, have a radius of 15 μm. And the white zones in images $a2$ and $b2$, representing the cross section of MMP, have a width of 10 μm. Reproduced from Ref. [18] with permission from Springer Nature.

were used. Figure 4.10(a) shows the acoustic streaming fields when the MMP's radius is 1, 8 and 12 μm, respectively, and Fig. 4.10(b) shows the change of $V_{s,max}$ and $V_{l,max}$ with the MMP's radius. The vertical axis of Fig. 4.10(b) represents the natural logarithm of $V_{s,max}/V_{s0}$ and $V_{l,max}/V_{l0}$, where V_{s0} (=0.05 μm/s) and V_{l0} (=0.66 μm/s) are the maximum acoustic streaming velocities on the substrate and along the z-direction for an MMP's radius of 1μm, respectively. From images $a1$ and $a2$, it is seen that when the MMP's radius is too small, a nanowire on the substrate may be sucked

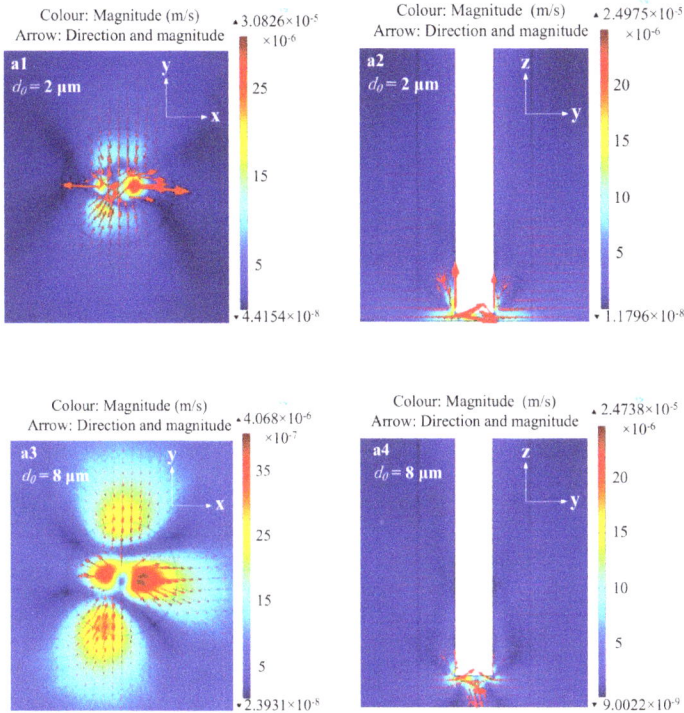

(a)

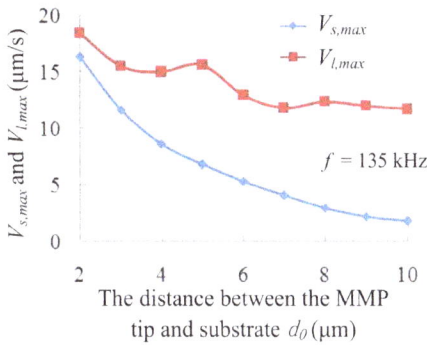

(b)

FIGURE 4.9 Acoustic streaming fields for different distance values between the MMP and substrate at 135 kHz. (a) Acoustic streaming patterns on the substrate surface and in the yz plane (the plane that is perpendicular to the substrate and vibration transmission needle, and passes the center axis of the MMP) at $d_0 = 2$ and 8 μm. (b) The maximum acoustic streaming velocities on the substrate surface and in the yz plane versus the distance between the MMP and substrate. The black circles in images $a1$ and $a3$, representing the boundary between the fine and coarse element regions, have a radius of 15 μm. And the white zones in images $a2$ and $a4$, representing the cross section of MMP, have a width of 10 μm. Reproduced from Ref. [18] with permission from Springer Nature.

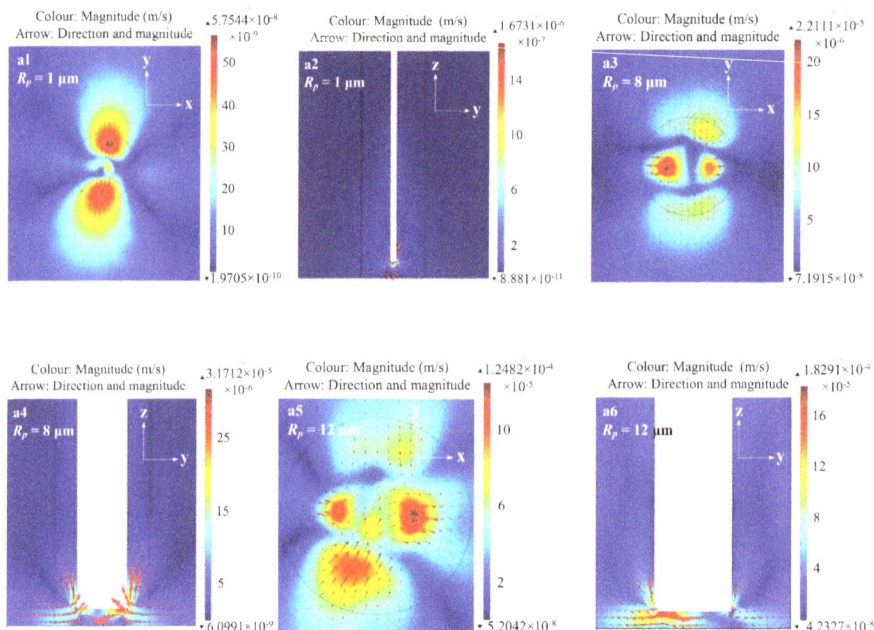

(a)

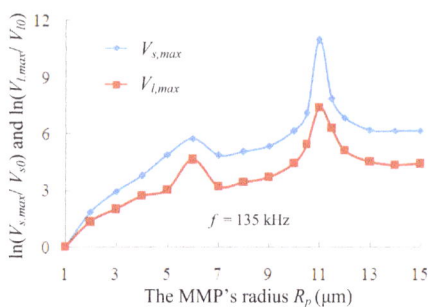

(b)

FIGURE 4.10   Acoustic streaming fields at different MMP's radii R_p at 135 kHz. (a) Acoustic streaming patterns on the substrate surface and in the yz plane (the plane that is perpendicular to the substrate and vibration transmission needle, and passes the center axis of the MMP) at $R_p = 1$, 8 and 12 μm. (b) The maximum acoustic streaming velocities on The substrate surface and in The yz plane versus the MMP's radius. The black circles in images $a1$, $a3$ and $a5$, representing the boundary between the fine and coarse element regions, have a radius of 10, 15 and 30 μm, respectively. And the white zones in images $a2$, $a4$ and $a6$, representing the cross section of MMP, have a width of 2, 16 and 24 μm, respectively. Reproduced from Ref. [18] with permission from Springer Nature.

into the region under the MMP, but it is difficult to lift the nanowire. Figure 4.10(b) shows that when the MMP's radius is about 11.5 μm, there is a large peak of acoustic streaming velocity. Our computation shows that this phenomenon is caused by a resonance of the MMP.

If the working point is at the peak or very close to the peak, a disturbance, such as the decrease of the water film thickness and an impact of micro/nano object onto the MMP, may cause a substantial change of the acoustic streaming velocity around the MMP. Thus, the MMP's radius should be so designed that the working point is shifted a bit from the MMP's resonance.

The effect of the MMP's length on the acoustic streaming field was computed, and the results are shown in Fig. 4.11. In the computation, the working frequency was 135 kHz, and the vibration excitation conditions listed in Table 4.2 were used. Figure 4.11(a) shows the acoustic streaming fields when the MMP's length is 0.3, 7 and 9.5 mm, respectively, and Fig. 4.11(b) shows the change of $V_{s,max}$ and $V_{l,max}$ with the MMP's length. The vertical axis of Fig. 4.11(b) represents the natural logarithm of $V_{s,max}/V_{s1}$ and $V_{l,max}/V_{l1}$, where V_{s1} (=6.84 μm/s) and V_{l1} (=16.71μm/s) are the maximum acoustic streaming velocities on the substrate and along the z-direction for the MMP's length of 0.3 mm, respectively. From images $a1$, $a2$ and $a3$, it is seen that the MMP's length affects the symmetry of the acoustic streaming field, which is caused by the change of phase difference Φ_{zy} and Φ_{zx}. Figure 4.11(b) shows that when the MMP's length is about 9.9 mm, there is a large peak of acoustic streaming velocity. The computation shows that this phenomenon is caused by a resonance of the MMP.

The dependency of the acoustic streaming field around the MMP on the water film's thickness was computed at 135 kHz, and the results are shown in Fig. 4.12. In the computation, all of the parameters other than the water film's thickness are kept constant. Figure 4.12(a) lists the acoustic streaming fields when the water film's thickness is 15, 50 and 300 μm, respectively. Figure 4.12(b) shows the maximum acoustic streaming velocities on the substrate $V_{s,max}$ and along the z-direction $V_{l,max}$ versus the water film's thickness. From Fig. 4.12(a), it is seen that the asymmetry of the acoustic streaming field increases as the water film's thickness decreases, which means that a thin water film may weaken the trapping capability. This is because as the water film's thickness decreases, the spatial asymmetry of driving force F_j (see Eq. 4.3) of the acoustic streaming is amplified. From Fig. 4.12(b), it is seen that when the water film is sufficiently thick, its thickness has little effect on the acoustic streaming velocities. This means that the acoustic streaming field for the nanowire trapping is mainly determined by the ultrasonic field near the substrate and MMP's tip when the water film is sufficiently thick.

The effects of the working and size parameters on the pattern and velocity of the acoustic streaming field are summarized in Table 4.3.

4.2.4 SUMMARY

The 3D acoustic streaming field in the probe-liquid-substrate system, in which a MMP is inserted into a layer of nano suspension film on a substrate, has been computed numerically and analyzed. The computational results can well explain why

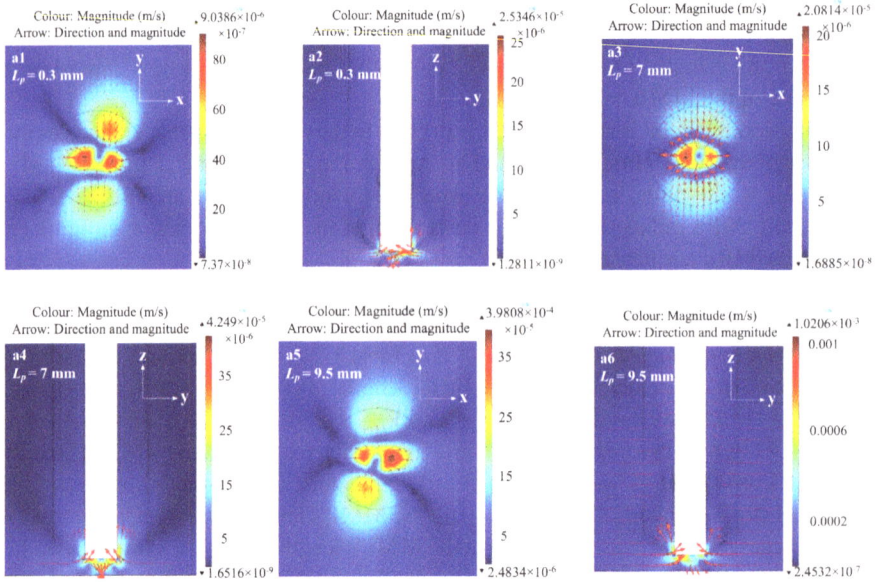

(a)

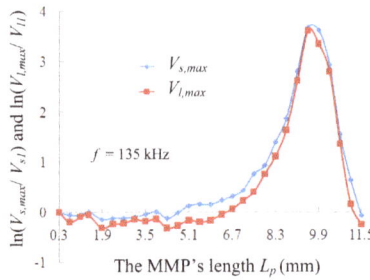

(b)

FIGURE 4.11   Acoustic streaming fields at different MMP's length L_p at 135 kHz. (a) Acoustic streaming patterns on the substrate surface and in the yz plane (the plane that is perpendicular to the substrate and vibration transmission needle, and passes the center axis of the MMP) at $L_p = 0.3$, 7 and 9.5 mm. (b) The maximum acoustic streaming velocities on The substrate surface and in The yz plane versus the MMP's length. The black circles in images $a1$, $a3$ and $a5$, representing the boundary between the fine and coarse element regions, have a radius of 15 µm. And the white zones in images $a2$, $a4$ and $a6$, representing the cross section of MMP, have a width of 10 µm. Reproduced from Ref. [18] with permission from Springer Nature.

the device works at some working frequencies and does not work at others, and the computed acoustic streaming field agrees with the observed one quite well. It is found that the phase difference and magnitude ratio among the orthogonal vibration components of the MMP root have a large effect on the symmetry of the acoustic streaming field, thus determining whether the acoustic streaming field is usable in

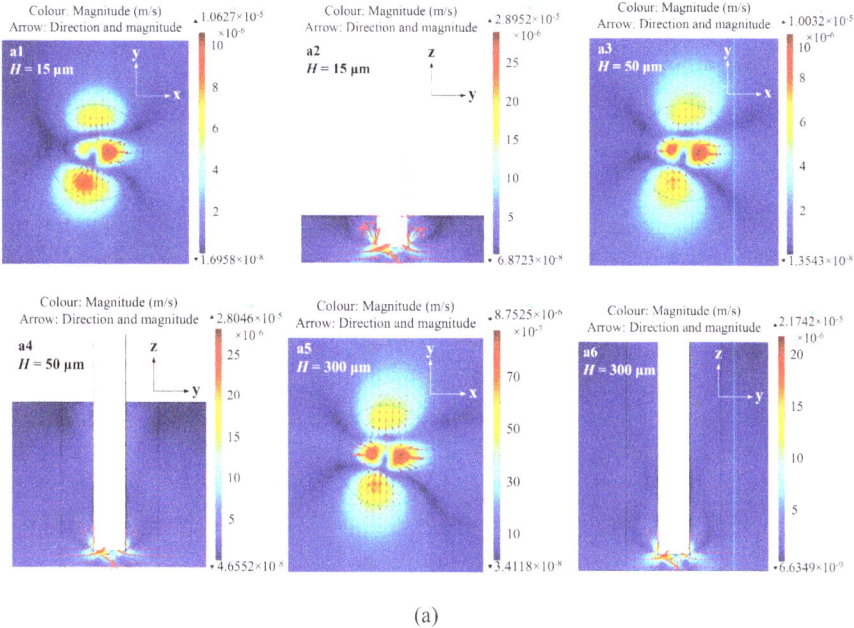

(a)

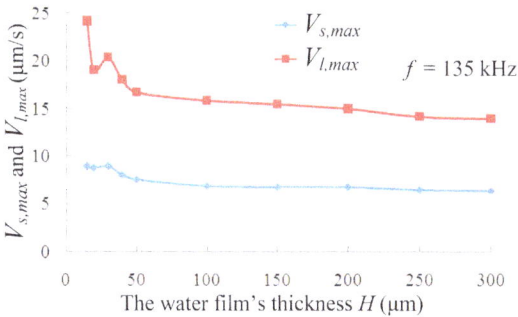

(b)

FIGURE 4.12 Acoustic streaming fields at different liquid film's thickness H at 135 kHz. (a) Acoustic streaming patterns on the substrate surface and in the yz plane (the plane that is perpendicular to the substrate and vibration transmission needle, and passes the center axis of the MMP) at $H = 15$, 50 and 300 μm. (b) The maximum acoustic streaming velocities on The substrate surface and in The yz plane versus the thickness of the liquid film. The black circles in images $a1$, $a3$ and $a5$, representing the boundary between the fine and coarse element regions, have a radius of 15 μm. And the white zones in images $a2$, $a4$ and $a6$, representing the cross section of MMP, have a width of 10 μm. Reproduced from Ref. [18] with permission from Springer Nature.

the contact type trapping of a single nanowire. The MMP root's elliptical vibration perpendicular to the substrate can generate an acoustic streaming field capable of trapping a single nanowire in the contact mode. It is also found that the velocity and pattern of the acoustic streaming change with the distance between the MMP and substrate, and the MMP's radius and length.

4.3 A DUAL-FUNCTIONAL ULTRASONIC NANO TWEEZER

In ultrasonic nano trapping, there are two working modes, that is, the noncontact and contact modes. In the noncontact mode, the trapped nano object is not in contact with the micro manipulating probe. While in the contact mode, the trapped nano object is in contact with the MMP. The noncontact mode has the merit that a trapped nano object does not stick to the micro manipulating probe, and releasing the trapped object is not a problem. However, in the noncontact mode, the trapped nano object can only be moved on the substrate surface. Other nano objects on its motional path will obstacle its transfer. The contact mode can overcome this problem. But sticky samples may adhere to the manipulating probe and do not fall even if the ultrasonic vibration is switched off.

This section gives the device structure, working principle and characteristics of a dual-functional ultrasonic nano tweezer, which has noncontact and contact-type nanowire trapping capability [4]. The device is simply made up of a piezoelectric component, a vibration transmission needle (VTN), and a micro manipulating probe (MMP). With the tweezer, individual silver nanowires with a 300 nm diameter and length up to 30 μm at the interface between a droplet and solid substrate can be trapped and transferred in contact with or without contact with the MMP. Analyses show that making the MMP oscillate linearly or elliptically at proper working frequencies is the key to integrating the noncontact and contact trapping functions in this device.

4.3.1 DEVICE STRUCTURE AND FUNCTIONS

Figure 4.13 shows the experimental setup for the noncontact and contact trapping of individual nanowires. The device is simply made up of a piezoelectric plate VTN made of steel, and MMP made of fiberglass.

The VTN is bonded along the narrow side of the piezoelectric plate. The MMP is bonded to the VTN's tip and parallel to the piezoelectric plate. The VTN, which is above the suspension film, is parallel to the substrate. The width, length and thickness of the piezoelectric plate are 5, 10 and 1 mm, respectively. The angle θ between the MMP and VTN is 95°. The electromechanical quality factor Q_m, piezoelectric coefficient d_{33}, and relative dielectric constant $\varepsilon_{33}^T/\varepsilon_0$ of the piezoelectric plate are 2000, 325×10^{-12} m/v and 1450, respectively. The VTN is 0.8 mm thick and 28 mm long out of the plate. The MMP is 10 μm thick and 2 mm long. The resonance frequency of the device is about 136 kHz, at which the VTN vibrates flexurally.

The water film thickness is about 0.3 mm. The suspension is formed by deionized water and dispersed silver nanowires. The distance between the MMP and substrate is 5 μm. The diameter and average length of the AgNW are about 300 nm and 10~30 μm, respectively.

Experimental results shown in Fig. 4.14 indicate that the device can trap a silver nanowire in the water film on the substrate surface in two working frequency ranges. In the figure, the vertical axis represents the measured average value of the z-directional vibration velocity magnitude of the VTN. The trapped nanowire is not in contact with the MMP in the frequency range from 131.2 to 132.2 kHz, and in contact with the MMP in the frequency range from 133.9 to 134 kHz. Figures 4.15 and 4.16 contain two groups of images to show the noncontact and contact trapping modes, respectively, and the transfer of a silver nanowire in these two modes.

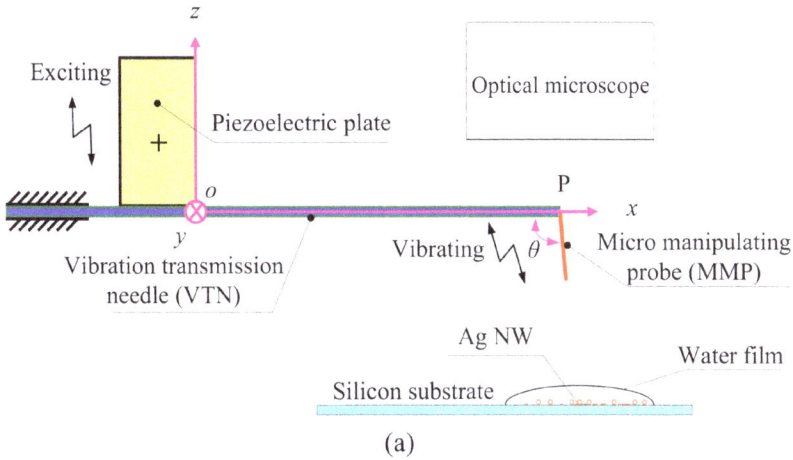

(a)

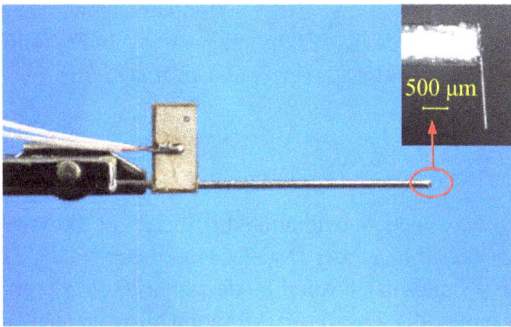

(b)

FIGURE 4.13   Experimental setup for the noncontact and contact-type trapping of a single silver nanowire in water film on a substrate. (a) Schematic diagram. (b) Construction of the ultrasonic transducer. Reproduced from Ref. [4] with permission from Elsevier.

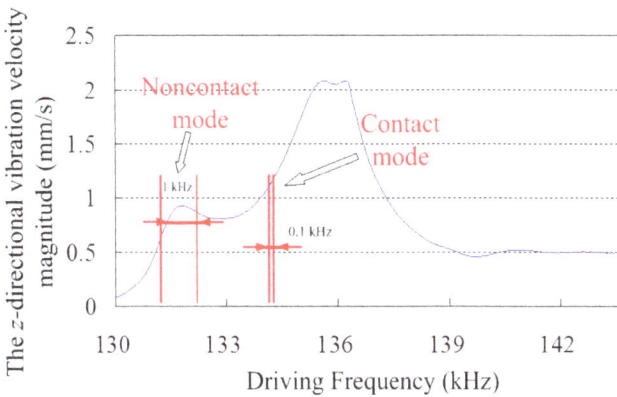

FIGURE 4.14   Measured z-directional vibration velocity magnitude of the vibration transmission needle (VTN) versus driving frequency. Reproduced from Ref. [4] with permission from Elsevier.

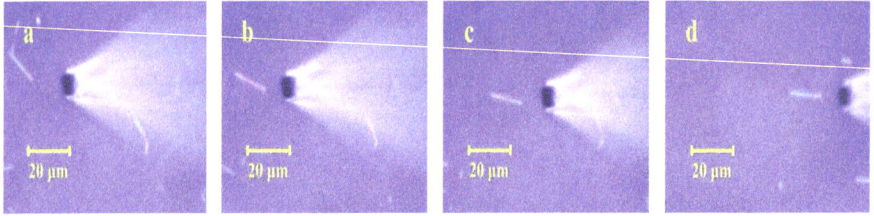

FIGURE 4.15 Noncontact trapping of a single AgNW under the micro manipulating probe's tip in water film on a substrate at a driving frequency of 131.2 kHz and Voltage of 40 V_{p-p}. Reproduced from Ref. [4] with permission from Elsevier.

In the noncontact mode, the trapped nanowire is on the substrate and can only be transferred to the substrate surface. In the contact mode, the trapped wire is in contact with the MMP. In both modes, the AgNW rotates while being sucked into the MMP. From image *b* to *d* in Fig. 4.15, the trapped nanowire is moved on the substrate surface by moving the device. From images *d* to *g* in Fig. 4.16, the trapped wire is transferred above the substrate surface, and in image *h* in Fig. 4.16, the trapped nanowire is released.

4.3.2 PRINCIPLE ANALYSES

Figure 4.17 shows the *z*-directional vibration distribution of the VTN at 133.9 kHz and 40 V_{p-p}, which is measured by a laser Doppler vibrometer (POLYTEC PSV-300 F). It shows that the VTN vibrates in a flexural mode with a 70 nm vibration amplitude.

Figure 4.18(a) shows the calculated vibration displacement distribution along the MMP at 131.2 kHz when the device operates in the noncontact trapping mode. For the noncontact trapping mode, the measured *x, y* and *z* components of the

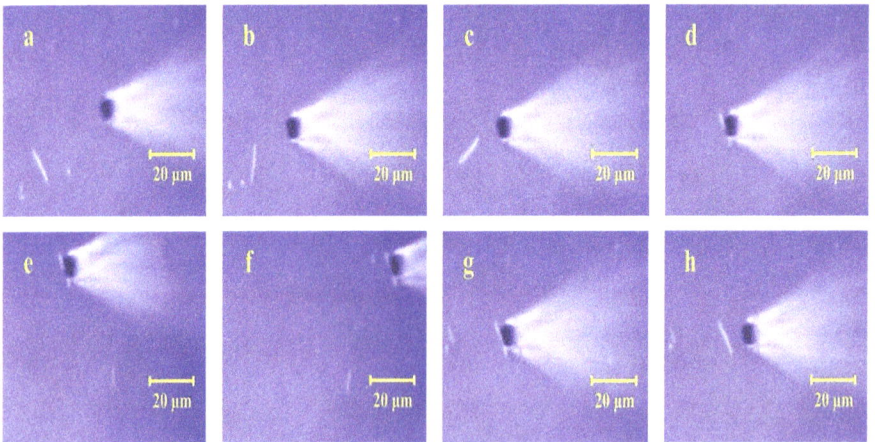

FIGURE 4.16 Contact trapping of a single AgNW under the micro manipulating probe's tip in water film on a substrate at a driving frequency of 133.9 kHz and Voltage of 40 V_{p-p}. Reproduced from Ref. [4] with permission from Elsevier.

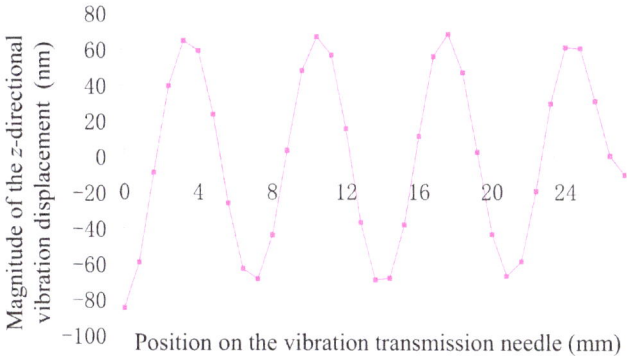

FIGURE 4.17   Distribution of measured z-directional vibration displacement of the vibration transmission needle along the x-direction at a driving frequency of 133.9 kHz and Voltage of 40 V_{p-p}. Reproduced from Ref. [4] with permission from Elsevier.

vibration velocity at its root (point P) are $5.67\times10^{-2}\angle-63°$m/s, $2.55\times10^{-2}\angle-71°$m/s, and $8.12\times10^{-2}\angle111°$m/s, respectively, at 131.2 kHz and 40 V_{p-p}.

Figure 4.18 (b) shows the vibration displacement distribution of the MMP at 133.9 kHz when the device operates in the contact trapping mode. For the contact trapping mode, the measured x-, y- and z- vibration velocity components at its root (point P) are $1.5\times10^{-2}\angle3°$m/s, $0.6\times10^{-2}\angle2°$m/s, and $3.14\times10^{-2}\angle89°$m/s, respectively, at 133.9 kHz and 40 V_{p-p}. In addition, Fig. 4.18(a, b) shows that in both trapping modes, the MMP vibrates with a 10~20 nm vibration amplitude.

The calculated acoustic streaming fields around the MMP's tip in the noncontact and contact trapping modes are shown in Figs. 4.19(a, b) and 4.20 (a, b), respectively, and the observed pattern of the acoustic streaming fields are shown in Figs. 4.19(c) and 4.20(c),

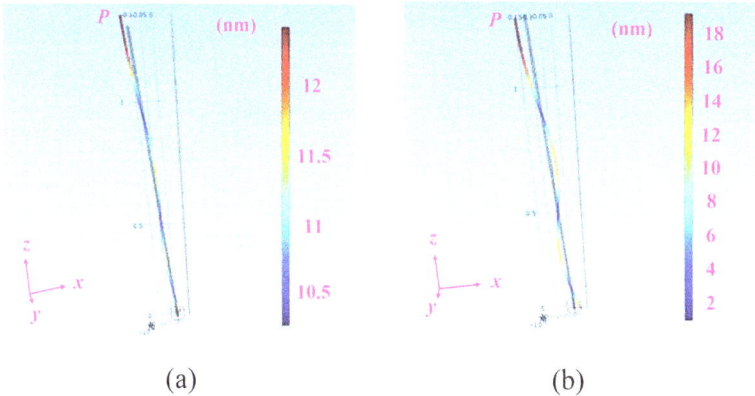

(a)   (b)

FIGURE 4.18   (a) Distribution of calculated vibration displacement of the micro manipulating probe at a frequency of 131.2 kHz and driving Voltage of 40 V_{p-p} at the noncontact trapping mode. (b) Distribution of calculated vibration displacement of the micro manipulating probe at a driving frequency of 133.9 kHz and Voltage of 40 V_{p-p} at the contact trapping mode. Reproduced from Ref. [4] with permission from Elsevier.

(a)   (b)

(c)

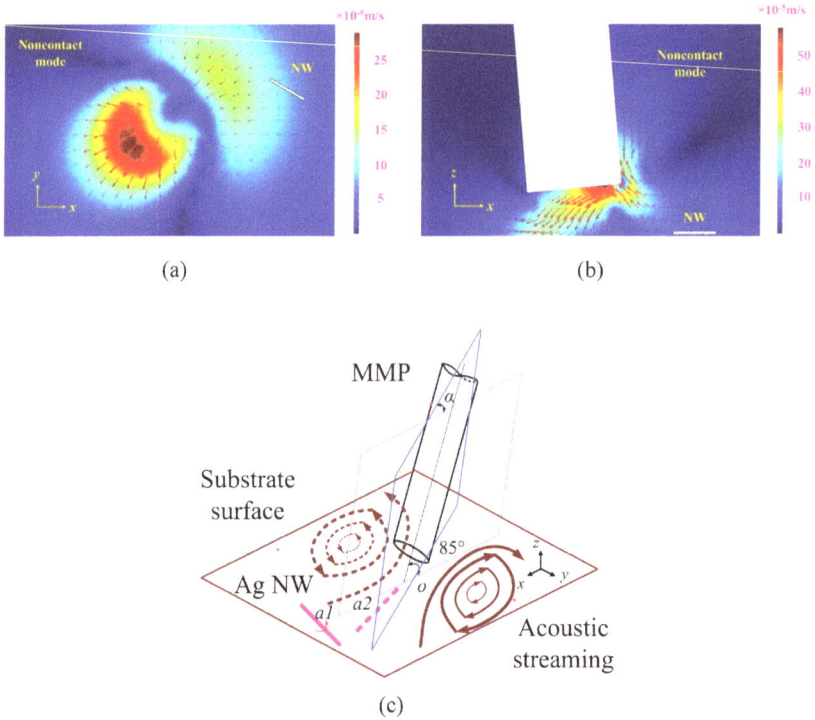

FIGURE 4.19   The calculated acoustic streaming field around the micro manipulating probe's tip in vibration in the noncontact trapping mode. (a) Acoustic streaming field on the substrate surface or in the xy plane. (b) Acoustic streaming field in the xz plane, which is the plane formed by the vibration transmission needle and micro manipulating probe. (c) Schematic diagram of the whole acoustic streaming field. Reproduced from Ref. [4] with permission from Elsevier.

respectively. In the calculation, only the system consisting of the MMP, droplet and substrate is considered; the driving frequency and vibration velocity components at the MMP's root, listed in the preceding paragraph, are used; the related parameters listed in the 1st paragraph of Sec. 4.3.1 and material constants in Table 4.4 are used; the acoustic streaming is calculated by the COMSOL Multiphysics software based finite element method. The acoustic streaming pattern can be visualized by using micro particles in the suspension.

Figures 4.19(a) and 4.20(a) show the flow patterns on the substrate surface for the noncontact and contact modes, respectively, and Figs. 4.19(b) and 4.20(b) show the flow pattern in the central plane of the MMP-VTN structure for the two modes, respectively.

The acoustic streaming on the substrate surface, as shown in Fig. 4.19(a), can well explain the noncontact trapping mode. As the acoustic streaming on the right side of the MMP flows towards the MMP's tip, it flushes the nanowire to the MMP's tip. When the sucking force which results from the acoustic streaming, is in balance with the friction force between the nanowire and substrate, the nanowire is trapped and can be transferred by moving the device. It was observed that when the vibration was

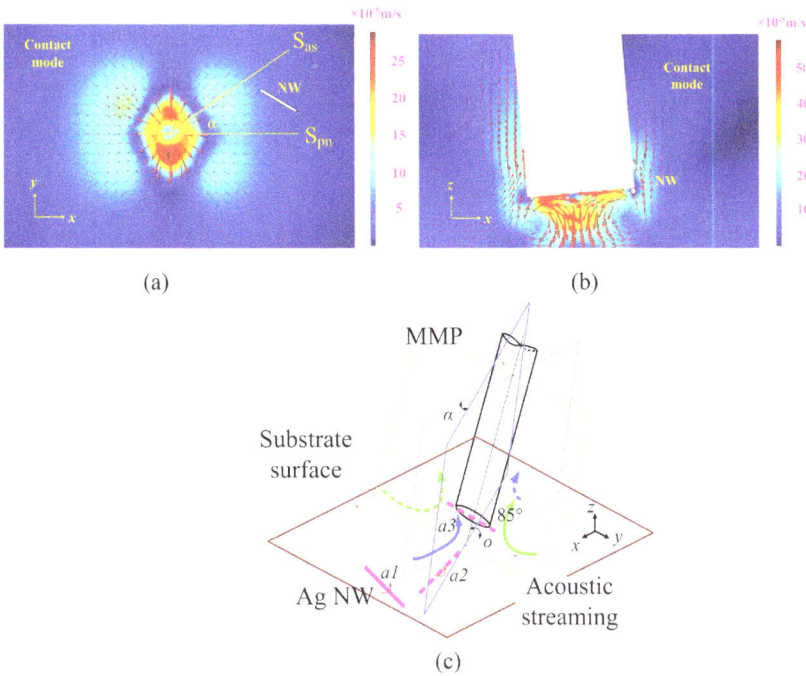

FIGURE 4.20   The acoustic streaming field around the micro manipulating probe's tip in vibration for the contact mode. (a) Acoustic streaming field on the substrate surface or in the *xy* plane. (b) Acoustic streaming field in the *xz* plane, which is the plane formed by the vibration transmission needle and micro manipulating probe. (c) Schematic diagram of the acoustic streaming field. Reproduced from Ref. [4] with permission from Elsevier.

too strong, the nanowire would be flushed away rather than being trapped at the tip of the MMP. This is because strong $-x$-directional acoustic streaming under the MMP's tip flushed the nanowire away along the $-x$ direction.

The acoustic streaming on the substrate surface, as shown in Fig. 4.20(a), can explain the contact trapping mode. As the acoustic streaming on the two sides of the MMP flows towards the MMP, a nanowire on either side of the MMP on the substrate is pushed towards the MMP tip. Under the MMP's tip, eddies shown in Fig. 4.20(b) can lift the sucked nanowire to the corner of the MMP's tip. For taking a balance at

TABLE 4.4

Material Constants Used in the Calculation

Material	Density kg/m³	Poisson's Ratio	Modulus of Elasticity Pa	Material Damping Coefficient	Velocity of Sound m/s	Viscosity Coefficient Pa s
Glass	2200	0.3	7.4×10^9	0.003	–	–
Water	1000	0.3	–	–	1500	0.0017

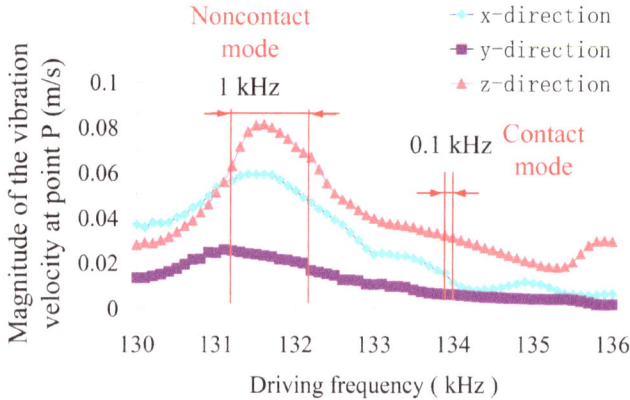

FIGURE 4.21 Measured magnitudes of the *x*-, *y*- and *z*-directional vibration velocities of the micro manipulating probe's root (point *P*) Versus driving frequency at 40 V$_{p-p}$. Reproduced from Ref. [4] with permission from Elsevier.

its final trapping position, the trapped nanowire must be perpendicular to the central plane S_{as} of the acoustic streaming field. From Fig. 4.20(a), it is known that the angle α made by the central plane S_{as} of the acoustic streaming field and the central plane S_{pn} of the MMP-VTN structure is 33°. This explains why the trapped nanowire is not perpendicular to the central plane of the MMP-VTN structure, shown in image *d* of Fig. 4.16.

The calculation conditions for Figs. 4.19 and 4.20 are the same except for the driving frequency and vibration velocity at the MMP's root. Thus, the possible factors causing the difference of acoustic streaming fields in Figs. 4.19 and 4.20 are the driving frequency and the vibration velocity at the MMP's root.

Figure 4.21 shows the measured frequency dependency of the magnitude of the *x*-, *y*- and *z*-directional vibration velocity components at point *P* (the root of the MMP) at 40 V$_{p-p}$. The frequency range is from 130 to 136 kHz, in which the noncontact and contact trapping modes occur, as well as the case without trapping capability. It is seen that the noncontact trapping mode has a larger vibration at point *P*, which means that a larger vibration velocity (or a large lifting force) does not necessarily cause a contact trapping mode. This indicates that the largeness of the vibration velocity components at point *P* is not the major factor in determining whether a trapped nanowire is in contact with the MMP.

To find the major factor that causes the difference in the acoustic streaming field in the noncontact and contact modes, the phase difference among the normal vibration components at point *P* in the contact and noncontact trapping modes was experimentally investigated. The phase difference between the *z*- and *y*-directional vibration components at point *P* is defined as $\Delta\varphi_{zy}$, and that between the *z*- and *x*-directional vibration components at point *P* is defined as $\Delta\varphi_{zx}$. Figure 4.22 shows the measured frequency dependency of $\Delta\varphi_{zy}$ and $\Delta\varphi_{zx}$ at 40 V$_{p-p}$. It is found that the phase difference values are around 180° in the noncontact trapping mode and 90° in the contact trapping mode. Mathematically, it is known that the resultant vibration of point *P* is a linear oscillation when $\Delta\varphi_{zy}$ and $\Delta\varphi_{zx}$ are around 180°, and it is an elliptical motion perpendicular to the substrate when $\Delta\varphi_{zy}$ and $\Delta\varphi_{zx}$ are around 90°.

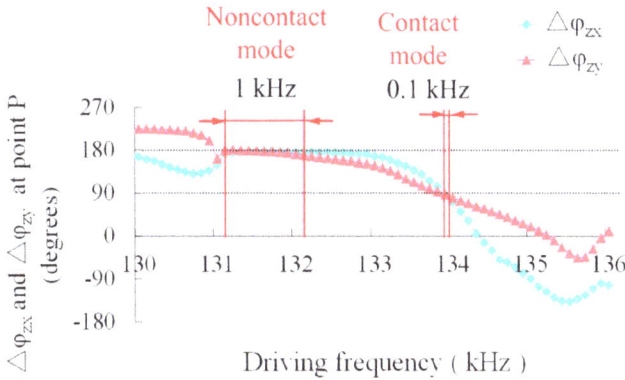

FIGURE 4.22 Measured distribution of the phase difference among the *x*, *y* and *z* components of the vibration velocity of the micro manipulating probe's root (point *P*) Versus driving frequency at 40 V_{p-p}. Reproduced from Ref. [4] with permission from Elsevier.

Therefore, it is concluded that the motional trajectory of the MMP root is linear in the noncontact trapping mode, and elliptical and perpendicular to the substrate in the contact trapping mode. This conclusion can be explained as follows. When the motional trajectory of the MMP root is linear, the *z*-directional flow is so weak that it cannot lift a sucked nanowire up. When the motional trajectory of the MMP root is elliptical and perpendicular to the substrate, a strong enough *z*-directional flow can be generated, which may lift the trapped nanowire onto the MMP tip. Next we try to explain why the devices reported in Refs. 1 and 2 only have one trapping mode, and the device in this work can possess the noncontact and contact trapping modes.

In Ref. [1], a VTN-MMP structure is excited by a Langevin transducer's end plate which has a larger size than the Langevin transducer's diameter. The longitudinal vibration of the Langevin transducer causes the roots of the VTN and MMP to vibrate in parallel to the substrate. Meanwhile, the in-plane flexural vibration of the Langevin transducer's end plate causes the VTN root and MMP to vibrate perpendicularly to the substrate. At the working frequencies at which the MMP's vibration components perpendicular and parallel to the substrate have a phase difference of 90°, the MMP root moves elliptically and perpendicularly to the substrate. This results in the contact trapping mode. The MMP's vibration component perpendicular to the substrate is generated by the radial vibration of the piezoelectric rings, and the component parallel to the substrate is generated by the longitudinal vibration of the piezoelectric rings. Due to the different transmission mechanisms of these two vibrations, it is impossible for the phase difference between the two vibration components of the MMP's root to be 0°or ±180°unless additional measures are taken in the structure design. Thus, the device reported in Ref. [1] only has contact trapping mode.

In Ref. [2], an acoustic needle with a micro beak, which is excited by a Langevin transducer, is used to trap a single nanowire. The micro beak and the acoustic needle are in the same plane perpendicular to the Langevin transducer's axis, and the micro beak's root vibrates along a linear trajectory parallel to the longitudinal vibration of the Langevin transducer. This vibration excitation structure cannot generate an elliptical motion at the micro beak's root, and thus, only the noncontact trapping mode can be realized.

In the device reported in this work, the width-directional vibration of the piezo-electric plate (in the x-direction) generates a flexural vibration in the VTN. This flexural vibration in the VTN causes the MMP's root (point P) to vibrate with its x-, y- and z-vibration components in phase or out of phase by ±180°. In this case, point P vibrates linearly, and this results in the noncontact trapping mode. Hence, the width-directional vibration of the piezoelectric plate causes the noncontact trapping mode. The length-directional vibration of the piezoelectric plate (in the z-direction) generates the flexural (or transverse) and longitudinal waves in the VTN. The flexural wave causes the MMP's root to vibrate in the z-direction perpendicular to the substrate. The longitudinal wave causes the MMP's root to vibrate in the x-direction (parallel to the substrate). Denoting the x- and z-directional vibration velocities at point O (Fig. 4.13) as V_{Ox} and V_{Oz}, respectively, and those at point P as V_{Px}, and V_{Pz}, respectively, vibration velocity phase difference $\angle V_{Px} - \angle V_{Ox} \approx 0$ or ±180°, vibration velocity phase difference $\angle V_{Pz} - \angle V_{Oz} \approx 90°$(see Fig. 4.17), and vibration velocity phase difference $\angle V_{Oz} - \angle V_{Ox} = 0$ (the x-directional vibration of the piezoelectric plate is caused by the Poisson's effect.). Thus, $\angle V_{Pz} - \angle V_{Px} = 90°$. Therefore, the MMP's root moves elliptically and perpendicularly to the substrate, which results in the contact trapping mode. Hence, the length directional vibration of the piezoelectric plate causes the contact trapping mode.

4.3.3 CHARACTERISTICS AND DISCUSSION

When the distance between an AgNW on the substrate surface and the MMP's tip is too large, the AgNW cannot be sucked by the MMP. The maximum distance d_m between the MMP's tip and an AgNW initially aligned at $y = 0$ along the x direction on the substrate surface (see Fig. 4.23(a)), at which the AgNW can be sucked, was measured. Figure 4.23(b, c) is the measured maximum distance d_m versus driving frequency at three driving voltages for the noncontact and contact modes, respectively. The AgNW used in the experiments is 300 nm thick and 28μm long. It is seen that the maximum distance d_m can be tuned by the voltage and reach several ten microns. It can also be tuned by the driving frequency, depending on the vibration response characteristic of the VTN. The peak in Fig. 4.23(b) is caused by the VTN's resonance, as shown in Fig. 4.21.

The effect of angle θ made by the VTN and MMP (see Fig. 4.13) on the device's trapping performance was experimentally investigated, and the results are shown in Table 4.5. The driving voltage was 40 V_{p-p}, and nanowires with a 300 nm diameter and 28 μm length were used. It was found that the noncontact mode can be realized in the frequency range from 131.2 to 132.2 kHz for the angle θ listed in the table, and the contact mode cannot be realized when angle θ is greater than 110°. At θ =120°and 135°, the MMP's tip still has the sucking capability but does not have the lifting function. Thus, increasing angle θ may decrease the upward acoustic streaming.

During the experiments, it was also seen that the vibration velocity components only have an order of magnitude of several centimeters per second in both

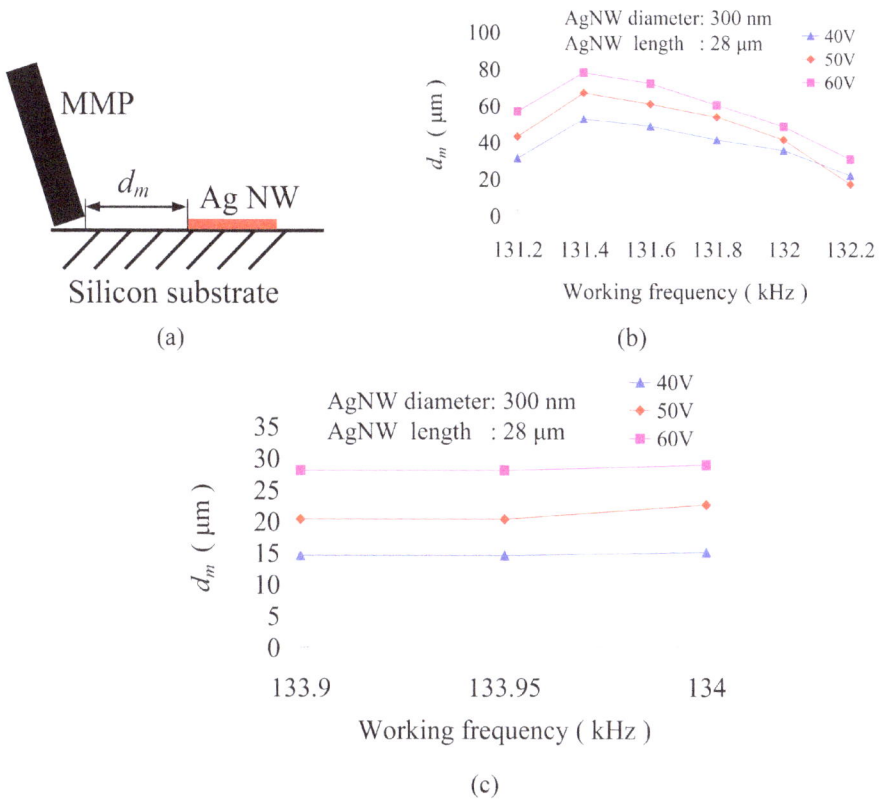

FIGURE 4.23 Measured maximum distance d_m between the micro manipulating probe's tip and an AgNW initially aligned in the plane formed by the vibration transmission needle and micro manipulating probe on the substrate surface, at which the AgNW can be sucked. (a) The micro manipulating probe and AgNW to be sucked. (b) d_m Versus the driving frequency at 40, 50 and 60 V_{p-p} at the noncontact mode. (c) d_m Versus the driving frequency at 40, 50 and 60 V_{p-p} at the contact trapping mode. Reproduced from Ref. [4] with permission from Elsevier.

the trapping modes. This means that it is possible to use non-resonance devices to implement the nano trapping, which can greatly enhance the stability of the device. In addition, it was seen that the frequency bandwidth for the contact trapping mode is narrower than that for the noncontact trapping mode. It needs to be improved by optimizing the structure of the device.

TABLE 4.5

The Effect of Angle θ Made by the VTN and MMP on the Performance of the Device

θ (degree)	95	105	120	135
Can the noncontact mode be realized?	yes	yes	yes	yes
Can the contact mode be realized?	yes	yes	no	no

4.3.4 SUMMARY

It has experimentally been demonstrated that the noncontact and contact trapping modes for individual nanowires can be integrated in one ultrasonic device by making the micro manipulation probe oscillate linearly and elliptically at proper working frequencies, respectively. The noncontact and contact trapping modes are realized by employing different acoustic streaming field patterns around the micro manipulating probe. The acoustic streaming field around the micro manipulating probe can be effectively controlled by the phase difference among the normal vibration components at the MMP root, which can be controlled by the driving frequency. The vibration displacement at the MMP root, needed for the trapping, is about several ten nanometers. In addition, it is found that the effective range of the acoustic streaming is several ten microns and can be tuned by the driving voltage and frequency.

The noncontact trapping mode enables the device to handle sticky nano samples, and the contact trapping mode makes the transfer of a trapped sample convenient. Thus, the proposed method makes the ultrasonic device more competitive in trapping and transferring nanowires.

4.4   ULTRASONIC NANO CLEANING FOR TARGETED POINTS ON A SUBSTRATE

The development of micro/nano manufacturing requires a technique of removing micro/nanoscale objects from a small area around a selectable point on the surface of a substrate in the medium such as liquid and air, which is useful in the defect control in nanomanufacturing, measurement of single nanoscale component, nano assembly, etc. The existing methods to clean the surface mainly include megasonics, spin rinse, liquid aerosol (i.e., spray), solid aerosol and brush scrubbing. Each of them has its own features and can be used to effectively clean a solid surface with a macro scale. However, their common shortcoming is that they only offer the nonselective removal function. Most of these cleaning methods suffer from a large amount of material loss and damage creation to the existing structures when removing contaminants generated in some steps of micro/nano manufacturing. It has been difficult to remove the micro/nanoscale objects from a small area around a selectable point on a solid substrate.

The acoustic micro/nano manipulation technology utilizes the physical effects of ultrasound to manipulate micro/nanoscale entities, such as solid particles, cells, organisms, tissues, crystals, bubbles and droplets. Diversified ultrasonic micro/nano manipulation functions have been implemented, which include the contact and noncontact trapping of single micro/nanoscale objects, 2D and 3D transferring of trapped micro/nanoscale objects, and controlled rotary driving of single nanowires. In this work [23], the author's group attempted to use acoustic streaming to remove micro/nanoscale particles in a small area at the interface between a substrate and water droplet. The experiments show that micro/nanoscale particles in a submillimeter-diameter area around a selectable point can be effectively removed by the proposed method. The physical mechanism of the cleaning process is analyzed, and characteristics of the cleaning method are measured. Also, the effect of working parameters on the acoustic streaming field is investigated by the FEM computation.

4.4.1 EXPERIMENTAL SETUP AND PHENOMENA

Figure 4.24 shows the experimental setup for the ultrasonic removal of micro/nanoscale particles in a submillimeter-diameter area. The device is composed of a piezoelectric plate, vibration transmission needle (VTN) and micromanipulation probe (MMP). The VTN is made of nickel-plated steel (Shanghai Dongfeng Co. Ltd, China), and the MMP is made of glass fiber (Nanjing Fiberglass Research & Design Institute Co. Ltd, China). The piezoelectric plate, made of lead zirconate titanate (P-81, Haiying Enterprise Group Co. Ltd, China), is bonded on the VTN along the narrow side of itself by the modified acrylate adhesive (Gleihow New Materials Co. Ltd, China). The

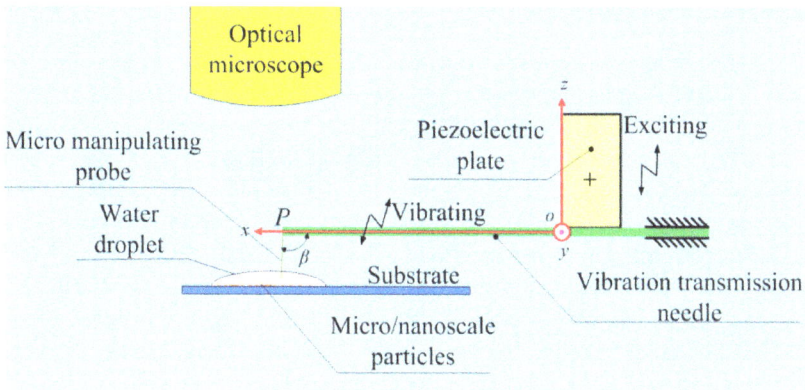

(a)

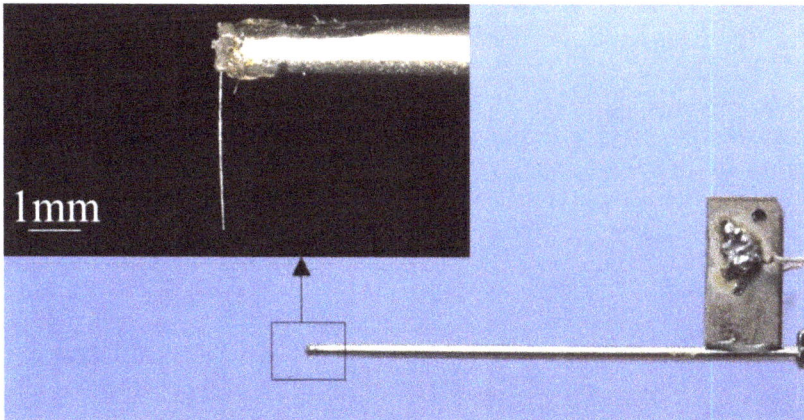

(b)

FIGURE 4.24 Experimental setup for the removal of micro/nanoscale particles in sub-millimeter-diameter area at the interface between a water droplet and substrate surface. (a) Schematic diagram. (b) Photograph of the device. The inset in (b) gives a magnified image of the micro manipulating probe and the tip of the vibration transmission needle. Reproduced from Ref. [24] with permission from AIP Publishing.

MMP is bonded to the VTN's tip (point P) by the super glue (Guangzhou Zhanba Adhesive Co. Ltd, China) and parallel to the piezoelectric plate. The width, length and thickness of the piezoelectric plate are 5, 10, and 1 mm, respectively. The angle β between the MMP and VTN is about 91°. The electromechanical quality factor Q_m, piezoelectric coefficient d_{33}, electromechanical coupling factor k_{33}, relative dielectric constant $\varepsilon_{33}^T/\varepsilon_0$, dielectric dissipation factor $\tan\delta$ and density of the piezoelectric plate are 800, 200×10^{-12} C/N, 0.6, 1000, 0.5% and 7450 kg/m³, respectively. The VTN has a uniform diameter of 0.9 mm and is 26 mm long out of the plate. The MMP is 20 μm thick and 3 mm long. The resonance frequency of the device, measured by a laser Doppler vibrometer (PSV-300F, Polytec GmbH, Waldbronn, Germany), is 124.5 kHz, at which the VTN vibrates flexurally, and the piezoelectric plate vibrates at its length vibration mode.

The VTN, which is above the suspension droplet, is parallel to the substrate. A layer of aqueous suspension droplet with a height of about 0.2 mm and a diameter of about 1 cm is dispersed on the silicon substrate by a dropper. Deionized water is used, and no bubbles are observed by the microscope (VHX-1000E, Keyence, Osaka, Japan) operating at an optical magnification of 500 times. The MMP is inserted into the suspension droplet, and the distance between the MMP and substrate surface is about 50 μm, which is controlled by the X-Y-Z moving stage (LD125-LM-2, Shengling Precise Machinery Co. Ltd, China). In the experiments, yeast cells (Angel Yeast Co. Ltd, China) with a diameter range of 3–5 μm (measured) and Si nanoparticles (NPs) (Beijing DK Nano Technology Co. Ltd, China) with a diameter range of 300–500 nm (provided by the manufacturer) are used as the experimental samples, and the concentration of the yeast cells and Si NPs is 1.38 mg/ml and 1.52 mg/ml, respectively. The experiments are conducted under an optical microscope.

Images $a1$–$a7$ and $b1$–$b7$ in Fig. 4.25 show the cleaning process for the yeast cells and Si NPs on the substrate by the MMP, respectively, at the sonication time of 0, 16, 32, 48, 60, 72 and 84 s.

In the experiments, the driving frequency and voltage are 124.5 kHz and 40 V_{p-p}, respectively. The particles are pushed outwards during the sonication, and as a result, the round-shaped cleaned areas with a quite clear boundary are formed. Images a8 and b8 show the distribution of the yeast cells and Si NPs on the substrate after the vibrating MMP is removed from the droplet. It is seen that the particles do not return back to the original location after the sonication is switched off.

As depicted in Fig. 4.25, the boundary of the cleaned area can be divided into two types, that is, stacking boundary (the left boundaries in images $b5$–$b8$) and non-stacking boundary (the left boundaries in images $a5$–$a8$). In the vicinity of the left boundary of the cleaned area in images $b5$–$b8$, there are stacks of NPs. This type of boundary is defined as the stacking boundary. In the vicinity of the left boundary of the cleaned area in images $a5$–$a8$, there are no stacks of yeast cells. The corresponding boundary is defined as the non-stacking boundary. One of the reasons for the difference in the boundaries is the size of the removed particles. Due to the smaller size, it is easier for NPs to overlap each other and stack up in the vicinity of the boundary. Thus, there is only the non-stacking boundary in Fig. 4.25(a). The boundary type also depends on the acoustic streaming pattern, which will be explained in the next section in details.

(a)

(b)

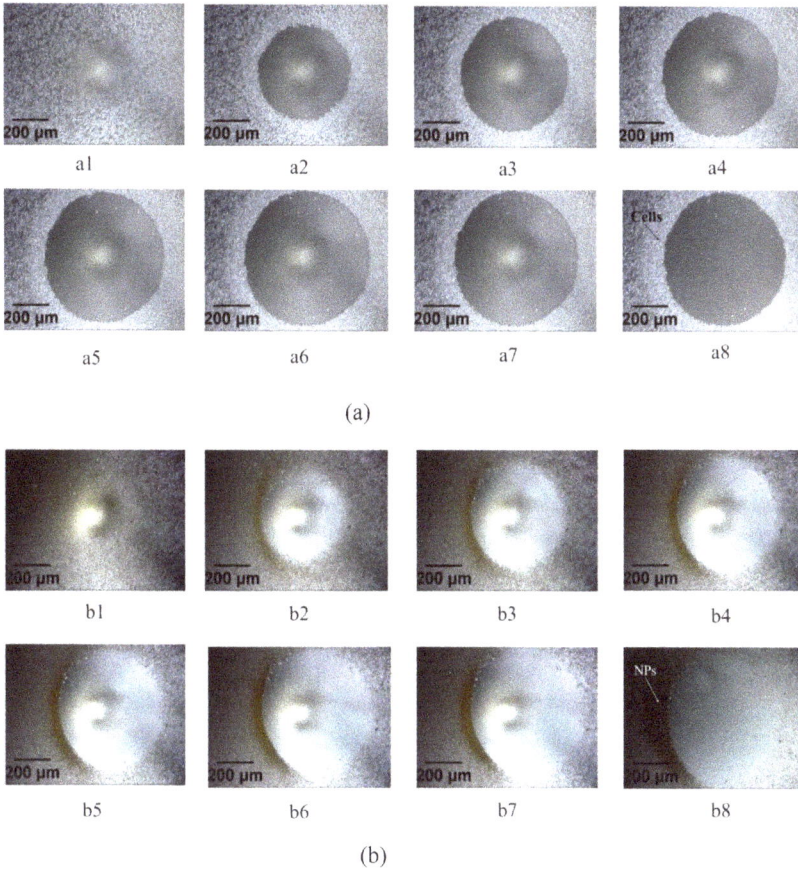

FIGURE 4.25   (a) Removal of the yeast cells under the micro manipulating probe at a driving frequency of 124.5 kHz and voltage of 40 V_{p-p}. (b) Removal of the Si NPs under the micro manipulating probe at a driving frequency of 124.5 kHz and voltage of 40 V_{p-p}. The sonication time for images *a1–a7* or *b1–b7* is 0, 16, 32, 48, 60, 72 and 84 s, respectively, and images a8 and b8 are taken after the micro manipulating probe is removed. Reproduced from Ref. [24] with permission from AIP Publishing.

4.4.2   PRINCIPLE ANALYSIS

Our measurement shows that the vibration amplitude of the piezoelectric plate in the z-direction (see Fig. 4.24) is about 14 and 5 times larger than those in the y-direction and x-direction, respectively. Thus, the VTN vibrates in the z-direction flexurally.

Figure 4.26(a) shows the z-directional vibration amplitude distribution of the VTN at 124.5 kHz and 40 Vp-p, which was measured by a laser Doppler vibrometer. The measured vibration amplitude at point P in the z-direction is 606.7 nm. Figure 4.26(b) shows the measured distribution of the z-directional in-plane vibration of the piezoelectric plate at 124.5 kHz and 40 Vp-p. The left and right insets in Fig. 4.26(b) show the deformation of the piezoelectric plate when it shrinks and extends, respectively.

(a)

(b)

(c)   (d)

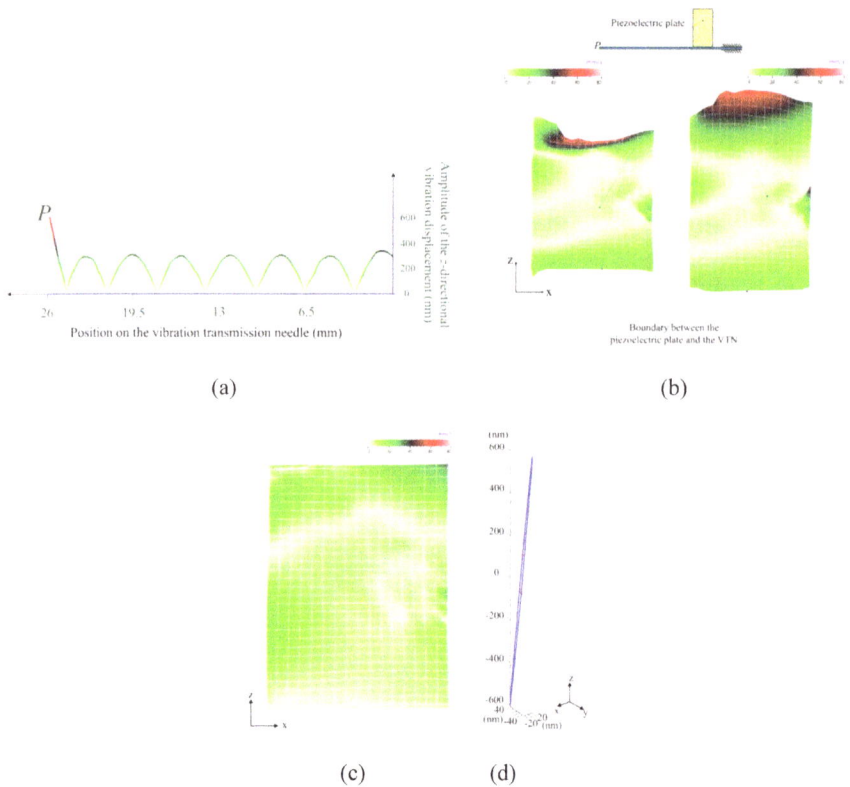

FIGURE 4.26 (a) Measured amplitude distribution of the z-directional vibration displacement of the vibration transmission needle along the x-direction at a driving frequency of 124.5 kHz and Voltage of 40 $V_{p\text{-}p}$. (b) Measured in-plane vibration mode of the piezoelectric plate at a driving frequency of 124.5 kHz and Voltage of 40 $V_{p\text{-}p}$. (c) Measured y-directional vibration velocity amplitude distribution of the piezoelectric plate at a driving frequency of 124.5 kHz and Voltage of 40 $V_{p\text{-}p}$. (d) Computed oscillation trajectory of point P at a driving frequency of 124.5 kHz and Voltage of 40 $V_{p\text{-}p}$. Reproduced from Ref. [25] with permission from AIP Publishing.

Figure 4.26(c) shows the measured y-directional vibration velocity amplitude distribution of the piezoelectric plate. Comparing Fig. 4.26(b, c), it is seen that the vibration of the piezoelectric plate in the z-direction is much larger than that in the y-direction. Thus, the piezoelectric plate operates at the length vibration mode at 124.5 kHz. The measured x, y and z components of the vibration velocity at point P are $1.598\times10^{-2}\angle35.1°$m/s, $3.863\times10^{-2}\angle-149.3°$m/s, and $0.4747\angle36.8°$m/s, respectively. The standard deviation of their phase angle is $1.6°$, $3.6°$ and $4.2°$, respectively, and that of their amplitude is 1.89×10^{-3} m/s, 2.21×10^{-3} m/s and 4.48×10^{-3} m/s, respectively, when the repetition measurement number for each working point is 15. It is seen that the z-directional vibration component at point P is about 12 and 30 times larger than the y- and x-directional vibration components, respectively. The phase difference between the z- and y-directional vibration components at point P is denoted as $\Delta\varphi_{zy}$, and that between the z- and x-directional vibration components at point P as $\Delta\varphi_{zx}$.

TABLE 4.6

Material Property Constants of the Probe-Droplet-Substrate System Used in the FEM Computation

Material	Density (kg/m³)	Poisson's Ratio	Modulus of Elasticity (Pa)	Acoustic Dissipation Factor* (Pa s)	Velocity of Sound (m/s)	Shear Viscosity Coefficient (Pa s)
Glass fiber (MMP)	2200	0.3	7.4×10^9	–	–	–
Water	1000	–	–	6.153×10^{-3}	1500	1.792×10^{-3}

* The acoustic dissipation factor b is computed by $b = \frac{4}{3}\eta + \eta'$, where η and η' are the shear and bulk viscosity coefficient of the acoustic medium, respectively.

Thus, $\Delta\varphi_{zy} = 186.1°$ and $\Delta\varphi_{zx} = 1.7°$, which are close to $180°$ and $0°$, respectively, and the oscillation trajectory of point P is approximately linear. The computed oscillation trajectory of point P is shown in Fig. 4.26(d), which indicates that the trajectory is very close to a linear line.

In order to explain the experimental phenomena, the 3D acoustic streaming field around the MMP tip was computed by the finite element method (FEM). The FEM computation model consists of the MMP, droplet and substrate. In the computation, the driving frequency was 124.5 kHz, and the measured x-, y- and z-directional vibration velocity components at the MMP root ($=1.598 \times 10^{-2} \angle 35.1°$ m/s, $3.863 \times 10^{-2} \angle -149.3°$ m/s and $0.4747 \angle 36.8°$ m/s, respectively) were used. The material property constants and size parameters of the probe-droplet-substrate system used in the computation are listed in Tables 4.6 and 4.7.

For the 3D acoustic streaming field, all of the boundaries are set to be the slip, which is deduced from our experimental phenomena [18]. The computation is implemented by COMSOL Multiphysics 4.3 software (COMSOL Inc., Stockholm, Sweden). Figure 4.27(a) shows the meshed 3D FEM computation model, and Fig. 4.27(b) shows the computed vibration displacement distribution of the MMP at 124.5 kHz and $40\ V_{p-p}$. The tetrahedral type of elements is used in the FEM computation model. The total element number in the probe is 813,600, and that in the liquid is 1,790,556. The relative tolerance factor of the FEM computation, which is used for the convergence

TABLE 4.7

Size Parameters of the Probe-Droplet-Substrate System Used in the FEM Computation

Water Doplet's Height (mm)	Water Droplet's Radius (mm)	MMP's Length (mm)	MMP's Radius (µm)	Distance between the MMP's Tip and Substrate (µm)	Angle between the MMP and Substrate (°)
0.15	1	3	10	50	89

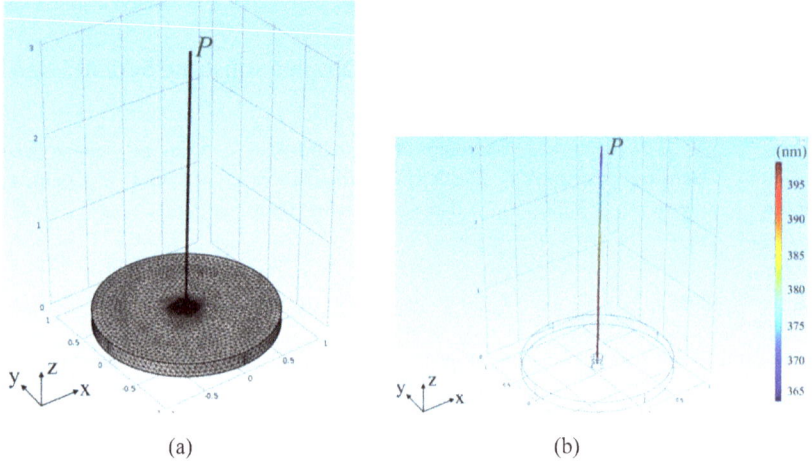

(a)   (b)

FIGURE 4.27 (a) Meshed 3D FEM computation model. (b) Computed distribution of the vibration displacement of the micro manipulating probe at a driving frequency of 124.5 kHz and Voltage of 40 V_{p-p}. Reproduced from Ref. [24] with permission from AIP Publishing.

judgment, is 0.001. The FEM model used in our computation is implicit, and only the steady states of the MMP's vibration, the acoustic field and the acoustic streaming field are computed.

Figure 4.28(a, b) shows the computed flow patterns on the substrate surface and in the central plane of the MMP-VTN structure, respectively, which agree with the observed one (Fig. 4.29) qualitatively. In Figs. 4.28(a) and 4.29, point O is the vertical projection of the center of the MMP's tip onto the substrate surface. The computed results can be used to explain the cleaning phenomenon quite well. As the acoustic streaming flows outwards from point O, it drags micro/nanoscale particles outwards. Due to the balance among the outward dragging force, inward dragging force and frictional force on the particles, a boundary is formed for the cleaned area.

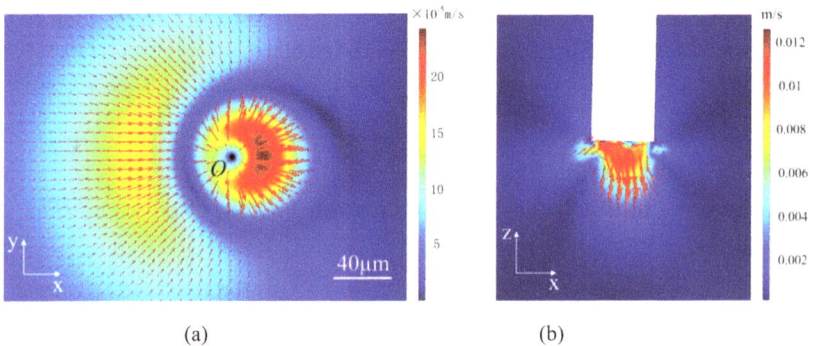

(a)   (b)

FIGURE 4.28 (a) Computed acoustic streaming field on the substrate surface. (b) Computed acoustic streaming field in the xz plane, which is the plane formed by the central axes of the vibration transmission needle and the micro manipulating probe. Reproduced from Ref. [24] with permission from AIP Publishing.

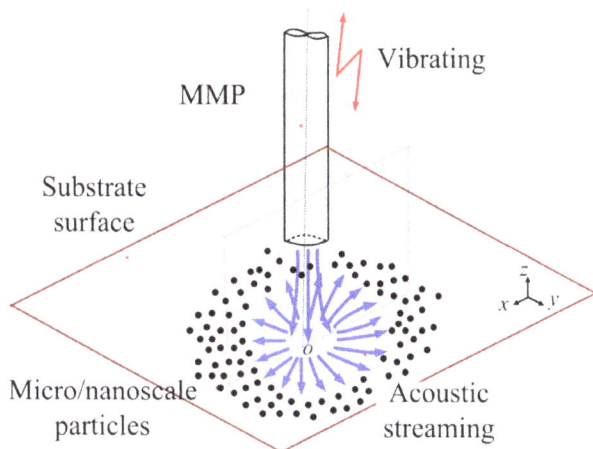

FIGURE 4.29 Observed pattern of the 3D acoustic streaming field around the micro manipulating probe. Reproduced from Ref. [24] with permission from AIP Publishing.

Comparing Figs. 4.25(b) and 4.28(a), it is known that another necessary condition to form the stacking boundary is that there are two flows with opposite directions at the boundary. At the right boundary of the cleaned area in Fig. 4.25(b), where there is only one directional flow (Fig. 4.28(a)), the flushed NPs scatter on the substrate rather than concentrating at the boundary. This is because there is no inward dragging force. At the left boundary of the cleaned area in Fig. 4.25(b), where the acoustic streaming has two opposite directions (Fig. 4.28(a)), the flushed NPs can stack up. Figure 4.28(b) indicates that the outward flow on the substrate comes from the downward flow under the MMP, which is part of the acoustic streaming eddies under the MMP. To form a cleaned area below the MMP, such a downward flow is essential.

4.4.3 CHARACTERISTICS AND DISCUSSION

The dependency of the acoustic streaming field on some working parameters was investigated by FEM computation. The working parameters included the distance between the MMP's tip and substrate, the angle between the MMP and substrate, and the ratio of the normal vibration components of the MMP. In the computation, unless otherwise specified, the driving frequency and the x-, y- and z-directional vibration velocity components at the MMP's root (point P) were identical to those in the preceding section, and the material property constants and size parameters of the probe-droplet-substrate system shown in Tables 4.6 and 4.7 were used.

Figure 4.30 shows the computed acoustic streaming fields at different distances between the MMP's tip and substrate (H). From images c1 and d1, it is seen that particles in range $A1$ may be flushed into range $A2$ owing to a relatively large speed of acoustic streaming in range $A1$ near boundary B, which is the curve on the substrate surface where the acoustic streaming velocity is zero. In image a1, the outward flow in range $A2$ flushes out the particles, and the particles flushed out are stopped by the inward flow near boundary B. Therefore, to implement the cleaning function, a large distance between the MMP's tip and substrate should be avoided.

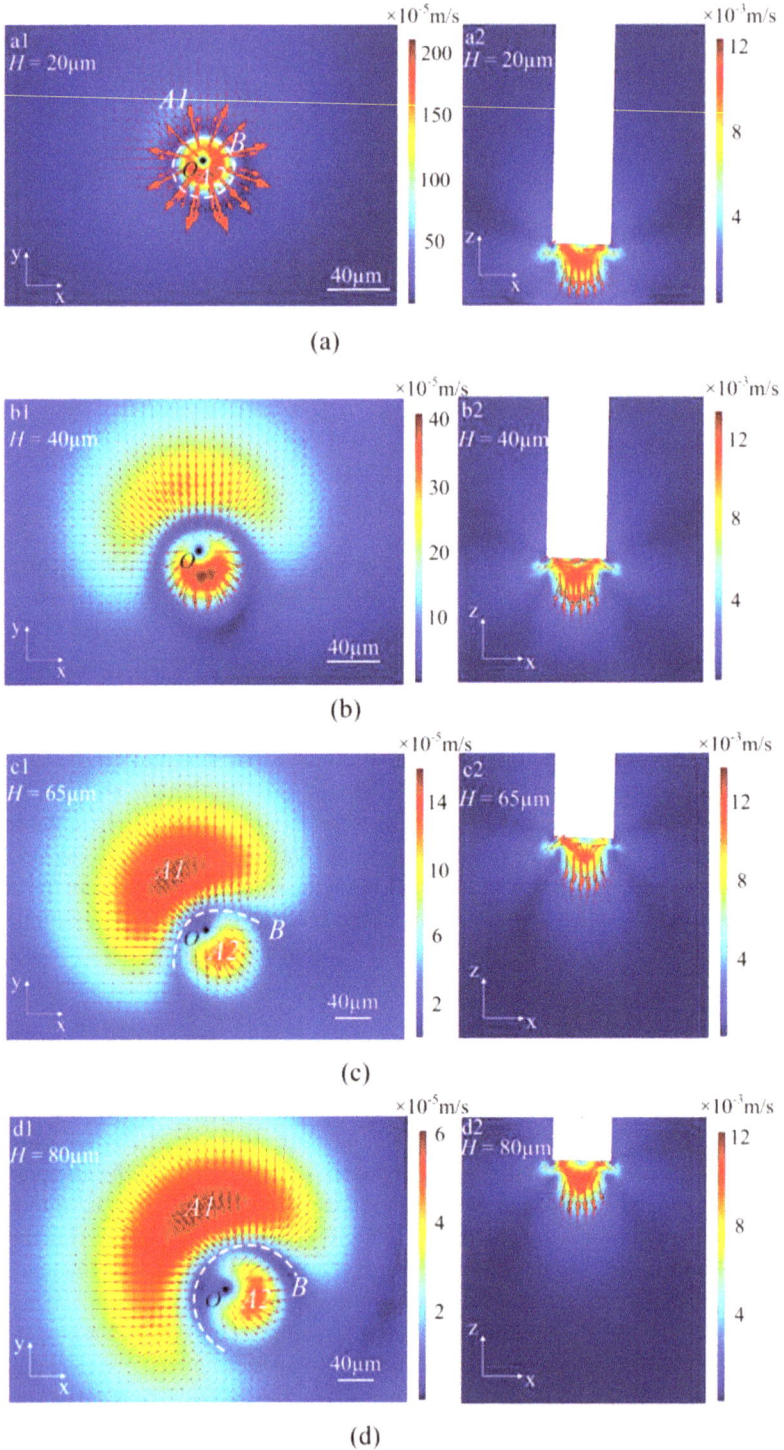

FIGURE 4.30 Acoustic streaming fields for different distance values H between the MMP tip and substrate. (a) $H = 20$ μm. (b) $H = 40$ μm. (c) $H = 65$ μm. (d) $H = 80$ μm. Reproduced from Ref. [24] with permission from AIP Publishing.

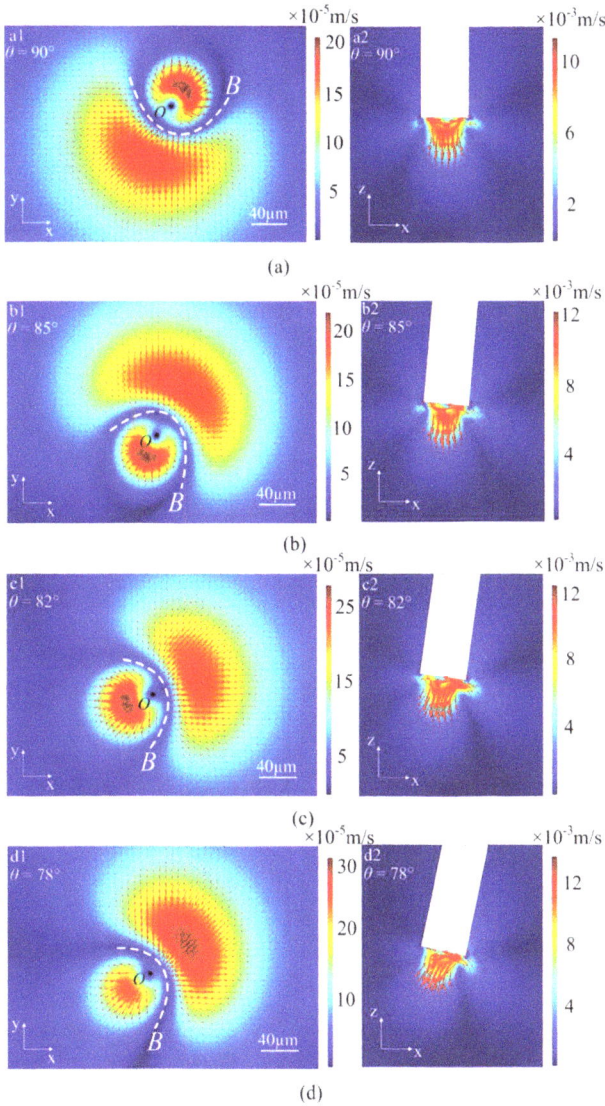

FIGURE 4.31 Acoustic streaming fields for different angles θ between the MMP and substrate. (a) $\theta = 90°$. (b) $\theta = 85°$. (c) $\theta = 82°$. (d) $\theta = 78°$. Reproduced from Ref. [24] with permission from AIP Publishing.

Figure 4.31 shows the computed acoustic streaming fields at different angles θ between the MMP and substrate. It is seen that as the angle θ deviates from $90°$, the acoustic streaming velocities on the two sides of boundary B become to have the same direction (toward the cleaned area), which means that the micro/nanoscale particles are more likely to be flushed into the cleaned area.

Figure 4.32 shows the computed acoustic streaming fields at different vibration velocity ratios of the z- and x-directional vibration velocity components V_z/V_x at $\Delta\varphi_{zx} = 0°$ and $V_y = 0$. In the computation, V_z/V_x decreased from 40 to 10, and V_z were

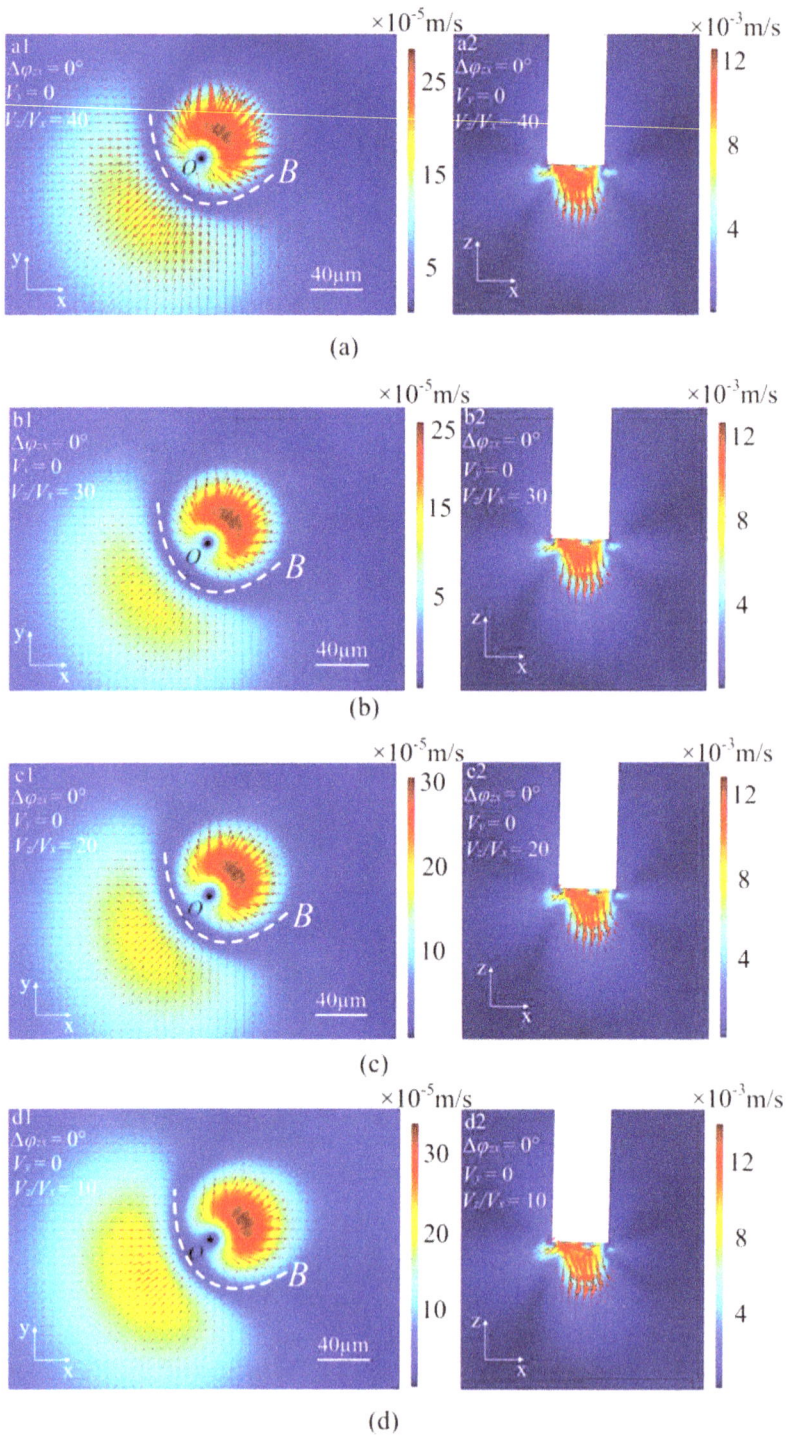

FIGURE 4.32 Acoustic streaming fields for different vibration velocity ratios between the z- and x-directional vibration velocity components V_z/V_x at $\Delta\varphi_{zx} = 0°$ and $V_y = 0$. (a) $V_z/V_x = 40$. (b) $V_z/V_x = 30$. (c) $V_z/V_x = 20$. (d) $V_z/V_x = 10$. Reproduced from Ref. [24] with permission from AIP Publishing.

kept constant (0.4747 m/s). Image *a1* indicates that the micro/nanoscale particles can be flushed out of the cleaned area from all directions on the substrate surface at $V_z/V_x = 40$. It is seen that as V_z/V_x decreases, the acoustic streaming velocities on the two sides of boundary B become to have the same direction (towards the cleaned area), which means that the micro/nanoscale particles are more likely to be flushed into the cleaned area.

To quantify the cleaning capability of the proposed method in this work, the equivalent diameter of the cleaned area D_e is used, which is defined as

$$D_e = 2\sqrt{\frac{S_c}{\pi}} \tag{4.7}$$

where S_c is the cleaned area, which can be measured by the images taken by the optical microscope. Figure 4.33(a, b) shows the measured diameter of the cleaned area

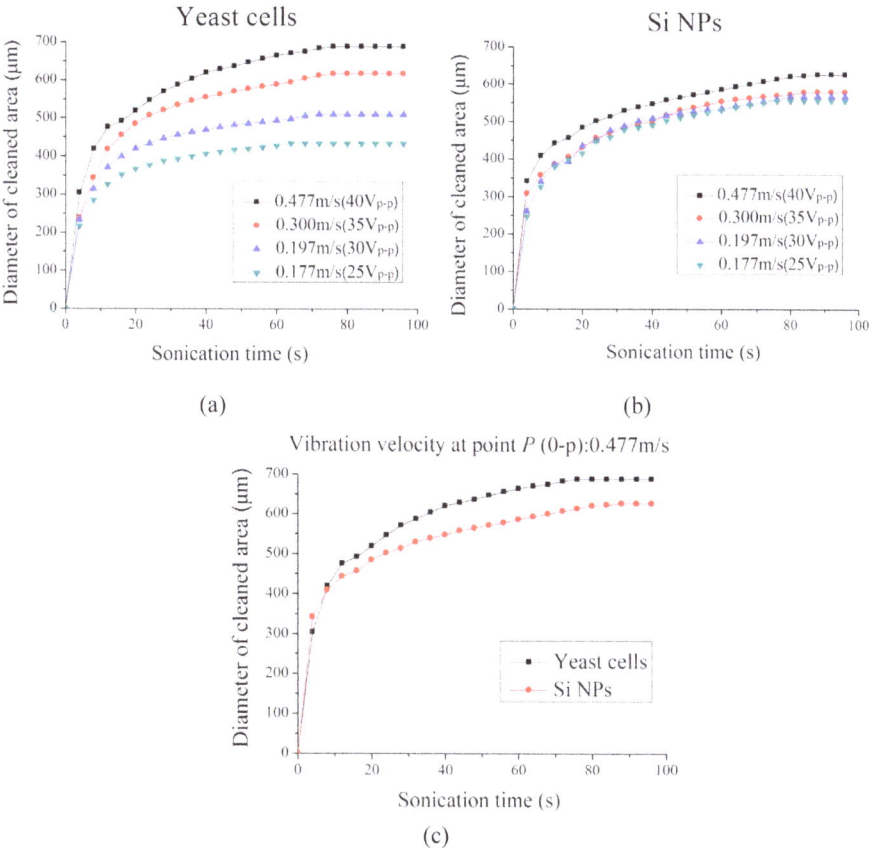

(a)   (b)

(c)

FIGURE 4.33   (a) Measured diameter of the cleaned area versus sonication time for the yeast cells at four different vibration velocities at Point P. (b) Measured diameter of the cleaned area versus sonication time for the Si NPs at four different vibration velocities at point *P*. (c) Measured diameter of the cleaned area versus sonication time for the yeast cells and Si NPs when the vibration velocity at point *P* is 0.477m/s. Reproduced from Ref. [24] with permission from AIP Publishing.

versus sonication time for the yeast cells and Si NPs, respectively, at four different vibration velocities at point *P*. It is seen that the cleaned area became larger with the increase of sonication time in the initial several ten seconds, and changed little after this time duration. Figure 4.33(c) compares the diameter of the cleaned area versus sonication time for the yeast cells and Si NPs when the vibration velocity at point *P* is constant (0.477 m/s). It is seen that the cleaned area diameter for the yeast cells is always larger than that for the Si NPs. The calculated mass of a yeast cell and Si NP is 3.7×10^{-14} kg and 7.8×10^{-17} kg, respectively. In the calculation, the density of yeast cells and Si NPs is 1.114×10^3 kg/m^3 and 2.33×10^3 kg/m^3, respectively. Thus, the yeast cells have a much larger inertia than the Si NPs. For this reason, the yeast cells driven by the acoustic streaming can go a longer distance than the Si NPs, which results in a larger cleaned area. It takes about 76 s and 88 s for the cleaned areas of yeast cells and Si NPs to reach stable diameters, respectively.

The stable diameter D_s of the cleaned area versus vibration velocity at point *P*, measured for the yeast cells and Si NPs, is shown in Fig. 4.34. Here, D_s is the diameter of the cleaned area sonicated for a sufficiently long time. It can be seen that when the vibration velocity is small, D_s for the Si NPs is larger than that for the yeast cells. This is because the Si NPs have lower inertia than the yeast cells and move more easily at lower acoustic streaming velocity. However, when the vibration velocity is larger, D_s for the yeast cells is larger than for the Si NPs. This is because the inertia of the yeast cells is larger than that of Si NPs, and the yeast cells can move farther when the acoustic streaming velocity is large. This analysis indicates that the particle inertia has different effects on the stable diameter of the cleaned area at low and high vibration velocity ranges.

The time constant τ of formation process of the cleaned area, which is defined as the time it takes for the cleaned area to reach 63% of the stable diameter, was measured for different vibration velocities at point *P*, and the results are shown in Fig. 4.35. It is

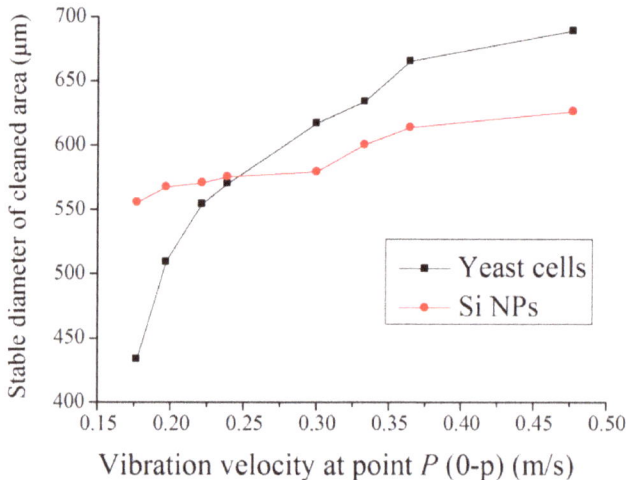

FIGURE 4.34 Measured stable diameter of the cleaned area versus vibration velocity at point *P* for the yeast cells and Si NPs. Reproduced from Ref. [30] with permission from AIP Publishing.

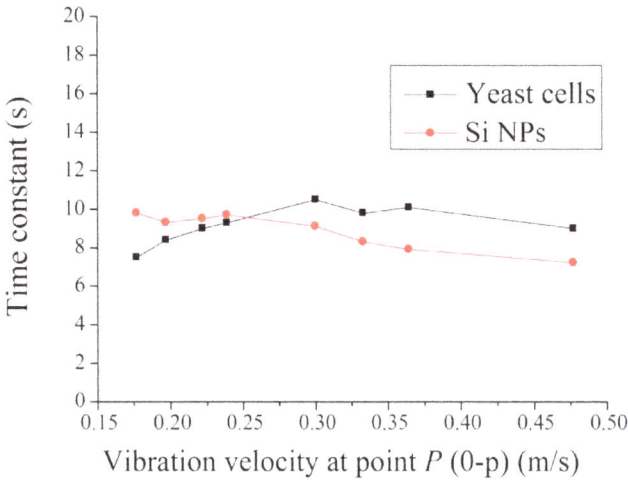

FIGURE 4.35   Measured time constant of the formation of the stable cleaned area versus vibration velocity at point P for the yeast cells and Si NPs. Reproduced from Ref. [30] with permission from AIP Publishing.

seen that the change of the time constant is less than 3 s for the vibration velocity range in the experiment. For the Si NPs, the time constant decreases as the vibration velocity increases. This is because the Si NPs move faster as the vibration velocity increases. For the yeast cells, with the vibration velocity increase, the time constant increases first and then decreases. The increase in the time constant is because of the larger stable diameter of the cleaned area (see Fig. 4.33(a)), and the decrease in the time constant is because the yeast cells move faster as the vibration velocity increases.

One merit of the method proposed and developed in this work is that the removal only occurs in a submillimeter-diameter range around a selectable point, and the removal process does not affect existing structures beyond this small range. Another merit is that the location of the cleaned area can be precisely controlled. Thus, it may be used to remove micro/nanoscale impurities away from the surface of interest in various nanomanufacturing processes, to prevent defects caused by the micro/nanoscale impurities. However, the removal effect depends on the bonding strength between the particles and substrate. This method cannot work if the bonding is very strong. Although the cleaned area in this paper is a submillimeter-diameter spot, the method proposed in this work can be used to remove micro/nanoscale particles along a narrow path on the substrate if the device is moved by a controlled X-Y-Z moving stage.

4.4.4   SUMMARY

A strategy to remove micro/nanoscale particles within the submillimeter-diameter area at the interface between a water droplet and substrate is demonstrated. The strategy is based on the utilization of the 3D flow field induced by a linearly vibrating micro manipulating probe (124.5 kHz). For the yeast cells with a diameter of 3–5 μm and Si NPs with a diameter of 300–500 nm, it takes about 1.5 min to form a round-shaped cleaned area with a stable diameter. The computation shows that the distance between the MMP tip and substrate, the angle between the MMP and substrate, and the ratio of the normal vibration components of the MMP have effects on whether the flow field can work. The cleaning process may be conducted around a selectable point. It has potential applications in nano sensor manufacturing, nano measurement, and nano assembly.

4.5   ULTRASONIC TWEEZERS FOR MICRO/NANO SAMPLES IN AIR

Controlled pickup and release of micro/nano objects in the non-aqueous environment has large potential applications in lab-on-chip systems, biotechnology, biomedicine, and micro device fabrication. There are already several effective methods to pick and release multiple micro/nano objects in a non-aqueous environment. However, the design of the devices, which are capable of manipulating single micro/nano objects at the air-substrate interface with a low-temperature rise, is still a big challenge mainly due to the limitation of working principle of the existing devices.

Several manipulation methods based on magnetic, optical, mechanical and acoustic principles have been reported for pickup and release of micro-objects in air. In the magnetic method [24], microscale materials with magnetism are trapped by the magnetic force, and manipulated microscale materials can be shifted by controlling the external magnetic field. Although it provides a method to manipulate a large number of micro objects, the manipulated materials need to be magnetized, and the device is usually bulky. Also, it is difficult to pick up a particular micro-object by this method. The optical method is based on the optical tweezers which were invented in 1986 by Arthur Ashkin [25–28]. It has been successfully applied to the manipulation of atoms, bacteria, viruses, cells and other organisms. The main drawback of this method is that heat generated by the laser at the manipulation spot is harmful to some biological samples [28, 29]. The mechanical method usually uses an AFM probe to pick up and release a micro-object [30]. It has a high requirement for the position control of the device, which makes the device bulky and costly. In this method, a captured sample may be contaminated by direct contact with the AFM probe and damaged by poor control of the clamping force. The acoustic method employs the acoustic radiation force to manipulate micro objects [31–43], with merits such as being little affected by the material properties of the manipulated objects and simple device structure. Devendran et al. proposed an optimized resonator system that could capture as small as 15 μm particles in air [35]. Prisbrey used the phased arrays of ultrasound transducers to create 3D dynamic patterns of micro particles in the air [37]. Koyama et al. designed a series of acoustic devices to manipulate microscale materials in air, which used a flexural vibrating circular disc [33] or flexural vibrating plate [34] as the ultrasonic source. However, all of these devices are not capable of picking up a selected single micro-object on a substrate owing to their structures and working principles. The authors' group once proposed a tapered metal plate system to trap micro-objects at the air-substrate interface [38–41]. With a simple and compact structure, it is still difficult to trap a selected micro-object. In 2015, the authors' group proposed a method to stably capture a selected micro copper wire at the air-substrate interface and transfer it through any 3D path. Although the method is already experimentally confirmed, it has a big disadvantage, that is, the local temperature rise of the manipulation spot may be as high as 40 degrees Celsius [42]. Thus, it is not suitable for the manipulation of biological samples.

In Ref. [43], the author's team reported a low temperature-rise and facile strategy to pick up and release single micro/nano objects in air on a substrate. Chlorella cells, pollens, glass fibers, micro copper wires and graphene sheets at the air-substrate interface could be successfully picked up by the Van der Waals force between the samples and a stationary micro manipulation probe and transported through an arbitrary three-dimensional path in the air to a desired position on the substrate by moving the device. The captured sample could be controllably released by ultrasonically exciting the micro manipulation probe. The effects of working parameters on the release position deviation were investigated and the design guidelines for the device were proposed. The measured temperature rise at the MMP tip, in contact with the captured sample, was lower than 0.4°C, which makes it competitive in the manipulation of heat-sensitive samples such as biological cells.

4.6 REMARKS

The examples in this chapter have indicated the potential of ultrasonic nano tweezers in diversified nano manipulation functions. For the manipulations at the liquid–solid interface, the manipulation functions are mainly determined by the pattern of acoustic streaming generated by the ultrasonic field around the micro manipulation probe (MMP), and controlling the pattern of the acoustic streaming is the key to implementing a particular nano manipulation function or generate a new manipulation function.

The pattern of acoustic streaming in probe-type ultrasonic tweezers can be affected by the following factors: vibration trajectory of the MMP's tip, the shape and size of the MMP, the distance between the MMP and substrate, acoustic properties of the droplet, the working frequency or wavelength, etc. Thus, one needs to pay attention to these factors when designing a probe-type ultrasonic tweezer or exploring its new functions.

The vibration trajectory of the MMP tip is mainly determined by vibration trajectory at the exciting location of the vibration transmission needle (VTN), the size, shape and material of the VTN, and the working frequency. As the VTN is bonded onto a piezoelectric component or the end plate of a sandwich-type ultrasonic transducer for vibration excitation, vibration trajectory at the exciting location may be changed by choosing different bonding point or location. The size, shape and material of the VTN affect its stiffness in the three orthogonal orientations, with which the vibration component(s) in some particular direction(s) can be suppressed and other component (s) may be amplified (by the cantilever effect). In addition, for the manipulation functions that make use of an elliptical vibration trajectory of the MMP's tip, one may use two piezoelectric components driven by AC voltages with a proper phase difference such as $\pi/2$.

The shape and size of the MMP affect the strength of the ultrasonic field around the MMP. Usually, properly increasing the MMP diameter can strengthen the ultrasonic field and generate stronger acoustic streaming. However, increasing the MMP diameter may decrease the positioning accuracy. Thus, one has to consider both the ultrasonic strength and positioning accuracy when deciding the MMP diameter. Choosing a proper MMP length can cause the MMP to resonate and hence effectively increase the manipulation force.

The distance between the MMP and substrate has a large effect on the acoustofluidic field under the MMP. For the same device and droplet, changing the distance may result in a totally different acoustic streaming field. Thus, the distance must be taken into consideration as an important parameter in the design of ultrasonic tweezers, and paid enough attention during the use.

The acoustic properties of the droplet, including the density, sound velocity and viscosity, affect the ultrasonic field and acoustic streaming in the droplet for a given vibration excitation. For a practical fluid in the droplet, the ultrasonic field becomes stronger as the density of the droplet increases in general, which may increase the acoustic streaming velocity. The effect of the viscosity of a fluid in the droplet on the acoustic streaming is a bit complicated, as the acoustic attenuation, the spatial gradient of Reynolds stress, and resistance to the micro flow of fluid are all affected by the viscosity. In most of the experiments on ultrasonic nano tweezers, water is the main ingredient of the droplet. Few experiments on ultrasonic nano tweezers utilize sticky fluid such as oil, as it is difficult to generate acoustic streaming in a sticky fluid. How to generate strong enough acoustic streaming in sticky fluid in a droplet and control it for manipulation of micro/nano objects is definitely an interesting and challenging research area.

As the MMP tip has different vibration trajectories at different vibration modes, tuning the working frequency may result in a different vibration trajectory of the MMP tip, which may generate a different manipulation function in many cases. So far, the probe-type ultrasonic nano tweezer has been mainly investigated in the low-frequency range (<200 kHz), and few reports have been on its manipulation behavior at a high frequency. At a higher working frequency, the size of the micro eddies of acoustic streaming under the MMP will be smaller because of a shorter wavelength. This will cause a change of the manipulation behavior, which may bring in new nano manipulation functions. However, to construct a high-frequency ultrasonic nano tweezer, one needs to design a vibration excitation structure that can excite a strong enough vibration velocity at the micro manipulation probe while keeping a high working frequency.

Apart from the manipulations at the droplet-substrate interface, the ultrasonic tweezers can also be employed to manipulate micro/nano samples at the air-substrate interface, using the Van der Waals force and ultrasonic vibration.

REFERENCES

1. N. Li, J. Hu, H. Li, S. Bhuyan and Y. Zhou, "Mobile acoustic streaming based trapping and 3-dimensional transfer of a single nanowire," *Appl. Phys. Lett.*, 101(9), 093113, 2012.
2. H. Li and J. Hu, "Noncontact manipulations of a single nanowire using an ultrasonic micro-beak," *IEEE Trans. Nanotechnol.*, 13(3), pp. 469–474, 2014.
3. N. Li and J. Hu, "Sound controlled rotary driving of a single nanowire," *IEEE Trans. Nanotechnol.*, 13(3), pp. 437–441, 2014.
4. X. Wang and J. Hu, "An ultrasonic manipulator with noncontact and contact-type nanowire trapping functions," *Sens. Actuator A Phys.*, 232, pp. 13–19, 2015.

5. X. Wang and J. Hu, "3D FEM analyses of the ultrasonic transducer for controlled nanowire rotary driving," *Appl. Acoust.*, 103, pp. 157–162, 2016.
6. Q. Tang, X. Wang and J. Hu, "Nano concentration by acoustically generated complex spiral vortex field," *Appl. Phys. Lett.*, 110, 104105, 2017.
7. X. Qi, Q. Tang, P. Liu, I. V. Minin, O. V. Minin and J. Hu, "Controlled concentration and transportation of nanoparticles at the interface between a plain substrate and droplet," *Sens. Actuators B Chem.*, 274, pp. 381–392, 2018.
8. Q. Liu, J. Hu, I. V. Minin and O. V. Minin, "High-performance ultrasonic tweezers for manipulations of motile and still single cells in a droplet," *Ultrasound Med. Biol.*, 45(11), pp. 3018–3027, 2019.
9. Q. Liu, K. Chen, J. Hu and T. Morita, "Double parabolic reflectors wave-guided high-power ultrasonic transducer (DPLUS) based ultrasonic tweezers for micro/nano manipulation," *Jpn. J. Appl. Phys.*, 59, SKKD12, 2020.
10. Q. Liu, K. Chen, J. Hu and T. Morita, "An ultrasonic tweezer with multiple manipulation functions based on the double-parabolic-reflector wave-guided high-power ultrasonic transducer," *IEEE Trans. Ultrason. Ferroelectr. Freq. Control*, 67(11), 2020.
11. P. Liu, Q. Tang, S. Su and J. Hu, "Principle analysis for the micromanipulation probe-type ultrasonic nanomotor," *Sens. Actuator A Phys.*, 318, 112524, 2021.
12. Q. Liu, Q. Tang, J. Hu, "An ultrasonic sweeper with micro/nano concentration, decorating, transmedium extraction and localized cleaning functions", *Rev. Sci. Instrum.*, 94, 073701, 2023.
13. Q. Tang and J. Hu, "Diversity of acoustic streaming in a rectangular acoustofluidic field," *Ultrasonics*, 58, pp. 27–34, 2015.
14. J. Hu, *Ultrasonic Micro/Nano Manipulations: Principles and Examples*, (World Scientific, New Jersey, 2014), pp. 30–32.
15. S. Ueha and Y. Tomikawa, *Ultrasonic Motors: Theory and Applications*, (Oxford University Press, Oxford, 1994), p. 32.
16. K. Uchino, *Piezoelectric Actuators and Ultrasonic Motors*, (Springer, New York, 1997).
17. C. Zhao, *Ultrasonic Motors: Technologies and Applications*, (Science Press Beijing & Springer, Beijing, 2011).
18. Q. Tang and J. Hu, "Analyses of acoustic streaming field in the probe-liquid-substrate system for nano trapping," *Microfluid Nanofluidics*, 19, pp. 1395–1408, 2015.
19. J. Lighthill, "Acoustic streaming," *Journal of Sound and Vibration*, 61(3), pp. 391–418, 1978.
20. R. T. Beyer, Nonlinear Acoustics, in W.P. Mason, Ed., *Physical Acoustics*, vol. 2B. (Academic Press, New York, 1965), pp. 231–263.
21. R. T. Beyer, The Parameter B/A, in M. F. Hamilton and D. T. Blackstock, Eds., *Nonlinear Acoustics: Theory and Applications*, (Academic Press, New York, 1997), pp. 25–39.
22. Y. Zhou, J. Hu and S. Bhuyan, "Manipulations of silver nanowires in a droplet on low-frequency ultrasonic stage," *IEEE Trans. Ultrason. Ferroelectr. Freq. Control*, 60(3), pp. 622–629, 2013.
23. P. Liu and J. Hu, "Controlled removal of micro/nanoscale particles in submillimeter-diameter area on a substrate," *Rev. Sci. Instrum.*, 88, 105003, 2017.
24. Q. Q. Wang, L. D. Yang, B. Wang, E. Yu, J. F. Yu and L. Zhang, "Collective behavior of reconfigurable magnetic droplets via dynamic self-Assembly," *ACS Appl. Mater. Inter.*, 11, pp. 1630–1637, 2019.
25. R. Omori, T. Kobayashi and A. Suzuki, "Observation of a single-beam gradient-force optical trap for dielectric particles in air," *Opt. Lett.*, 22, pp. 816–818, 1997.
26. D. G. Grier, "A revolution in optical manipulation," *Nature*, 424, pp. 810–816, 2003.
27. A. H. J. Yang, S. D. Moore, B. S. Schmidt, M. Klug, M. Lipson and D. Erickson, "Optical manipulation of nanoparticles and biomolecules in sub-wavelength slot waveguides," *Nature*, 457, pp. 71–75, 2009.

28. J. Castillo, M. Dimaki and W. E. Svendsen, "Manipulation of biological samples using micro and nano techniques," *Integr. Biol-UK*, 1, pp. 30–42, 2009.

29. H. Xie and S. Regnier, "Three-dimensional automated micromanipulation using a nanotip gripper with multi-feedback," *J. Micromech. Microeng.*, 19, 075009, 2009.

30. M. H. Korayem, M. Taheri, H. Badkoobehhezaveh and H. Khaksar, "Simulating the AFM-based biomanipulation of cylindrical micro/nanoparticles in different biological environments," *J. Braz. Soc. Mech. Sci.*, 39, pp. 1883–94, 2017.

31. R. Yamamoto, D. Koyama and M. Matsukawa, "On-chip ultrasonic manipulation of microparticles by using the flexural vibration of a glass substrate," *Ultrasonics*, 79, pp. 81–6, 2017.

32. D. Koyama and K. Nakamura, "Noncontact ultrasonic transportation of small objects in a circular trajectory in air by flexural vibrations of a circular disc," *IEEE. Trans. Ultrason. Ferroelectr.*, 57, pp. 1434–42, 2010.

33. D. Koyama and K. Nakamura, "Noncontact ultrasonic transportation of small objects over long distances in air using a bending vibrator and a reflector," *IEEE. Trans. Ultrason. Ferroelectr.*, 57, pp. 1152–9, 2010.

34. R. Kashima, D. Koyama and M. Matsukawa, "Two-dimensional noncontact transportation of small objects in air using flexural vibration of a plate," *IEEE. Trans. Ultrason. Ferroeletr.*, 62, pp. 2161–8, 2015.

35. C. Devendran, D. R. Billson, D. A. Hutchins and A. Neild, "Optimization of an acoustic resonator for particle manipulation in air," *Sensor Actuators B*, 224, pp. 529–38, 2016.

36. H. Tanaka, Y. Wada, Y. Mizuno and K. Nakamura, "Effect of holed reflector on acoustic radiation force in noncontact ultrasonic dispensing of small droplets," *Jpn. J. Appl. Phys.*, 55, 067302, 2016.

37. M. Prisbrey and B. Raeymaekers, "Ultrasound noncontact particle manipulation of three-dimensional dynamic user-specified patterns of particles in air," *Phys. Rev. Appl.*, 10, 034066, 2018.

38. J. Hu and A. Santoso, "A Pi-shaped ultrasonic tweezers concept for manipulation of small particles," *IEEE. Trans. Ultrason. Ferroelectr.*, 51, pp. 1499–507, 2004.

39. J. Hu, J. Xu, J. Yang, J. Du, Y. Cai and C. Tay, "Ultrasonic collection of small particles by a tapered metal strip," *IEEE. Trans. Ultrason. Ferroelectr.*, 53, pp. 571–578, 2006.

40. Y. Liu and J. Hu, "Analysis of the ultrasonic collection of small particles by a tapered metal strip," *Sensor Actuator A*, 141, pp. 321–327, 2008.

41. Y. Liu and J. Hu, "Trapping of particles by the leakage of a standing wave ultrasonic field," *J. Appl. Phys.*, 106, 034903, 2009.

42. G. Chen, N. Li and J. Hu, "Capture of individual micro metal wires in air by ultrasonic tweezers," *IEEE ASME Trans. Mechatron.*, 20(6), pp. 3053–3059, 2015.

43. X. Qi, P. Liu and J. Hu, "A low temperature-rise and facile manipulation method for single micro objects at the air-substrate interface," *J. Micromech. Microeng.*, 29, 105007, 2019.

5 Ultrasonic Driving of Gas Molecules and Its Applications

With the recent 40 years' research and development on ultrasonic actuation technology, the physical effects of ultrasonic vibration, such as the linear and elliptical vibration trajectories of a solid surface, acoustic radiation force and acoustic streaming, have been successfully employed in the driving of macro, micro and nano scale solid objects and microfluid. The ultrasonic device that can drive a macro scale object is termed ultrasonic motor, and the one which can drive a micro/nano scale object is termed ultrasonic manipulator. Ultrasound can also be used to drive gas molecules, which have a diameter in sub-nm order. In this case, the ultrasonic nano drive may bring in the so-called ultrasonic catalysis effect, that is, the ultrasonic wave speeds up the chemical reaction at a solid-gas interface under proper sonication. In this chapter, the principle of ultrasonic driving of gas molecules is explained first in Section 5.1. Then an application example of ultrasonic driving of gas molecules in a gas sensing system is described in Section 5.2, and a further investigation of the ultrasonic driving process of gas molecules near the sensing surface of a metal oxide semiconductor (MOX) gas sensor is given in Section 5.3. After that, an application example of ultrasonic driving of gas molecules in the metal-air battery is demonstrated in Section 5.4.

5.1 INTRODUCTION TO ULTRASONIC DRIVING OF GAS MOLECULES

It is well known that small particles in liquid or at the liquid-solid interface can be driven by the physical effects of ultrasound [1–8]. As driving smaller particles such as the sub-nm particles is also one of the important aims of manipulation technology, the author's research team has been exploring the feasibility of manipulating gas molecules by ultrasound. From their experiments of ultrasonic driving of gas molecules which have a diameter in the sub-nm scale, it was found that ultrasonic waves can drive gas molecules onto a gas-solid interface [9, 10]. In the experiments, gas sensors such as MOX and catalytic combustion sensors were employed to detect the driving effect. Figure 5.1 shows one possible principle of ultrasonic driving of gas molecules near a gas-solid interface, which was proposed by the author's group.

On the gas-solid interface in ultrasonic field or at the boundary of ultrasonic field, there exists an acoustic boundary layer in which the viscosity of gas has a very large effect on the ultrasonic field [1, 11]. For this reason, the spatial gradient of ultrasonic field inside the acoustic boundary layer is very large, compared to that outside the

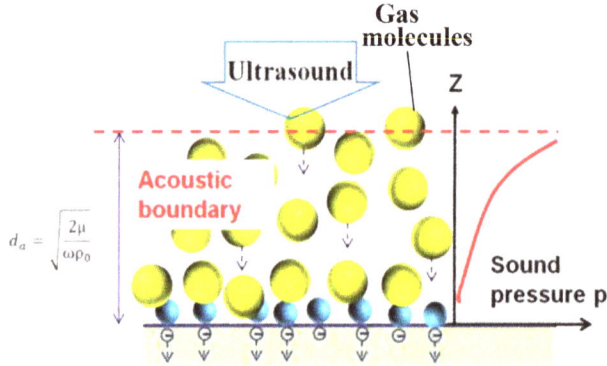

FIGURE 5.1 Principle of ultrasonic driving of gas molecules.

acoustic boundary layer. The acoustic potential energy E_p of a gas molecule with a radius R at a position with sound pressure p in the boundary layer is

$$E_p = \frac{\langle p^2 \rangle}{2\rho_0 c^2} V = \frac{2\pi \langle p^2 \rangle R^3}{3\rho_0 c^2}, \tag{5.1}$$

where ρ_0 is the density of the gas when there is no ultrasound, c is the sound speed in the gas and $< >$ represents the time average in one or multiple time periods. The acoustic radiation force F_z acting on the gas molecule is

$$F_z = -\frac{dE_p}{dz} \tag{5.2}$$

where the z-axis is perpendicular to the gas-solid interface, pointing from the gas-solid interface to the gas. When the solid is not in vibration, $dp/dz > 0$ in the acoustic boundary layer. Thus, acoustic radiation force F_z is negative, which means that the gas molecule experiences an acoustic radiation force pointing to the gas-solid interface. This force (F_z) will drive the gas molecules in the acoustic boundary layer onto the gas-solid interface. At 25 kHz and room temperature in air, the thickness of acoustic boundary layer is 14.3 μm, which is much larger than the diameter of target gas (usually in sub-nm range). This means that the acoustic boundary layer contains large amounts of gas molecules, and gas molecules exerted by force F_z are massive.

Herein, the author would like to present another possible principle model for the ultrasonic driving of gas molecules near a gas-solid interface [9, 10]. During the positive half-cycles of sound pressure on the interface, target gas near the interface is compressed and its molecules are ejected onto the interface. Although the negative half-cycles of the sound pressure cause the desorption effect on target gas molecules on the interface, the desorption effect is weak if there is strong physical/chemical bonding between the target gas and solid material, or a quick reaction between the target gas molecules and substances at the interface. Thus, in each time period of sound pressure, target gas molecules transferred onto the interface by the positive sound pressure are more than that desorbed from the interface. This causes the phenomenon of ultrasonic driving of gas molecules. At the present stage, the mechanism of ultrasonic driving of gas molecules is still a bit illusive, and molecular dynamics modeling and more experiments are needed to reveal the mechanism.

5.2 APPLICATION OF ULTRASONIC DRIVING OF GAS MOLECULES IN GAS SENSING

As an application example of ultrasonic driving of gas molecules, the method to ultra-sonically enhance the metal oxide (MOX) gas sensor's sensitivity to low-concentration volatile organic compound (VOC) gases is given in this section [9]. The system uses an ultrasonic transducer to produce a standing wave field above the sensing layer. Ultrasonic field near the sensing layer drives the VOC molecules to the sensing layer and promotes the oxidation-reduction reaction between the VOC molecules and oxygen species on the sensing surface, which increases the resistance change of the sensor. The experiments show that this method can greatly enhance the gas sensor's sensitivity. With the assistance of ultrasound, the sensitivity can be increased up to 30 times.

5.2.1 INTRODUCTION

The VOCs which include benzenes, alcohols, ethers, ketones, etc., are one of the major air pollutants. They are generated in petroleum refining, petrochemical processing, natural gas processing, water treatment, human daily activity, etc. Improper exposure of human beings and animals to VOCs may cause various health problems such as central nervous system anesthetization, skin and eyes irritation and vomiting. Also, many VOCs are inflammable and susceptible to combustion. Detection of VOC gases is very important and has wide applications in air quality monitoring, leakage detection of poisonous gas and fuel, pesticide residue testing and so on. The sensors for detecting various VOCs mainly include metal oxide, electrochemical and optical gas sensors, and each of them has its own features.

The metal oxide gas sensor is an important gas sensor. Its operation is based on the oxidation-reduction reaction between the target gas molecules and reactive oxygen species on the surface of a metal oxide sensing layer. The oxidation-reduction reaction changes the resistivity of the sensor resistor and causes the output voltage to change. Metal oxide gas sensors have been widely used in the detection of various gases, as they have the merits such as good durability and long working life, low-cost, good sensitivity, fast response and excellent portability.

An improvement of the sensitivity of metal oxide gas sensors will definitely widen their application range. The efforts to improve the sensitivity have been mainly in the optimization of the sensing material and fabrication process, which is to enhance the resistance change of the sensing layer. A method to enhance the sensitivity of metal oxide gas sensors by ultraviolet (UV) light irradiation was once proposed and investigated. The reported relative resistance change of the sensing resistor before and after UV irradiation is 900%. It can be combined with the sensing material optimization to increase the sensitivity further. The demerit of this method is that it is only effective for sensing materials with photo-catalytic effects such as WO_3.

In this work, an ultrasonic method to enhance the metal oxide gas sensor's sensitivity to VOCs is proposed and the ultrasonic effect on the sensing performance is experimentally demonstrated. Also, the working principle, dependence of the sensing response on ultrasonic parameters, and effects of target gas concentration and vibration velocity on the sensing response and response time under sonication are investigated. The experimental results show that ultrasound can enhance the gas sensor's sensitivity by one order of magnitude. To the best of our knowledge, this work is the first attempt to employ gas-borne ultrasound to enhance the metal oxide gas sensors' sensitivity.

5.2.2 EXPERIMENTAL METHOD

Figure 5.2(a) shows the experimental setup to investigate the ultrasonic effect on the sensitivity of a MOX gas sensor, and to measure the characteristic change of the sensor system before and after the sonication. It is mainly composed of a sealed plexiglass box (700 mm × 500 mm × 500 mm), a Langevin ultrasonic transducer (HNC-4AH-2560, Hainertec Co. Ltd), a SnO_2 gas sensor (MQ-6, Zhengzhou Winsen

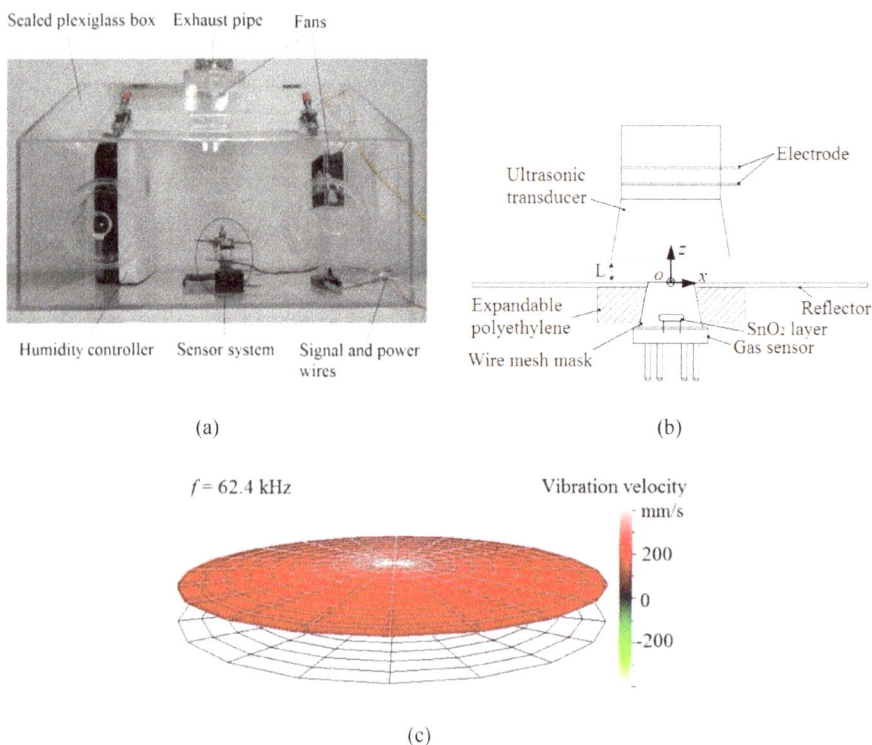

(a)

(b)

(c)

FIGURE 5.2   Experimental setup and vibration mode. (a) Photo of the chamber for testing the ultrasonic effect on the sensitivity of a metal oxide gas sensor. (b) Schematic diagram of the sensor system. (c) Vibration mode of the radiation surface at 62.4 kHz. Reproduced from Ref. 9 with permission from Elsevier.

Electronics Technology Co., Ltd.) and a polyvinyl chloride (PVC) plate reflector. The ultrasonic transducer, SnO_2 gas sensor and reflector form the sensor system, as shown in Fig. 5.2(b). The top surface of the gas sensor is covered with the reflector, which is parallel to the transducer's radiation surface. A circular hole is disposed at the center of the reflector to make ultrasonic energy transmit into the gas sensor. The gas sensor and reflector are placed on a micro-displacement platform, for adjusting the gap thickness between the radiation surface and reflector. The strength of ultrasonic field in the air gap between the radiation surface and reflector can be adjusted by the gap thickness and ultrasonic transducer's vibration velocity. Also, the wire mesh mask of the sensor is sealed by expandable polyethylene on its side in order to enhance the ultrasonic effect. In practical applications, the micro-displacement platform is not necessary, because the distance between the reflector and transducer is fixed.

In the experiments, the transducer was driven by an electrical system containing a signal generator, a power amplifier and an oscilloscope. Prior to assembling the sensor system, vibration characteristics of the transducer's radiation surface were measured by a laser Doppler Vibrometer (PSV-500, POLYTEC). It shows that the radiation surface has piston-like and flexural vibration modes. Our experiments indicated that the piston-like vibration mode at 62.4 kHz has a better effect of the sensing performance enhancement. The measured vibration mode of the radiation surface at 62.4 kHz is shown in Fig. 5.2(c).

VOC samples used in the experiments include ethanol (CH_3CH_2OH), methylbenzene (C_7H_8), formaldehyde (HCHO), acetone (CH_3COCH_3) and xylene (C_8H_{10}) (Sinopharm Chemical Reagents Co., China). They were obtained by evaporation of the corresponding organic solvents. A microliter syringe (LC-10 μL, Shanghai High Pigeon Industry & Trade Co., Ltd.) was used to measure the volume of the organic solvents, which was used to compute the concentration of the VOC gas in the test chamber after they evaporated.

Figure 5.3(a) shows the measurement circuit for the sensor. It mainly consists of a tin dioxide (SnO_2) sensing layer with a resistance R_S, a resistive heater, a DC voltage source V_S ($= 5.0 \pm 0.1$ V), an adjustable resistor R_L and a read-out unit to give the output voltage V_O. The SnO_2 layer is coated on a hollow aluminum oxide (Al_2O_3) ceramic tube. The resistive heater, passing through the ceramic tube, is used to raise the working temperature of SnO_2 layer to activate its sensing capability. Figure 5.3(b) shows the configuration of the sensor. The sensing resistor is protected by a stainless steel wire mesh and mechanically supported by six hard electrodes. Figure 5.3(c) shows photos of the sensor. The sensor is connected to a single-chip microcomputer (SCM) system (DFRduino UNO R3 and LCD12864 shield, Arduino) for control, data acquisition and reading. The SCM system may be connected to a personal computer for further data processing.

During the measurement, temperature in the test chamber was 20°C ± 1°C and the relative humidity was 65% ± 5%. Liquid VOC with controlled amount was injected into the test chamber to obtain the VOC target gas for measuring the effect of ultrasound on the sensitivity and sensing response. There were two fans in the test chamber. One was used to disperse the gas inside the chamber, and another to exhaust gas from the chamber after each measurement.

In the experiments, the output voltage V_O of the sensor that was exposed to the VOC target gases with different concentrations was measured. Based on the output

(a)   (b)

(c)

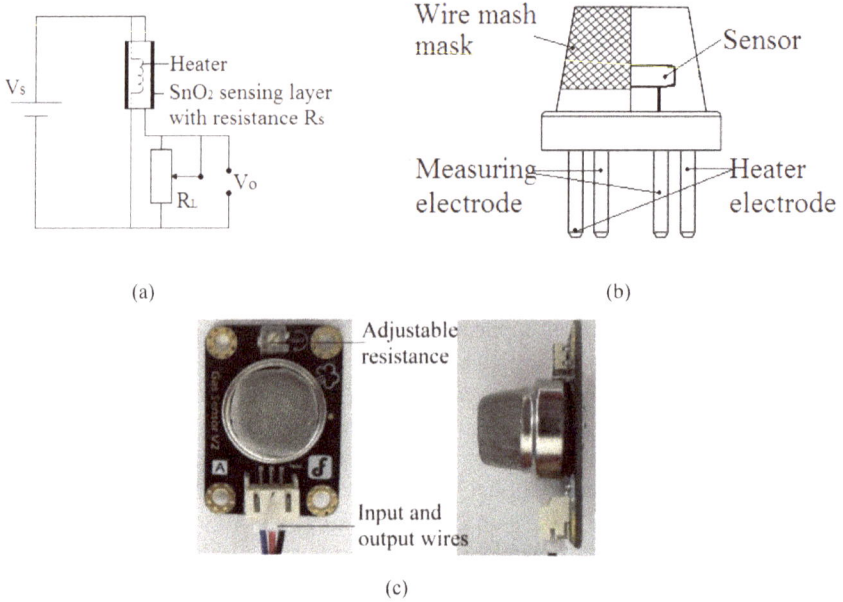

FIGURE 5.3 The SnO_2 gas sensor used in the experiments. (a) Measurement circuit. (b) Schematic diagram of the sensor configuration. (c) Photos. Reproduced from Ref. 9 with permission from Elsevier.

voltage, the resistance R_S of the sensing layer (denoted as sensor resistance) was deduced by the following equation:

$$V_O = V_S \frac{R_L}{R_L + R_S},$$ (5.3)

where $R_L = 2.5$ kΩ in this work.

From the deduced resistance values of the sensing layer exposed to the VOC gas and to pure air ($R_{S,g}$ and $R_{S,0}$), the sensing response γ was determined by the following definition:

$$\gamma = \frac{R_{S,0} - R_{S,g}}{R_{S,g}}.$$ (5.4)

Based on the sensing response γ, the sensitivity S was determined by the following definition:

$$S = \frac{d\gamma}{dC},$$ (5.5)

where C was the VOC gas concentration to be measured. The sensitivity determined by Eq. (5.5) represents the slope of the γ-C curve.

In this work, unless otherwise specified, the working frequency f was 62.4 kHz, the air gap thickness L was 2.78 mm (half of the wavelength) and the ultrasonic transducer operated in resonance. The vibration velocity u at the center of the radiation surface was used to represent the strength of vibration excitation, and it was 270 mm/s (0–p) unless otherwise specified. The measured working temperature was 210°C when there was no sonication. The measurement was carried out by a thermal coupler (GM1312, k-type, Shenzhen Jumaoyuan Science and Technology Co. Ltd, China), which was inserted into the space covered by the wire mesh mask through a small hole on the side of the mask, and was in contact with the sensing layer.

5.2.3 EXPERIMENTAL PHENOMENA AND PRINCIPLE ANALYSES

Figure 5.4(a) shows the measured output voltage V_O and sensor resistance R_S versus acetone gas concentration. In the experiment, the mixture of acetone and air was used as the target gas. It is seen that sonication can effectively increase the output voltage and decrease the sensor resistance for a given target gas concentration. Figure 5.4(b, c) shows the measured sensing response and sensitivity versus acetone gas concentration, respectively, under the same experimental conditions as Fig. 5.4(a). From Fig. 5.4(b), it is seen that the sonication can greatly enhance the sensing response. From Fig. 5.4(c), it is seen that the sonication can enhance the sensitivity.

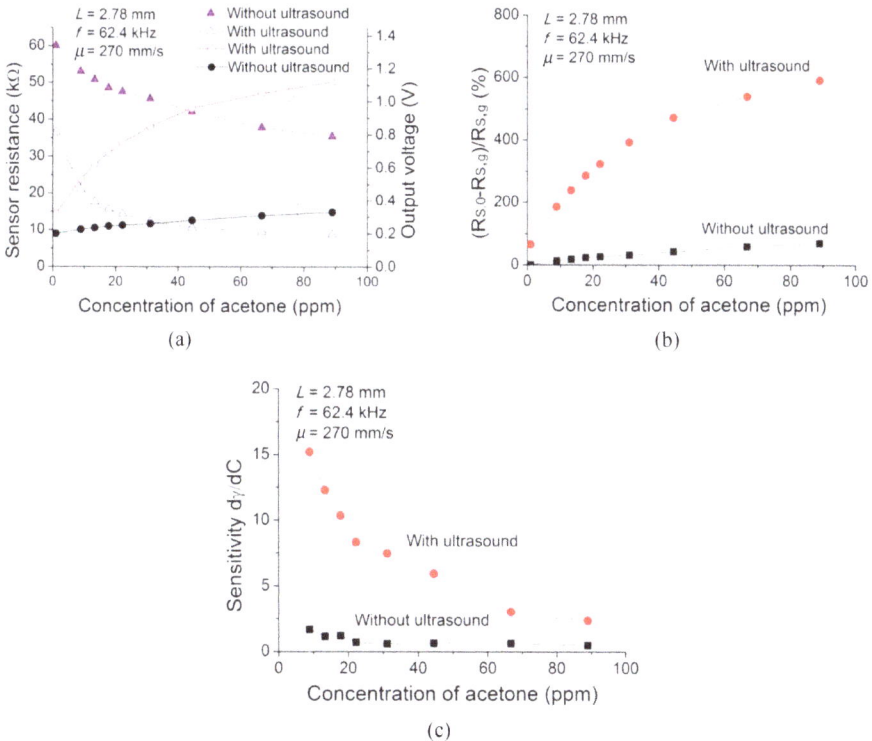

FIGURE 5.4 Measured effects of ultrasound on the sensor performance at different concentrations of acetone. (a) Output voltage and sensor resistance. (b) Sensing response. (c) Sensitivity. Reproduced from Ref. 9 with permission from Elsevier.

(a)

(b)

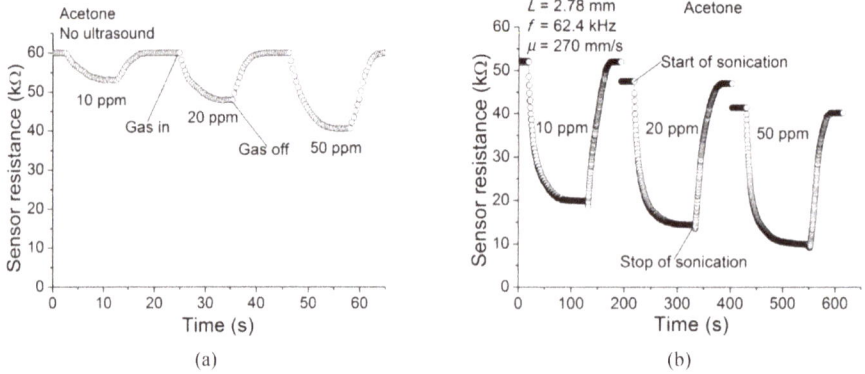

FIGURE 5.5   The transient processes of sensor resistance when acetone gases with different concentrations are measured. (a) Without sonication. (b) With sonication. Reproduced from Ref. 9 with permission from Elsevier.

Figure 5.5(a, b) shows the measured transient processes of the sensor resistance without sonication and under the ultrasonic condition, respectively. It is seen that the sonication decreases the sensor resistance greatly. To explain the experimental phenomenon stated above, the ultrasonic field in the air gap between the radiation surface and reflector was analyzed by the finite element method (FEM). COMSOL Multiphysics software was used in the calculation. Figure 5.6(a) shows the sound pressure distribution in the air gap right above the wire mesh mask. In the computation, a_z (the acceleration of the radiation surface) $= 1.06 \times 10^5$ m/s^2, f (working frequency) $= 62.4$ kHz, L (the distance between the radiation surface and wire mesh mask) $= 2.78$ mm, r_a (the ultrasonic field's radius) $= 15$ mm, and c (sound speed in air) $= 341$ m/s. It is seen that there is a standing wave in the air gap, which generates sound pressure at the top surface of the wire mesh mask. The sound pressure at the top surface of the wire mesh mask will generate an ultrasonic field in the space enclosed by the wire mesh mask, as shown in Fig. 5.6(b).

For the reason described in Section 5.1.1, ultrasound near the sensing surface makes more target gas molecules transported onto the sensing layer surface per unit time. As a result, ultrasound increases the number of free electrons that are

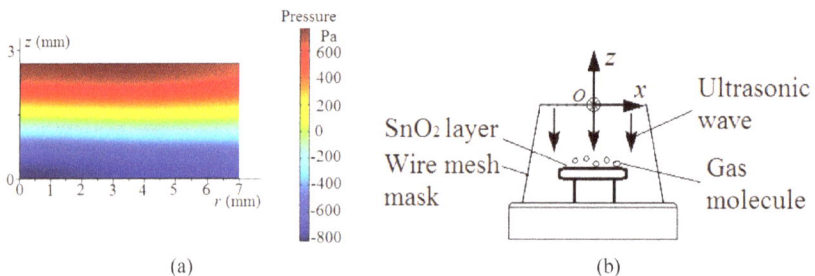

(a)

(b)

FIGURE 5.6   Ultrasonic field. (A) Calculated acoustic pressure at $L = 2.78$ mm, $f = 62.4$ kHz, $a_z = 1.06 \times 10^5$ m/s^2. (b) A model to show the gas molecules driven by ultrasonic field. Reproduced from Ref. 9 with permission from Elsevier.

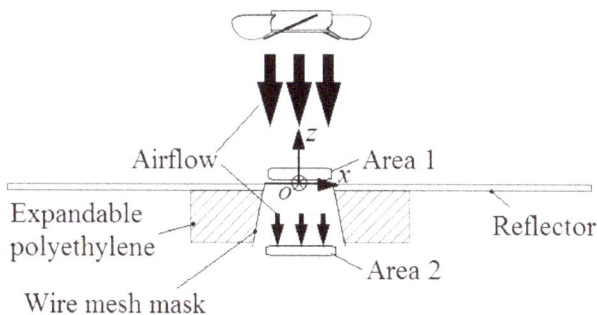

FIGURE 5.7 Experimental setup for measuring the relationship between the airflow velocities inside and outside the wire mesh mask. Reproduced from Ref. 9 with permission from Elsevier.

generated in the oxidation-reduction reaction and released into the sensing resistor in unit time, and decreases the sensor resistor R_S. The ultrasound-induced vibration velocity of the target gas molecules is less than 1–2 m/s, which is much lower than the velocity of the thermal motion (at least several hundred m/s). Thus, the target gas molecules' kinetic energy increased by the ultrasound is much smaller than that of their thermal motion. This indicates that the ultrasound affects little on the collision energy between the target gas molecules and oxygen species.

Theoretically, the acoustic streaming toward the sensing layer, generated by the ultrasonic field, can also affect the sensing response by enhancing the molecular transport and lowering the working temperature. To investigate the effects of the acoustic streaming, an experimental simulation was carried out, as shown in Fig. 5.7. In the experiment, the airflow toward the top surface of the wire mesh mask was generated by a small fan suspended right above the mask, and acetone gas with different concentrations was used as the target gas. The measured sensing response versus target gas concentration when the estimated airflow velocity w in the mask is 0.02, 0.05 and 0.2 m/s, respectively, is shown in Fig. 5.8.

The airflow velocity inside the wire mesh mask was controlled not to exceed 0.2 m/s in the experimental simulation, because the acoustic streaming usually has a velocity less than 0.2 m/s. For comparison, the experimental results when there is ultrasound only and when there is no ultrasound and airflow are also presented in Fig. 5.8. By a comparison of the curves in Fig. 5.8, it is known that the acoustic streaming in the vicinity of the sensing layer can contribute to the sensing response enhancement, but its contribution is less than that of the sound pressure. This is reasonable because the sound pressure is a first-order physical effect of acoustic field and the acoustic streaming is a second-order one. From Fig. 5.8, it is also seen that as the airflow velocity (no ultrasound) increases, the sensing response to the acetone gas increases. This indicates that the change of sensing response which results from the molecular transport enhancement by the acoustic streaming is larger than that which results from the cooling effect of acoustic streaming.

The air flow velocity inside the wire mesh mask in the above experiments was estimated by the following method. The relationship between the air flow velocities outside and inside the wire mesh mask (areas 1 and 2) was measured first. The

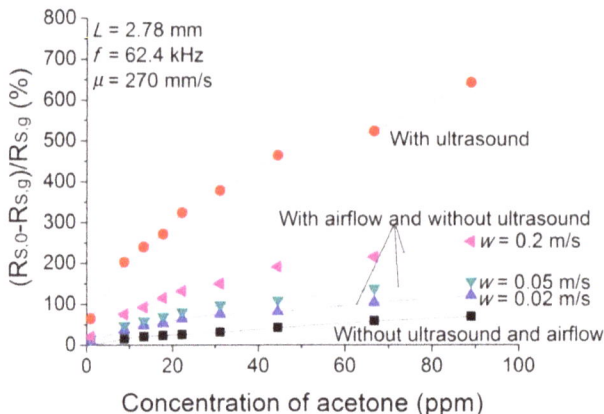

FIGURE 5.8 Sensing response versus acetone gas concentration under different airflow and ultrasonic conditions. Reproduced from Ref. 9 with permission from Elsevier.

experimental setup was obtained by removing the ultrasonic transducer, sensing resistor and base from the structure shown in Fig. 5.2(b), and then suspending a small fan above the mask. The air flow velocity was measured by an anemograph meter (HT-9829, XINTEST). After obtaining this relationship, the air flow velocity inside the wire mesh mask could be estimated by measuring the air flow velocity outside the wire mesh mask at the top surface.

The temperature decrease at the sensing layer surface, caused by the cooling effect of acoustic streaming, was measured and confirmed by the thermal coupler for different vibration velocities, and the result is shown in Fig. 5.9. In the experiments, three different distance values between the ultrasonic transducer's radiation face and wire mesh mask's top were used ($L = 1/2$, 1 and 3/2 wavelengths). It is seen that the surface temperature of the sensor decreases as the vibration velocity increases. This is because the acoustic streaming becomes stronger as the vibration velocity

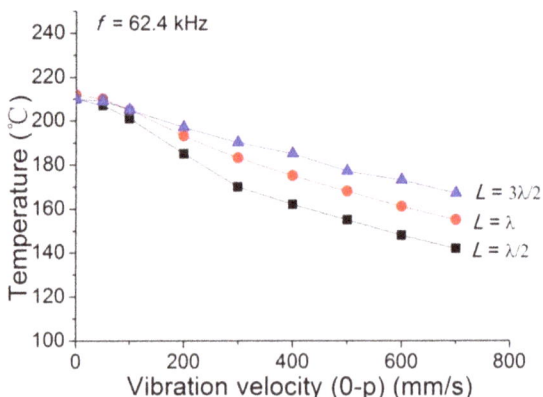

FIGURE 5.9 The effect of vibration velocity on the temperature at the sensing layer surface. Reproduced from Ref. 9 with permission from Elsevier.

increases. Figure 5.9 can well explain the experimental phenomenon that the base-line sensor resistance (Rs when there is no target gas) with the ultrasound is lower than that without ultrasound, which is indicated in Fig. 5.4(a). When the sensing layer is sonicated in air without the target gas, temperature at the sensing layer surface decreases due to the acoustic streaming. This enhances the desorption process of the oxygen species on the sensing layer, which increases the number of free electrons in the sensing layer. Figure 5.9 shows that temperature decreases at the sensing layer due to the acoustic streaming can be up to 60°C.

Thus, it is concluded that the ultrasonic effect on the sensing response results from two factors, that is, molecular transportation enhancement by the sound pressure, the acoustic streaming and cooling effect of the acoustic streaming. The increase of the sensing response is mainly caused by the molecular transportation caused by the sound pressure. Acoustic streaming has two effects on the sensing process. It not only brings the target gas molecules onto the sensing layer but also lowers the sensing layer temperature. The cooling effect may strengthen the desorption process of oxygen species on the sensing layer.

5.2.4 CHARACTERISTICS AND DISCUSSION

Figure 5.10(a) shows the measured sensing response versus the concentration of five different VOC gases with and without sonication. The target gases include ethanol (CH_3CH_2OH), methylbenzene (C_7H_8), formaldehyde (HCHO), xylene (C_8H_{10}) and acetone (CH_3COCH_3). In the experiments, the working voltage $V = 100$ V_{p-p} and the vibration velocity $u = 690$ mm/s (0–p). It is seen that the sonication can increase the sensing response by one order of magnitude. Figure 5.10(b) shows the measured ultrasound-induced sensitivity change when the concentration of target VOC gas is about 11 ppm. It is seen that the sonication can increase the sensitivity by one order of magnitude. The ultrasonic effect on the sensitivity is defined by S_u/S_0, where S_u and S_0 are the sensitivity with and without sonication. This quantified ultrasonic effect may also be viewed as the ultrasonic catalysis effect.

Figure 5.10(c) shows the measured ultrasonic effect for different target gases at different concentrations. It is seen that the sensitivity can be increased up to 30 times with the ultrasonic catalysis. Also, it indicates that the ultrasonic effect depends on the concentration of target VOC gases, and it increases as the concentration decreases. This phenomenon is explained as follows. The variation of acoustic medium density $\Delta\rho$ due to sound pressure p is

$$\Delta\rho = p\rho_0/\beta, \tag{5.6}$$

where ρ_0 is the medium density without sonication, and β is the adiabatic bulk modulus which is

$$\beta = \gamma_s P_0, \tag{5.7}$$

where γ_s is the ratio of specific heats and P_0 is the gas pressure without sonication. Assuming the experimental has little effect on the acoustic parameters ρ_0 and β, the change of target gas density near the sensing layer is proportional to the sound

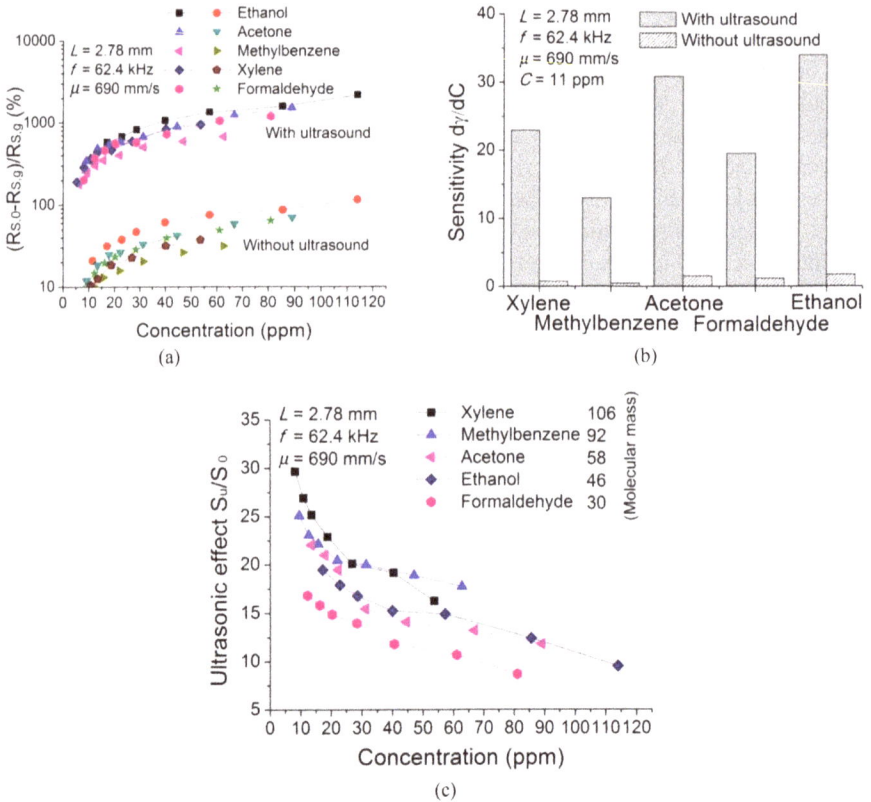

FIGURE 5.10 Ultrasonic enhancement of the sensing response and sensitivity to various VOC gases. (a) Sensing response versus gas concentration with and without ultrasound. (b) Sensitivity comparison at a constant gas concentration. (c) Ultrasonic effect versus gas concentration. Reproduced from Ref. 9 with permission from Elsevier.

pressure. Thus, for a given ultrasonic field with a constant sound pressure amplitude at the sensing surface, the ultrasound caused change of target gas density or concentration has a constant amplitude approximately. Therefore, ultrasound has a relatively large effect when the target gas concentration is low.

Figure 5.10(c) also indicates that the molecular mass of target gases may affect the ultrasonic effect on the sensitivity, and a large molecular mass tends to strengthen the ultrasonic effect. To confirm the effectiveness of this deduction, the ultrasonic effect for hydrogen gas (molecular mass = 2) was measured with the following experimental conditions: Working frequency = 62.4 kHz, vibration velocity = 690 mm/s (0–p), target gas concentration = 10 ppm and gap thickness = 2.78 mm. According to this experiment, it is known that S_u/S_0 is 5 for the hydrogen gas, which is much less than the values of the VOC gases. This phenomenon can be well explained by the above deduction. As the molecular mass increases, the sound pressure in the positive half-cycles makes the target gas molecules travel a longer distance because the inertia of the target gas molecules increases. Thus, for a given ultrasonic field, more

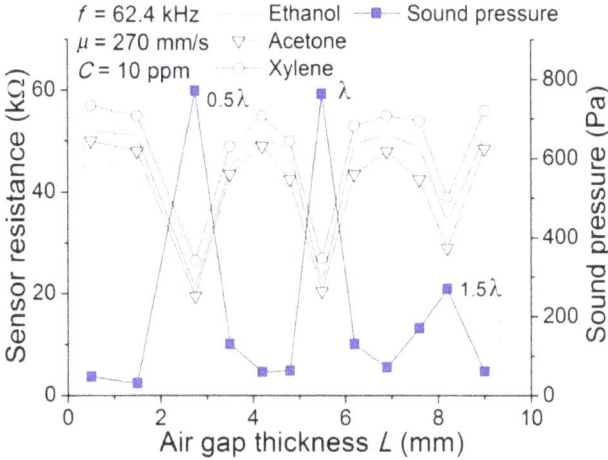

FIGURE 5.11   The influence of the air gap thickness L on the sensor resistance and sound pressure. Reproduced from Ref. 9 with permission from Elsevier.

target gas molecules are transported onto the sensing layer surface as the molecular mass increases.

Figure 5.11 shows the measured relationships between the sensor resistance R_S and air gap thickness L for ethanol, acetone and xylene gases with an identical concentration of 10 ppm. In the experiments, the working voltage V was 30 V_{p-p}. It is seen that when L is 2.78, 5.5 and 8.24 mm, the sensor resistance R_S is the minimum. As the wavelength of ultrasound in the air gap λ is 5.46 mm (= c/f), 2.78, 5.5 and 8.24 mm are quite close to $\lambda/2$, λ, $3\lambda/2$, respectively. Thus, the sensor resistance R_S is the minimum when the ultrasonic field in the air gap is in resonance. The resonance of the air gap is confirmed by an FEM computation, and the result is also shown in Fig. 5.11. The sound pressure at $r = 0$ and $z = 0$ (the central point of the reflector) is presented in the figure.

To understand the effect of working frequency on the sensing response, the sensing response versus vibration velocity for 10 ppm acetone gas at 62.4, 40.0 and 21.8 kHz was measured, and the results are shown in Fig. 5.12. In the experiment, the air gap thickness L was half-wavelength. It indicates that the working frequency does affect the sensing response, and a higher working frequency is beneficial to the improvement of the sensing response. The latter phenomenon is attributed to the following two reasons: There are more positive half-cycles of sound pressure in unit time as the working frequency increases; the sound pressure increases as the working frequency increases when the shape, size and vibration excitation of an ultrasonic field are fixed.

Response time of the gas sensor t_{ur} is defined as the time taken by the sensor to complete 90% change of the sensor resistance under an ultrasonic condition. Figure 5.13(a) shows the measured response time versus vibration velocity at four different concentrations of acetone gas. It is seen that the response time increases with the increase of vibration velocity. As the vibration velocity increases, the sound pressure near the sensing layer also increases. This results in more target gas molecules transported onto the sensing layer surface per unit time, which can be equivalently

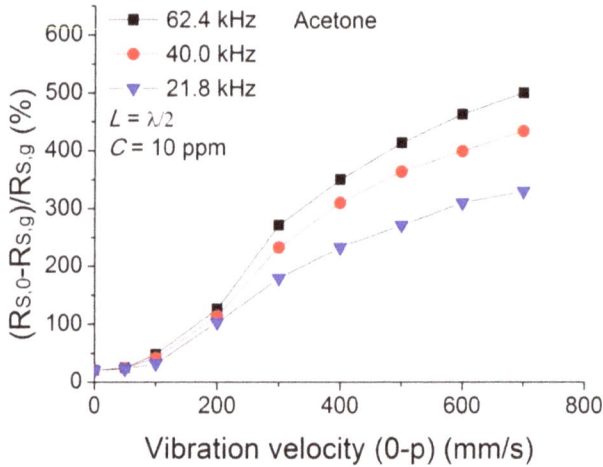

FIGURE 5.12   The influence of the working frequency and vibration velocity on the sensing response. Reproduced from Ref. 9 with permission from Elsevier.

viewed as an increase of the target gas concentration in the vicinity of the sensing layer. Figure 5.13(b) shows one example of measured transient responses of the sensor resistance after the sonication is started and stopped.

In all of the above experiments, sensor MQ-6 was used. The sensing responses of sensor MQ-135 (Zhengzhou Winsen Electronics Technology Co., Ltd.) to acetone and of 2M010 (Beijing Guo-Tai-Heng-An Technology Co., Ltd.) to gasoline versus vibration velocity was also measured, and the results are shown in Fig. 5.14. MQ-6 and MQ-135 have the identical main sensing material (SnO_2) but different additives. The main sensing material of 2M010 is WO_3. The results show that the sensing response of MQ-135 and 2M010 increases by 21 and 27 times, respectively, when the

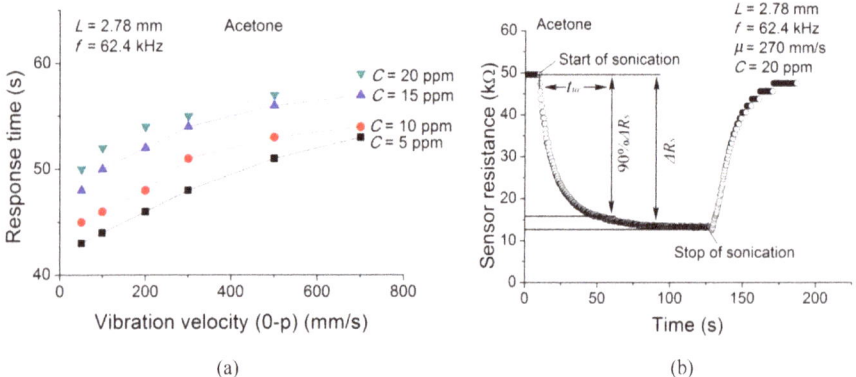

FIGURE 5.13   Response time to ultrasound for acetone gas. (a) Response time versus vibration velocity at different gas concentrations. (b) A transient response of the sensor resistance. Reproduced from Ref. 9 with permission from Elsevier.

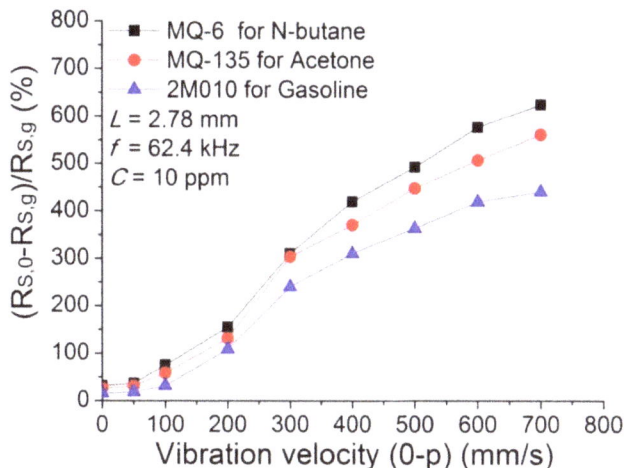

FIGURE 5.14   The sensing responses of sensor MQ-6 to n-butane, MQ-135 to acetone and 2M010 to gasoline versus vibration velocity. Reproduced from Ref. 9 with permission from Elsevier.

vibration velocity increases from 0 to 700 mm/s (0–p), and the target gas concentration is 10 ppm. This shows that the ultrasonic method proposed in this work is also effective to other MOX gas sensors.

5.2.5 CONCLUSIONS

In this work, a method that employs ultrasound to greatly enhance the sensitivity of a metal-oxide VOC gas sensor has been demonstrated. The sensitivity can be increased by one order of magnitude by ultrasound. The working principle is analyzed, and the characteristics of the ultrasound-assisted sensing system are experimentally clarified. The method utilizes ultrasonic standing wave field to enhance the transport rate of target gas molecules onto the sensing layer. It becomes more effective as the target gas concentration decreases and the molecular mass of target gas increases. Apart from the ultrasonic field strength, the sensing response is also affected by the working frequency, and the response time is affected by the vibration velocity and target gas concentration. This work provides a new and effective way to improve the performance of MOX gas sensors. It can be used in a high-sensitivity gas sensing system, together with other methods such as the use of porous and nano sensing materials, and UV irradiation.

5.3   MORE MECHANISM ANALYSES OF ULTRASOUND-ASSISTED GAS SENSING

The gas-borne ultrasound provides a new and universal physical method to greatly improve the sensitivity of gas sensors. In this work [10], to obtain deeper insights of its physical principle, ultrasonic physical effects on the sensing performance of the ultrasound-assisted MOX gas sensor are investigated experimentally and theoretically, and the effects of sound pressure and acoustic streaming on the sensing process are directly verified. It indicates that the transportation of target gas molecules onto the sensing surface can be enhanced by sound pressure on the sensing surface, which results in a significant increase in both the sensing response and sensitivity. Also, it is found that the sensing surface may be cooled down by the acoustic streaming, which causes a sensing response change opposite to that caused by the sound pressure. It is predicted and experimentally verified that when both of the acoustic streaming and sound pressure exist on the sensing surface, the sensing characteristics should be between those of the two extreme working modes in which there is only sound pressure or acoustic streaming on the sensing surface.

5.3.1 INTRODUCTION

The author's team once proposed a new physical assistance method to effectively enhance the sensitivity of MOX gas sensors to low-concentration VOC gases and lower their lower detection limit (LDL), in which an acoustic standing wave was employed (Section 5.2). With the assistance of ultrasound, the sensitivity of a commercial MOX gas sensor was improved by at least one order of magnitude. For example, for a SnO_2 gas sensor MQ-6 (Zhengzhou Winsen Electronics Technology Co., Ltd) exposed to 5 ppm acetone gas, its sensitivity could be increased by up to 30 times by ultrasound. Sections 5.3.2–5.3.4 will give the details of an endeavor of the author's team to investigate the working mechanism of gas-borne ultrasound assisted MOX gas sensors.

5.3.2 EXPERIMENTAL METHOD

The experimental gas-borne ultrasound assisted MOX gas sensor system was mainly composed of a SnO_2 gas sensor (MP-4, Zhengzhou Winsen Electronics Technology Co., Ltd.) and a Langevin ultrasonic transducer (HNC-4SH-3840, Hainertec Co., Ltd.), as shown in Fig. 5.15. The SnO_2 gas sensor was right below the center of the transducer's radiation face which was parallel to the sensing surface. The center of the sensor component was always in the z-axis. The sensor component had a plate shape with dimensions of 1.5 mm × 1.5 mm × 0.2 mm, with its sensing layer coated on the front side of an Al_2O_3 substrate, and the resistive heater pasted on the back

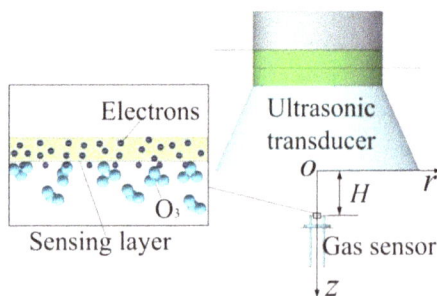

FIGURE 5.15 Experimental setup to investigate the physical principle of BAW-assisted gas sensors. Reproduced from Ref. 10 with permission from AIP Publishing.

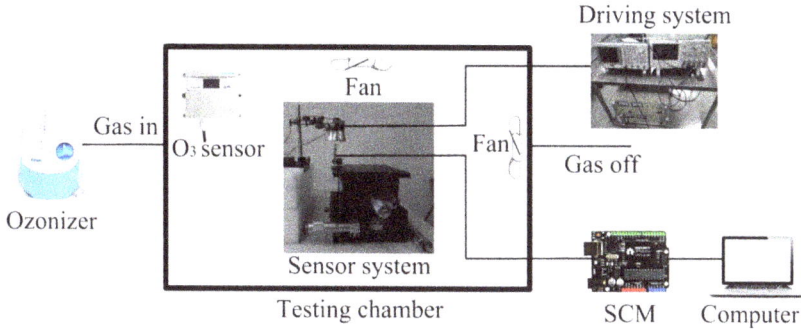

FIGURE 5.16   Schematic diagram of the measurement system. Reproduced from Ref. 10 with permission from AIP Publishing.

side. The sensor plate was supported by a metal support via four metal pins, with a 2 mm separation between the sensor plate and metal support. The measured LDL of the SnO_2 gas sensor for O_3 was 2 ppm. The ultrasonic transducer had a resonance at 40.2 kHz when its driving voltage was 24 V_{p-p}. Based on the vibration measurement by a laser Doppler Vibrometer (PSV-500, POLYTEC), it was known that the radiation face had a piston-like vibration mode at the resonance. The distance between the sensing surface and transducer's radiation face was adjusted by a 3D micro-displacement platform (LD125-LM-2, Shengling Precise Machinery Co., Ltd.) in order to tune the ultrasonic field on the sensing surface. Ozone gas generated by an ozonizer (FC558, Xiamen Laisen Electronics Co., Ltd.) was used as the target gas, and its concentration was also monitored by a commercial ozone gas sensor (JXBS-3001-O3, Weihai JXCT Technology Co., Ltd.).

Figure 5.16 shows the system to measure the resistance of the sonicated SnO_2 sensor exposed to O_3 gas under different ultrasonic conditions. The test chamber had dimensions of 700 mm × 500 mm × 500 mm, and its walls were made of plexi-glass. The system consisting of the ultrasonic transducer and SnO_2 sensor component, as shown in Fig. 5.15, was placed in it. During the experiments, the metal protection cover of the gas sensor was removed for the convenience of experiments and analyses. The sensor component was electrically connected to a microcomputer (DFRduino UNO R3, Arduino) for data processing. The ultrasonic transducer was driven by an electrical system containing a signal generator, a power amplifier and an oscilloscope. In all experiments, the ultrasonic transducer was kept resonance by tuning the working frequency, and the transducer vibration velocity was varied by controlling the driving voltage. There were two fans in the test chamber. One was used to disperse the target gas inside the chamber and another to exhaust target gas after each measurement. Working temperature of the sensor was measured by a thermal coupler (GM1312, K-type, Shenzhen Jumaoyuan Science and Technology Co., Ltd.), which was in contact with the sensing layer.

In this work, the sensing response γ is defined by $\gamma = (R_{S,g} - R_{S,0})/R_{S,0}$, where $R_{S,g}$ is the resistance of the gas sensor exposed to ozone gas (with or without sonication) and $R_{S,0}$ is the resistance of the gas sensor exposed to pure air (with or without sonication).

Unless otherwise specified, the experimental and computation conditions were as follows: The working frequency f was around 40.2 kHz; the working temperature of the gas sensor was 195°C when there was no sonication; the ambient temperature in the test chamber was 20°C ± 1°C and the relative humidity 65% ± 5% RH (relative humidity). Also, vibration velocity u at the center of the radiation face of the transducer was used to represent the strength of vibration excitation for the ultrasonic field.

5.3.3 COMPUTATIONAL METHOD

The factors which may affect the sensing process under sonication were searched from the major first- and second-order physical effects of ultrasound. The major first-order physical effect includes the sound pressure and linear vibration velocity/displacement/acceleration of target gas. The major second-order physical effect includes the acoustic streaming and acoustic radiation force [12–14].

As the sensing surface is acoustically hard, the linear vibration velocity (or displacement or acceleration) of the target gas at the sensing surface is zero, which means that it has no chance to affect the sensing process. Meanwhile, for a particle with a radius smaller than several hundred nanometers, the effect of acoustic radiation force on the particle motion is much less than that of acoustic streaming. This is because the acoustic radiation force is proportional to the cubic of the particle radius, whilst the Stokes' force caused by the acoustic streaming is linearly proportional to the particle radius. Thus, the effect of acoustic radiation force on the motion of target gas molecules can be ignored. Therefore, the sound pressure and acoustic streaming are supposed to be the physical factors that may affect the sensing process. In this work, the maximum sound pressure p_s and acoustic streaming velocity v_{as} on the sensing surface were used to represent the strength of the sound pressure and acoustic streaming, respectively.

As the distance between the sensor component and radiation face is only several millimeters, an accurate measurement of sound pressure and acoustic streaming on the sensing layer is almost impossible. For this reason, we used the FEM to obtain the sound pressure and acoustic streaming. The acoustic/solid coupling module of COMSOL Multiphysics software (R4.3) was used in the computation, with the parameters listed in Tables 5.1 and 5.2.

TABLE 5.1
Property Constants of Acoustic Medium in the Ultrasonic Field

Property Constants	Value
Dynamic viscosity (kg/(m·s))	1.812×10^{-5}
Ratio of specific heats	1.4
Heat capacity at constant pressure (J/(kg·K))	1005.6
Density (kg/m³)	1.204
Thermal conductivity (kg·m/(s³·K))	2.573×10^{-2}
Speed of sound (m/s)	343.2
Bulk viscosity (kg/(m·s))	5.436×10^{-6}

TABLE 5.2

Dimensions and Vibration Excitation Conditions of the Ultrasonic Field

Parameters	Value
Radius of the ultrasonic field r (mm)	24
Height of the ultrasonic field h (mm)	16
Excitation frequency f (kHz)	40.2
Measured acceleration of the radiation face a_z (km/s²)	0–198.5

5.3.4 PHYSICAL PRINCIPLE ANALYSES AND DISCUSSION

To verify the effect of sound pressure on the sensing surface when the sensor is exposed to O_3 gas, the position of the sensing surface in the computational model is varied along the z-direction, together with its substrate, in order to find out the position where acoustic streaming velocity on the sensing surface is close to zero and sound pressure on the sensing surface is sufficiently large. The computation shows that there exist several such positions in the ultrasonic field, and one of them is at $H = 4.1$ mm under the radiation surface. Here, H is the distance between the sensor and radiation surface, as shown in Fig. 5.15.

Figure 5.17(a, b) shows the computed acoustic streaming and sound pressure fields around the sensor, respectively, when the sensor is at $H = 4.1$ mm with vibration velocity u of 600 mm/s (0–p). It is seen that at this position, the sound pressure on the sensing surface is 453 Pa (0–p), whereas the acoustic streaming velocity on the sensing surface is 7 mm/s, which can be neglected. The working mode at which there is sound pressure, but no acoustic streaming on the sensing surface is defined as the sound pressure mode. The measured sensor resistance versus sound pressure p_s on the sensing surface at different O_3 concentrations when the sensor is at $H = 4.1$ mm is shown in Fig. 5.17(c). The sound pressure p_s was deduced from the measured vibration velocity u and computed ratio of the sound pressure p_s to vibration velocity u. It is seen that the sensor resistance increases monotonously as the sound pressure p_s increases. This gives a direct experimental verification that the sound pressure is capable of affecting the sensing process and sensitivity. For the 5 ppm O_3 gas, the sensing response γ and sensitivity ($d\gamma/dC$) are 2.2% and 3.7×10^{-3}/ppm at $p_s = 0$, respectively, and become 8.5% and 1.3×10^{-2}/ppm at $p_s = 670$ Pa (0–p), respectively. In addition, measured LDL of the sensor system for O_3 is about 0.5 ppm at 650 mm/s (0–p), which means that the LDL is lowered to about 1/4 of that without sonication. The working mode at which there is sound pressure but no acoustic streaming on the sensing surface is defined as the sound pressure mode (*P* mode)

During the positive half-cycles of sound pressure p_s on the sensing surface, O_3 gas on the sensing surface is compressed, and its molecules are ejected onto the sensing surface. During the negative half-cycles of sound pressure, the desorption of target gas molecules, which is caused by the rarefaction of target gas on the sensing surface, is weak due to the chemical bonding between the target gas species and sensing materials. Thus, in each time period of sound pressure p_s, target gas

(a)   Acoustic streaming

(b)   Sound pressure field

(c)

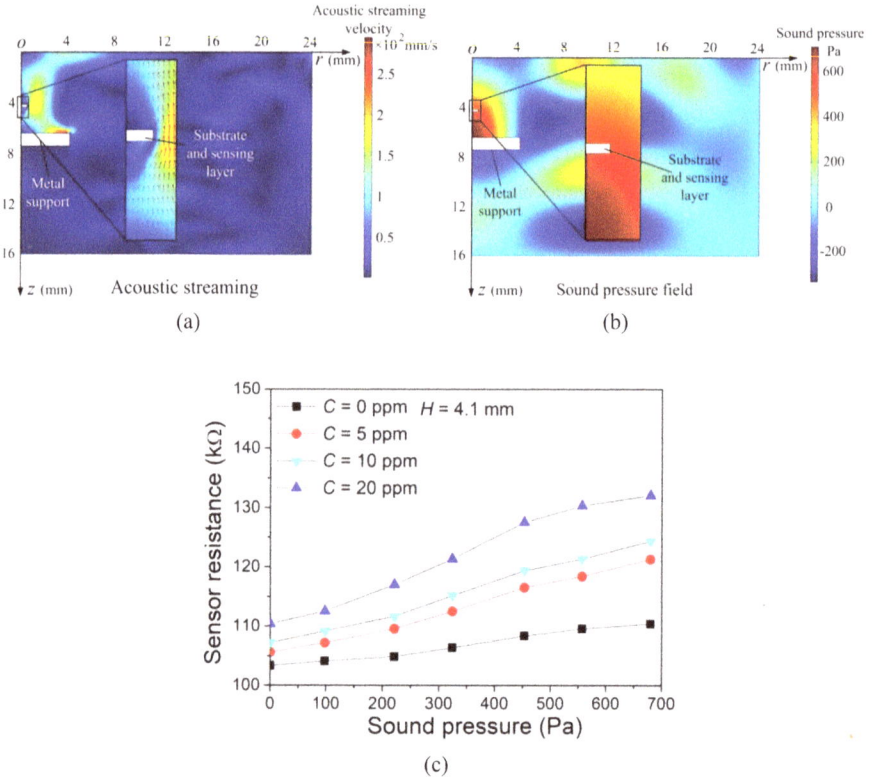

FIGURE 5.17   Ultrasonic field and sensor resistance when the sensor is located at the position ($H = 4.1$ mm), where the acoustic streaming velocity on the sensing surface is close to zero and sound pressure on the sensing surface is sufficiently large. (a) Computed acoustic streaming field. (b) Computed sound pressure field. (c) Measured relationship between the sensor resistance and sound pressure at different O_3 gas concentrations. In the computation for (a) and (b), $u = 600$ mm/s (0–p). Reproduced from Ref. 10 with permission from AIP Publishing.

molecule number transported onto the sensing layer by the positive sound pressure is larger than that desorbed from the sensing layer by the negative sound pressure. This causes an ultrasound-induced enhancement of target gas molecule transportation onto the sensing layer. Therefore, as the sound pressure increases, more O_3 molecules are transported onto the sensing surface per unit time. As a result, more electrons in the N-type semiconductor material are attracted by the O_3 molecules which adsorb on the sensing surface, as the sound pressure increases. This causes the sensor resistance to increase. The sensor resistance increases at zero concentration of target gas O_3 with the vibration velocity, which is caused by the enhanced O_2 molecules transfer onto the sensing layer.

To investigate the effect of acoustic streaming on the sensing surface when the sensor is exposed to O_3 gas, the position of the sensing surface is varied along the z-direction in the FEM computation, in order to find out the position where sound pressure on the sensing surface is close to zero and acoustic streaming velocity v_{as} on

(a)

(b)

(c)

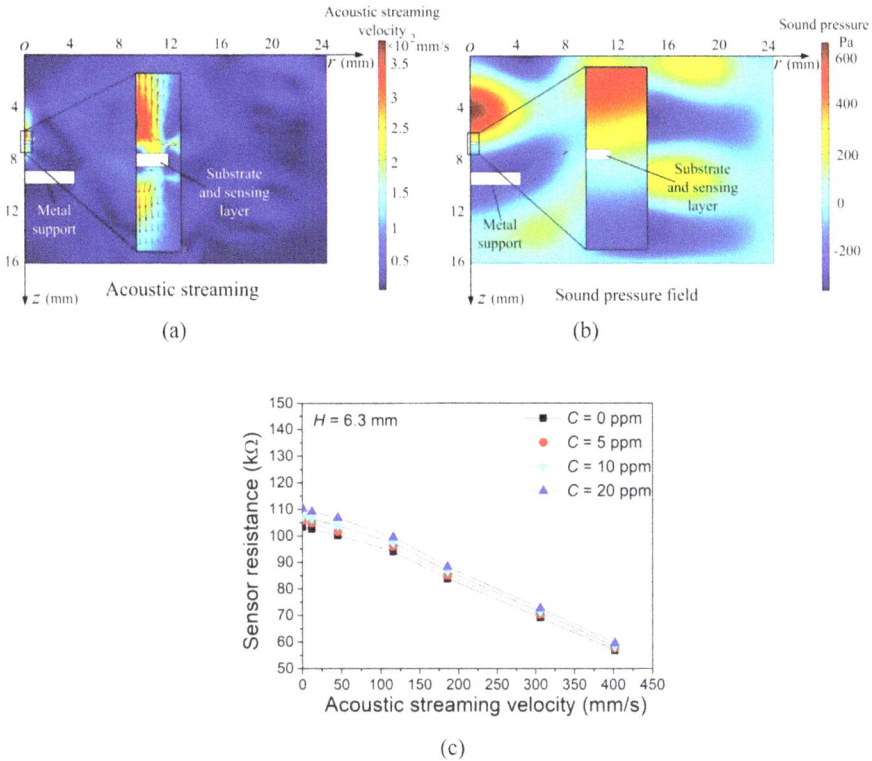

FIGURE 5.18   Ultrasonic field and sensor resistance when the sensor is located at the position ($H = 6.3$ mm), where the sound pressure on the sensing surface is close to zero and acoustic streaming velocity on the sensing surface is sufficiently large. (a) Computed acoustic streaming field. (b) Computed sound pressure field. (c) Measured relationship between the sensor resistance and acoustic streaming velocity at different O_3 gas concentrations. In the computation for (a) and (b), $u = 600$ mm/s (0−p). Reproduced from Ref. 10 with permission from AIP Publishing.

the sensing surface is sufficiently large. The computation shows that such positions do exist, and one of them is at $H = 6.3$ mm. The working mode at which there is acoustic streaming but no sound pressure on the sensing surface is defined as the acoustic streaming mode (A mode). Figure 5.18(a, b) shows the computed acoustic streaming and sound pressure fields, respectively, when $H = 6.3$ mm with $u = 600$ mm/s (0−p). It is seen that when the sensor is at this position, the acoustic streaming velocity v_{as} on the sensing surface is 176 mm/s and the sound pressure on the sensing surface is 18 Pa (0−p) which can be neglected. For the sensor at $H = 6.3$ mm under the radiation surface, the sensor resistance versus acoustic streaming velocity v_{as} at different O_3 concentrations was measured, and the result is shown in Fig. 5.18(c). The acoustic streaming velocity v_{as} was deduced from the measured vibration velocity u and computed ratio of the acoustic streaming velocity v_{as} to vibration velocity u.

It is seen that as the acoustic streaming velocity increases, the sensor resistance decreases monotonously, which is different from the characteristic shown in Fig. 5.17(c). This gives a direct experimental verification that the acoustic streaming is capable of

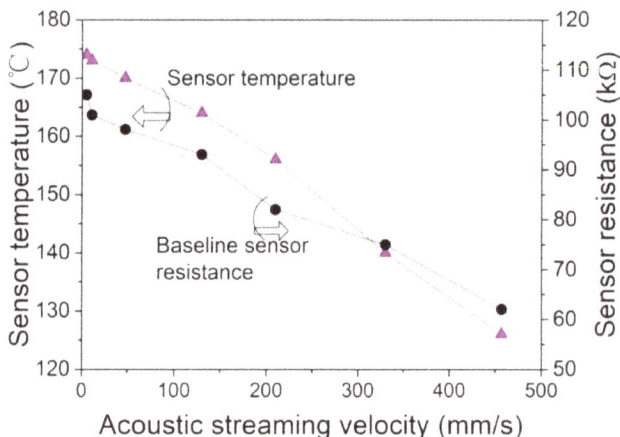

FIGURE 5.19   Measured relationship among the working temperature, acoustic streaming velocity and sensor resistance. Reproduced from Ref. 10 with permission from AIP Publishing.

affecting the sensing process. For the 5 ppm O_3 gas, the sensing response and sensitivity ($d\gamma/dC$) are 2.2% and 3.7×10^{-3}/ppm at $v_{as} = 0$, respectively, and become 1.7% and 2.6×10^{-3}/ppm at $v_{as} = 402$ mm/s, respectively. It seems that the acoustic streaming decreases the sensing response and sensitivity a bit. In addition, measured LDL of the sensor system for O_3 is about 3 ppm at 650 mm/s (0–p), which means that the LDL is increased by about 50% from that without sonication.

To find the mechanism of the above phenomena, working temperature of the sensing surface was measured at different acoustic streaming velocity v_{as}, and the result is shown in Fig. 5.19. In the measurement, the distance H was fixed at 2.1 mm, and the acoustic streaming velocity v_{as} was varied by changing the driving voltage of ultrasonic transducer in resonance. It is seen that as the acoustic streaming velocity increases, the working temperature decreases due to the cooling effect. The baseline resistance of the sensor (not exposed to the target gas) was also measured at different working temperatures, and the result is also shown in Fig. 5.19. It is seen that the sensor resistance decreases as the working temperature decreases, which indicates that the working temperature decrease enhances the desorption process of the oxygen species from the sensing surface. This result agrees very well with the conclusions given in Refs. [15, 16]. Thus, the phenomenon shown in Fig. 5.18(c) is caused by the cooling effect resulting from the acoustic streaming.

The P mode and A mode are two extreme working modes for the gas-borne ultrasound assisted gas sensor, which cause the sensing resistance to change in the opposite directions. The working mode in which both acoustic streaming and sound pressure exist on the sensing surface is defined as the complex mode (C mode). The measured dependency of sensitivity ($d\gamma/dC$) on the vibration velocity u at the sound pressure, acoustic streaming and complex modes is listed in Fig. 5.20. It is seen that as the vibration velocity increases, the sensitivity of the sound pressure mode increases but that of the acoustic streaming mode changes little. As the sound pressure and acoustic

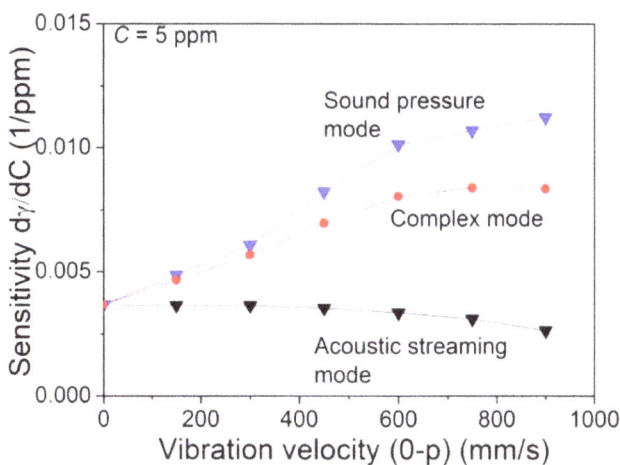

FIGURE 5.20   Sensitivities of the sound pressure, acoustic streaming and complex modes at different vibration velocities. Reproduced from Ref. 10 with permission from AIP Publishing.

streaming modes are two extreme working modes, it may be predicted that the sensing performance of the complex mode should be between those of the sound pressure and acoustic streaming modes for the same vibration velocity and target gas concentration. The measured sensitivity of the complex mode shown in Fig. 5.20 provides an experimental verification for this prediction. The slope of the curves of P and C modes has a decrease when the vibration velocity is larger than some critical value. This means that the cooling effect, which is caused by the acoustic streaming, starts to affect the sensing process when the vibration velocity is larger than the critical value.

In experiments of the complex mode, the sensor was at $H = 2.1$ mm under the radiation surface, where there were sufficiently large sound pressure and acoustic streaming on the sensing surface ($v_{as} = 162$ mm/s and $p_s = 390$ Pa (0–p) at $u = 600$ mm/s (0–p), based on the FEM computation). The sensor resistance versus vibration velocity at different O_3 concentrations was measured, and the result is shown in Fig. 5.21. The measured LDL of the sensor system for O_3 is about 1 ppm at 650 mm/s (0–p), which means that the LDL is lowered to about 1/2 of that without sonication. Figure 5.21 also indicates that the critical vibration velocity at which the cooling effect starts to affect the sensing process, is little affected by the target gas concentration, which is in expectation.

Figure 5.22 shows the measured transient sensing response of the three working modes for O_3 target gas of 20 ppm when the vibration velocity is 450 mm/s (0–p). In the measurement, the ultrasonic field and sensing response were already in the steady state before applying the target gas, and the discharge of target gas and switch-off of ultrasound were carried out simultaneously. The response time of the A mode, C mode and P mode is 11, 39 and 43 s, respectively. Here, the response time is defined as the time taken by the sensor to complete 90% change of the sensor resistance. Thus, the measured response time of the C mode is between those of the A and P modes, which verifies the above prediction again.

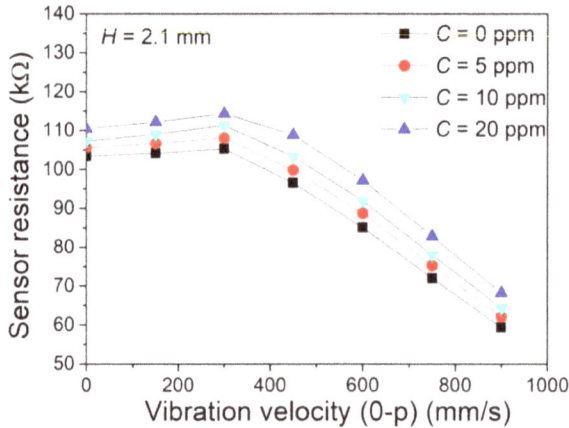

FIGURE 5.21   Measured relationship between the sensor resistance and vibration velocity at different O_3 gas concentrations at the complex mode. Reproduced from Ref. 10 with permission from AIP Publishing.

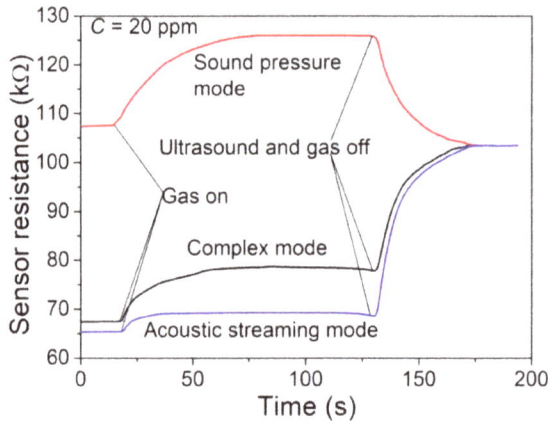

FIGURE 5.22   Transient sensing response processes of the sound pressure, acoustic streaming and complex modes when the target gas is 20 ppm O_3 gas and vibration velocity of the ultrasonic transducer is 450 mm/s (0–p). In the measurement, the ultrasonic field and response are already in a steady state before applying the target gas, and the discharge of the target gas and switch-off of ultrasound are carried out simultaneously. Reproduced from Ref. 10 with permission from AIP Publishing.

5.3.5   CONCLUSIONS

The author's team verified the individual effect of sound pressure and acoustic streaming on the sensing process of the gas-borne ultrasound assisted MOX gas sensor. Based on the experimental result analyses and FEM computation, the following conclusions are achieved. Enhanced transportation of target gas molecules onto the sensing surface, caused by the sound pressure, results in a sensitivity enhancement; The acoustic streaming may cool down the sensing surface, and decrease the sensing response and sensitivity a bit; The gas-borne ultrasound assisted MOX gas sensor has three working modes, that is, P mode, A mode and C mode. It is predicted and experimentally confirmed that sensing characteristics of the complex mode should be between those of the P mode and A modes. It is understandable that the sensitivity enhancement mechanism can also work for other types of gas sensors such as the catalytic combustion, electrochemical and diode gas sensors. Moreover, It is predicted that ultrasonically induced transportation of gas molecules onto a solid surface may enhance other types of gas-solid reaction processes.

5.4   AIRBORNE ULTRASOUND CATALYZED SALTWATER AL/MG-AIR FLOW BATTERIES

5.4.1   INTRODUCTION

Metal-air batteries (MABs) are regarded as a type of clean and sustainable power source for applications in next-generation electronics and energy storage systems due to their high theoretical energy density, high safety, eco-friendliness and low cost. In recent years, various MABs with different metal anode materials have been developed and investigated, such as Li-air batteries, Zn-air batteries, Al-air batteries and Mg-air batteries. Among them, the Li-air battery has the highest theoretical energy density (11.6 kW h kg^{-1}), which can even rival the gasoline engine (13 kW h kg^{-1}). However, there are still many challenges in the practical application of Li-air batteries, such as safety and stability issues due to the lithium dendrites, low coulombic efficiency and inefficient catalytic capacity. Zn-air batteries also suffer from similar safety and stability problems due to the formation of zinc dendrite during operation. However, Al-air and Mg-air batteries can avoid such safety and stability problems and are able to work well in neutral electrolytes such as saltwater. Although Al/Mg-air batteries are usually electrochemically non-rechargeable, they are safe, stable, cost-effective and environmentally friendly, and their anode materials are earth-abundant, which makes them attractive green power sources. Over recent years, Al/Mg-air batteries have received widespread attention and have been developed rapidly in the energy field.

Despite many advantages, there are still some scientific and technical challenges in the development of Al/Mg-air batteries, such as inefficient oxygen reduction reaction (ORR) rate at the air cathode, which restricts power output of the batteries. Till now, researchers have put great efforts to improve the discharge performance of Al/Mg-air batteries, which include the development of metal alloys as the anodes, use of gel electrolytes or corrosion inhibitors for preventing erosion and designs of highly efficient ORR catalysts for air cathodes and so on. In addition, through the means of electrolyte flow, the continuous supply of fresh electrolytes makes batteries effectively avoiding the negative influence of the reaction by-products, which can further improve the battery discharge performance. Thus, Al/Mg-air batteries incorporating electrolyte flows, which are termed as Al/Mg-air flow batteries, are attracting more and more attention recently.

Ultrasonic methods have been employed to enhance the performance of batteries. They include the acoustic streaming induced electrolyte agitation to improve the discharge performance of a button zinc-air battery, ultrasonic capillary effect-based electrolyte circulation to enhance the discharge performance of an Al-air flow battery, ultrasound-enhanced diffusion through the membrane separator to improve the performance of zinc-alkaline batteries and SAW-driven electrolyte flow for improving the charging rate of lithium-air batteries. In these works, ultrasound was used to prompt the ion conduction process in electrolytes.

In this work, we have proposed and developed an airborne ultrasound-based catalysis method for promoting the ORR and improving the discharge performance of saltwater Al/Mg-air flow batteries, in which a focused airborne ultrasound (FAU) is applied onto the cathode surface. Enhancement of the battery power is experimentally confirmed, and the working principle is analyzed. To the best of our knowledge, this is the first attempt to utilize the gas-borne ultrasonic catalysis effect to promote the ORR of metal-air flow batteries (MAFBs).

5.4.2 EXPERIMENTAL

5.4.2.1 Principle and Structural Design

Figure 5.23 shows a schematic of the saltwater Al/Mg-air flow batteries catalyzed by the FAU generated by a focused ultrasonic transducer, in which the cathode

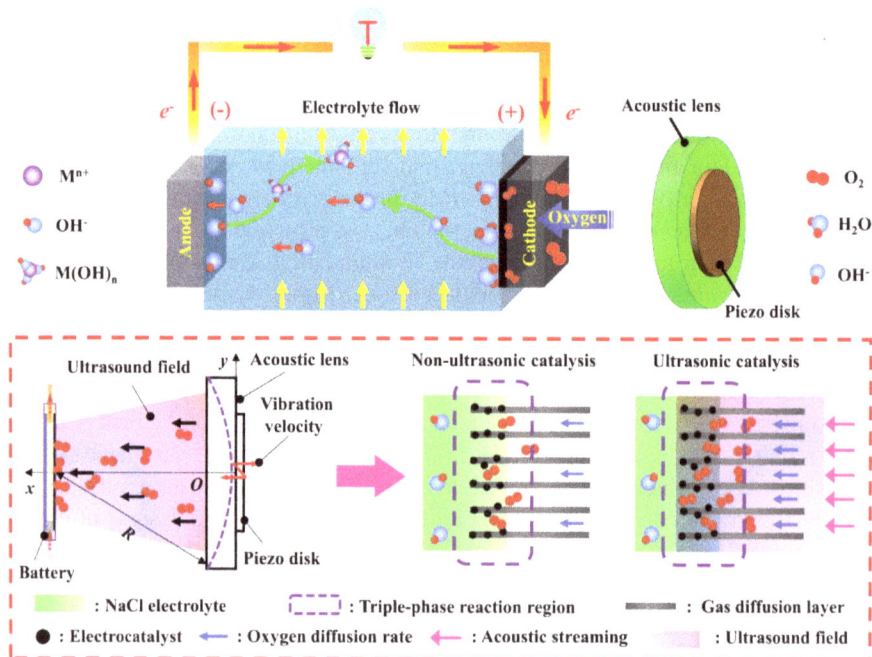

FIGURE 5.23 Principle schematic of the saltwater Al/Mg-air flow battery catalyzed by a focused airborne ultrasound. Reproduced from Ref.17 with permission from Elsevier.

is sonicated by the FAU. The electrochemical reactions of Al-air batteries are as follows:

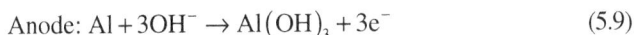

$$\text{Cathode: } O_2 + 2H_2O + 4e^- \rightarrow 4OH^- \tag{5.8}$$

$$\text{Anode: } Al + 3OH^- \rightarrow Al(OH)_3 + 3e^- \tag{5.9}$$

$$\text{Overall: } 4Al + 3O_2 + 6H_2O \rightarrow 4Al(OH)_3 \tag{5.10}$$

The electrochemical reactions of Mg-air batteries are as follows:

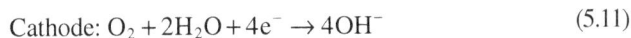

$$\text{Cathode: } O_2 + 2H_2O + 4e^- \rightarrow 4OH^- \tag{5.11}$$

$$\text{Anode: } Mg \rightarrow Mg^{2+} + 2e^- \tag{5.12}$$

$$\text{Overall: } 2Mg + O_2 + 2H_2O \rightarrow 2Mg(OH)_2 \tag{5.13}$$

Near the three-phase reaction interface (TPRI) on the cathode, oxygen molecules are ejected onto the cathode surface by the strong acoustic pressure generated by the FAU, and the oxygen diffusion rate is enhanced by the acoustic streaming generated by the spatial gradient of the FAU. As a result, the ORR rate at the TPRI is enhanced greatly, compared to the case without sonication. The existence of the acoustic pressure and acoustic streaming near the cathode is confirmed by the FEM computation in Section 5.4.2.4.

Figure 5.24(a, b) shows the structure and the photo of the saltwater MAFB catalyzed by the FAU. The focused ultrasonic transducer consists of a concave spherical acoustic lens made of aluminum alloy and a 2-mm thick piezoelectric disk with a diameter of 30 mm (P-8, Wuxi Haiying Co., Ltd.). The diameter, thickness and focal radius of the concave spherical shape are 48, 7.93 and 48 mm, respectively (Fig. 5.24(c)). The piezoelectric disc is bonded onto the back of the acoustic lens by epoxy (UHU PLUS ENDFEST 300) at the central area. Based on our measurement, the focused ultrasonic transducer has multiple resonance modes at 122.1, 409.4, 584.7 and 608.4 kHz. All of them have served as the working frequencies in the experiments.

The battery has a sandwich structure, composed of an anode electrode, a cathode electrode and a PMMA plate (polymethyl methacrylate, 35 mm × 35 mm × 2.8 mm) in the middle. The PMMA plate is fabricated by cutting a raw PMMA plate by a five-axis vertical machining center (DMU 60 monoBLOCK, DMG). To form an electrochemical reaction chamber, a square reaction channel of 30 mm × 30 mm is cut out in the middle of the PMMA plate, and thus the electrochemical reaction area of each electrode is 9 cm^2 (Fig. 5.24(c)). To install the two electrodes conveniently and increase the tightness between the electrode and PMMA plate, two square cavities with a size of 34 mm × 34 mm × 0.2 mm are cut out on both sides of the PMMA plate. Thus, the distance between the two electrodes is 2.4 mm. Also, the electrolyte inlet and outlet are set at the bottom and top of the PMMA plate, respectively, for the inflow and outflow of saltwater.

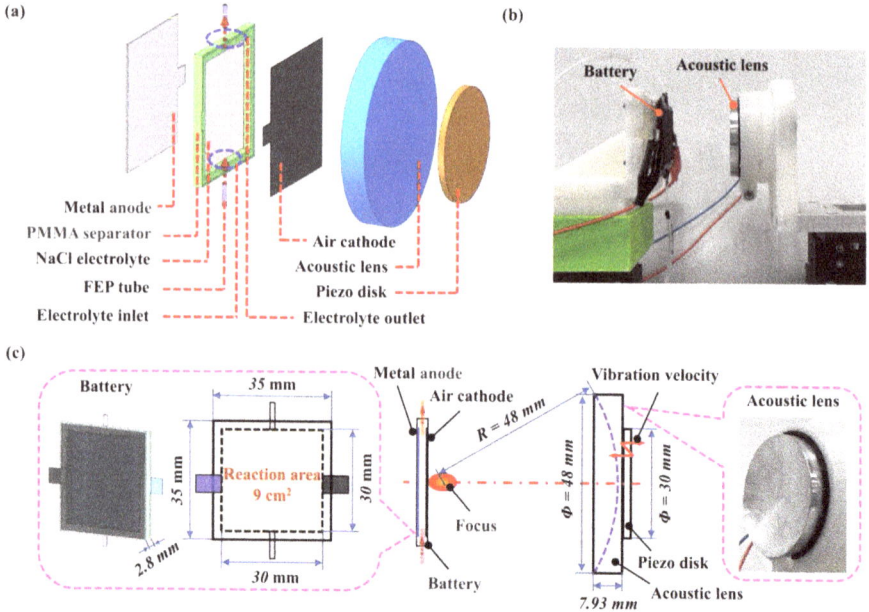

FIGURE 5.24 (a) Structure diagram and (b) photo of a saltwater MAFB catalyzed by the FAU. (c) Sizes of the battery and acoustic lens. Reproduced from Ref.17 with permission from Elsevier.

5.4.2.2 Materials and Chemicals

The saltwater was prepared with a deionized water preparation system (JCL-20-15, Ultrapure Water Co., Ltd.) and analytical pure sodium chloride ($\geq 99.8\%$, Sinopharm Chemical Reagent Co., Ltd.). In the experiments, the saltwater electrolyte solution concentrations were 1, 2, 3, 4 and 5 M. Pure aluminum (Al, purity of ~99.99%) and AZ31B magnesium alloy (Mg, purity of ~96%) were used as the anodes, respectively. Carbon paper (HCP 120, Shanghai Hesen Electric Co., Ltd.) was used as the cathode electrode, which also served as the gas-diffusion electrode (GDE).

Pt/C suspension was sprayed onto the surface of the GDE to form a catalytic layer, in order to improve the intrinsic discharge performance of the experimental batteries. The Pt/C catalyst suspension was prepared by mixing catalyst powder (40 wt.% Pt/C, Johnson Matthey), ethanol-deionized water solvent and Nafion solution (5 wt.%, 45 ± 3 mg cm^{-3}, DuPont). The mass ratio of the catalyst powder, the ethanol-deionized water solvent and the Nafion solution was 1:40:10, and the volume ratio of ethanol to deionized water was 1:1. The catalyst powder and the Nafion solution were dispersed in the ethanol-deionized water solvent by a high-power sonication system (UH600, Shanghai OuHor Machinery Equipment Co., Ltd. 600 W) for 60 min to obtain the catalyst suspension. Then, the catalyst suspension was sprayed onto the carbon paper with an overall Pt/C load of 2 mg cm^{-2} and dried at 90°C for 6 h in a drying oven (DHG500-00, Supo Instrument Co., Ltd.).

During the battery assembly, the cathode and anode electrodes were bonded on the two opposite sides of the PMMA plate by the fuel cell sealant (K-704NB, Kafuter).

Furthermore, a copper foil, which was bonded to the carbon paper using conductive silver paste, was used as the current collector for the link to the external circuit.

5.4.2.3   Electrochemical Measurements

After the MAFBs were assembled, the saltwater electrolyte was pumped into the electrochemical reaction chamber at a flow rate of 1 ml min^{-1} by an external mechanical pump (LM60A, Nanjing Runze Fluid Control Equipment Co., Ltd). The electrochemical measurements were carried out by an electrochemical workstation (CHI 760E, Shanghai Chenhua Instrument Co., Ltd.) at room temperature (298 ± 2 K). The open-circuit voltages (OCVs) of batteries were recorded, and the polarization curves were obtained by the potentiostatic measurement method. The testing voltage value ranged from 0.55 to 0 V with a step of 0.05 V for the Al-air flow batteries and from 1.1 to 0 V with a step of 0.1 V for the Mg-air ones. Each current measurement lasted for 30 s, and current values in each 30s duration were averaged to obtain the average current in the 30 s. In the polarization curve measurement, the power density (PD) of MAFBs was calculated as follows:

$$\text{Power density}\,(PD) = (V \times I)/A, \tag{5.14}$$

where V is the battery voltage, I is the battery current and A is the active electrode area. The single electrode potential was also measured using a reference electrode (Ag/AgCl in saturated KCl, Shanghai Chenhua Instrument Co., Ltd.), as well as the polarization curves of single electrode potential. In the measurement, the cathode or anode was connected to the reference electrode through a Fluke digital multimeter (Fluke 17B+). Measurement of each polarization curve was repeated three times to take the average.

In addition, to obtain the battery impedance, the electrochemical impedance spectroscopy (EIS) measurement method was used with a testing frequency range from 100 kHz to 0.1 Hz and A.C. (alternative current) amplitude of 5 mV at 0.55 V for the Al-air flow batteries and 1.10 V for the Mg-air ones, respectively.

5.4.2.4   Finite Element Analyses

To clarify the principle of the discharge performance enhancement of MAFBs sonicated by the FAU, the acoustic pressure field and the steady-state acoustic streaming field in the air around the battery were computed by the FEM (COMSOL Multiphysics 6.0, MA, Burlington). We employed a 2D physical model that coupled the piezoelectric disc, acoustic lens, battery structure and surrounding air domain. A perfectly matched layer (PML) was adopted to mimic the far-field air domain. The entire simulation process comprises two steps. First, by applying the measured working frequency and vibration velocity to the piezoelectric disc, the vibration mode of the acoustic lens and the acoustic pressure field in the air domain are computed. Then, the steady acoustic streaming field in the air domain is computed, using the computational results of the acoustic field.

The vibration measurement result (Fig. 5.25(a)) shows that the piezoelectric disc at 122.1 kHz has a thickness vibration pattern with a vibration velocity of 100 mm s^{-1} (rms or root mean square). Using the vibration excitation conditions, vibration pattern

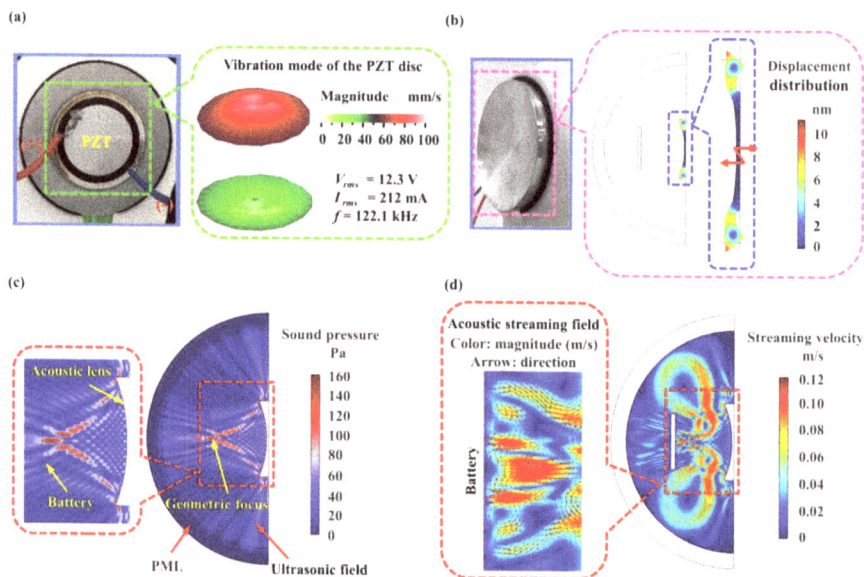

FIGURE 5.25 (a) Measured vibration mode of the PZT (lead zirconate titanate) disc. (b) Simulated vibration of the acoustic lens. (c) Sound pressure distribution in the air domain around the cathode. (d) Acoustic streaming distribution around the cathode. Reproduced from Ref. 17 with permission from Elsevier.

of the acoustic lens was simulated, and the result is shown in Fig. 5.25(b), from which it is seen that there is a flexural vibration in the concave surface of the acoustic lens. Computed sound pressure distribution in the air domain is shown in Fig. 5.25(c). It is seen that the diameter of the focal region of acoustic pressure is about 2.5 mm. The strong acoustic pressure near the cathode can eject the oxygen molecules onto the electrode and thus catalyze the ORR. Furthermore, computed acoustic streaming distribution around the battery is shown in Fig. 5.25(d), in which the color and arrow represent the velocity and direction of the acoustic streaming, respectively. It can be seen that the acoustic streaming flows to the cathode, and multiple vortices are formed on the electrode surface. As the acoustic streaming induced by airborne ultrasound can speed up the convective flow rate of air, the acoustic streaming around the cathode may improve the oxygen diffusion rate, and thus enhance the ORR too.

5.4.3 RESULTS AND DISCUSSION

5.4.3.1 Battery Performance Enhancement by the FAU

Unless otherwise specified, the experimental saltwater electrolyte concentration, electrolyte flow rate and operating voltage and frequency for exciting the acoustic lens are 5 M, 1 ml min^{-1}, 12.5 V$_{rms}$ and 122.1 kHz, respectively.

Figure 5.26 shows the measured electrochemical performance of the Al/Mg-air flow batteries catalyzed by the FAU.

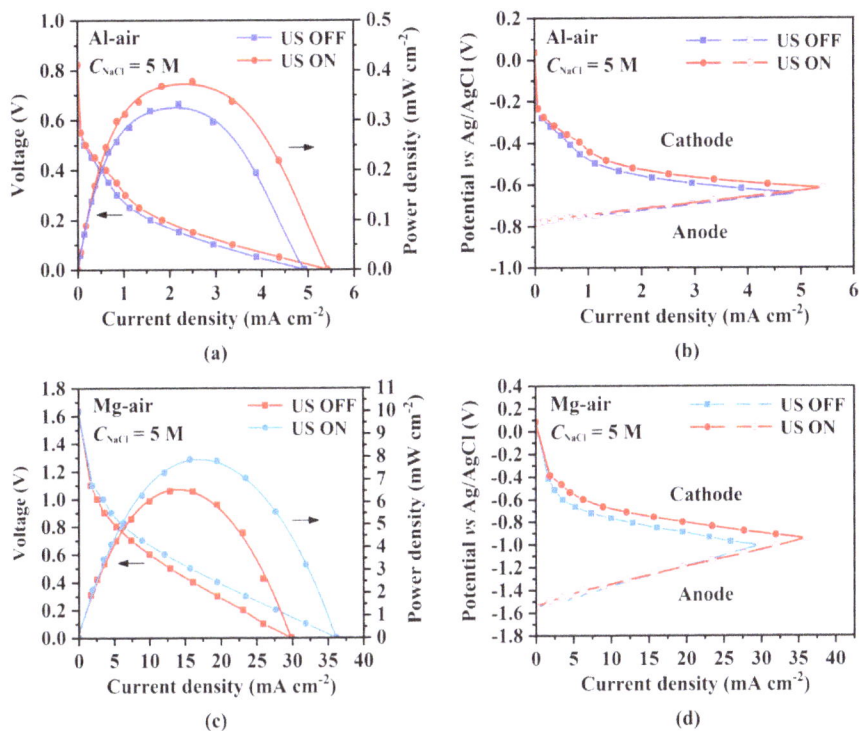

FIGURE 5.26 Discharge performance of the Al/Mg-air flow batteries catalyzed by the FAU. (a) Polarization and power density curves of the Al-air flow battery. (b) Single electrode potential curves of the Al-air flow battery. (c) Polarization and power density curves of the Mg-air flow battery. (d) Single electrode potential curves of the Mg-air flow battery. Reproduced from Ref. 17 with permission from Elsevier.

For the Al-air flow battery, as shown in Fig. 5.26(a), both the peak power density and the short-circuit current density are improved by the FAU. The peak power density increases from 0.332 to 0.378 mW cm^{-2} (13.86%), and the short-circuit density increases from 4.922 to 5.427 mA cm^{-2} (10.26%). For the Mg-air one, as shown in Fig. 5.26(c), the peak power density increases from 6.461 to 7.859 mW cm^{-2} (21.64%) and the short-circuit density increases from 29.893 to 36.134 mA cm^{-2} (20.90%).

Figures 5.26(b) and 5.26(d) show the single electrode potential curves for the Al-air and Mg-air flow batteries, respectively. The cathodic and anodic potentials are around 0.035 V and −0.787 V for the Al-air flow battery, respectively, and around 0.083 V and −1.547 V for the Mg-air one, respectively. It is seen that the improvement of battery performance mainly arises from the cathode side, while the anode's potential almost has no change before and after the sonication. This is because the airborne ultrasonic catalysis effect happens on the cathode, rather than on the anode.

The EIS was also measured for the Al/Mg-air flow batteries with and without FAU, and the results are shown in Fig. 5.27. The equivalent circuits used to fit the EIS plots are shown in the insets and the equivalent parameters are listed in Table 5.3. In the

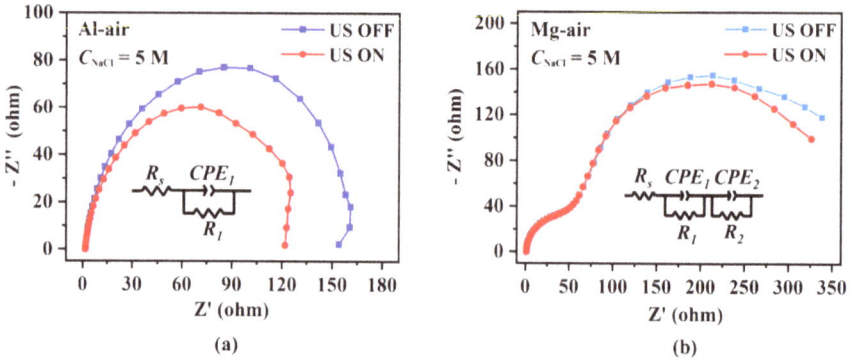

(a)   (b)

FIGURE 5.27   Measured EIS of the MAFBs. (a) Al-air flow battery. (b) Mg-air flow battery. Reproduced from Ref. 17 with permission from Elsevier.

equivalent circuits, R_s represents the total ohm resistance of the battery, contributed by the electrolyte, contact and lead wires, CPE_1 and CPE_2 are the constant phase elements of the cathode and anode, respectively, and R_1 and R_2 represent the polarization resistance of the cathode and anode, respectively, which include the charge transfer and diffusion layer resistances.

The EIS in Fig. 5.27 gives the total impedance of each battery, characterized by a high-frequency semicircle (the smaller one) and a low-frequency semicircle (the larger one). The high- and low-frequency semicircles represent the polarization impedance of the anode and cathode, respectively, and the measured R_1 and R_2 are summarized in Table 5.3. Due to the anode impedance being much smaller than the cathode one for the Al-air battery, the anode impedance can be ignored for the Al-air battery. From Table 5.3, it can be seen that for the Al-air flow battery, the polarization resistance values of the cathode (R_1) are 162.5 $\Omega \cdot cm^2$ without sonication and 128.2 $\Omega \cdot cm^2$ with the sonication, respectively, which indicates that the FAU can decrease of the polarization resistance of the cathode. This is because a faster electrochemical reaction rate at the cathode causes more hydroxide ions generated per unity time and fast consumption of electrons at the cathode. For the Mg-air one, there is a similar result as shown in Fig. 5.27(b).

TABLE 5.3

Fitting Parameters of the Equivalent Circuit for the Measured EIS of the Al/Mg-Air Batteries

Element		R_s ($\Omega \cdot cm^2$)	R_1 ($\Omega \cdot cm^2$)	CPE_1 ($\Omega^{-1} \cdot S^{n1} \cdot cm^{-2}$)	n_1	R_2 ($\Omega \cdot cm^2$)	CPE_2 ($\Omega^{-1} \cdot S^{n2} \cdot cm^{-2}$)	n_2
Al-air	US OFF	1.845	162.5	1.19×10^{-4}	0.945	/	/	/
	US ON	1.849	128.2	1.21×10^{-4}	0.940	/	/	/
Mg-air	US OFF	1.437	44.79	2.58×10^{-4}	0.981	340.6	0.0019	0.920
	US ON	1.438	44.03	2.65×10^{-4}	0.978	320.4	0.0019	0.924

5.4.3.2   Effect of the Vibration Velocity

In this work, measured vibration velocity at the back side of the piezoelectric disk was used to represent the FAU strength. The polarization and power density curves under different vibration velocities were measured with 5M NaCl electrolyte, and the results for the Al-air flow battery are shown in Fig. 5.28(a, c), respectively. It is observed that when the vibration velocity increases from 0 to 100 mm s^{-1}, the peak power density increases from 0.332 to 0.378 mW cm^{-2}. Compared with the peak power density without sonication, the percentage increase of the peak power density at vibration velocities of 25, 40, 70 and 100 mm s^{-1} (RMS) are 1.81%, 6.02%, 9.64% and 13.86%, respectively. This performance enhancement is due to an increase of sound pressure at the TPRI on the cathode (from 40.30 to 161.18 Pa), as shown in Fig. 5.28(d). The larger the sound pressure, the stronger the ultrasonic catalysis effect for the ORR. For the Mg-air one, the peak power density increases from 6.461 to 7.859 mW cm^{-2} as the vibration velocity increases from 0 to 100 mm s^{-1}, as shown in Fig. 5.28(b, c). Additionally, the percentage increase of peak power density of battery is 3.31%, 7.27%, 14.04% and 21.64%, respectively.

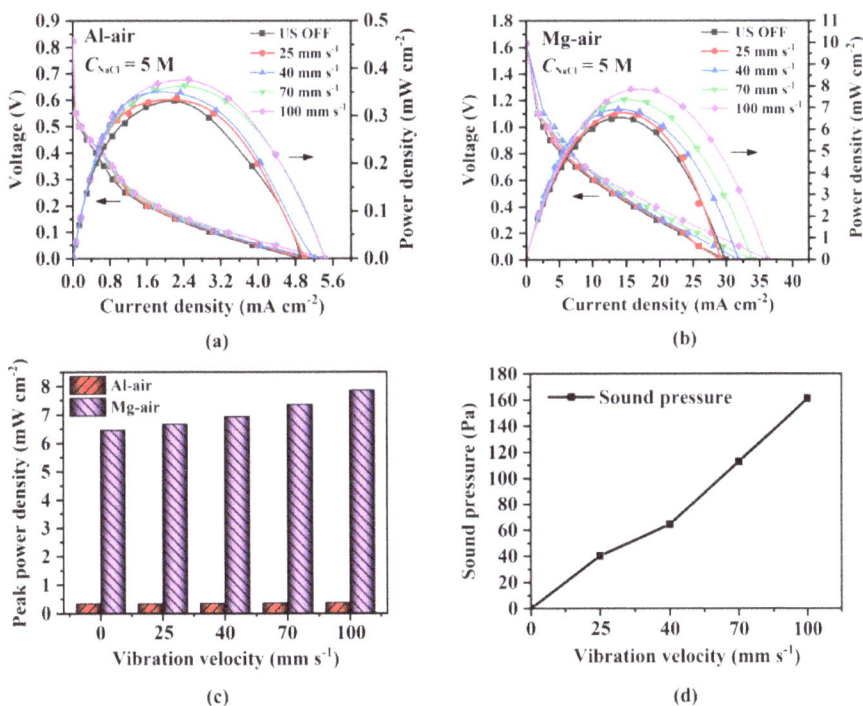

FIGURE 5.28   Discharge performance of the Al/Mg-air flow batteries catalyzed by the FAU at different ultrasonic vibration velocities (RMS). (a) Polarization and power density curves of the Al-air flow battery. (b) Polarization and power density curves of the Mg-air flow battery. (c) Peak power density of the Al/Mg-air flow battery. (d) Sound pressure at the three-phase reaction interface on the cathode under different ultrasonic vibration velocities. Reproduced from Ref. 17 with permission from Elsevier.

5.4.3.3   Effect of Operating Frequency

Based on our experimental measurement, it was found that the ultrasonic transducer has multiple resonance points (122.1, 409.4, 584.7 and 608.4 kHz). In this work, we measured the battery performance under these resonance frequencies at 12.5 V_{rms} driving voltage. For the Al-air flow battery, as shown in Fig. 5.29(a, c), when the operating frequency is 122.1, 409.4, 584.7 and 608.4 kHz, the peak power density is 0.378, 0.387, 0.395 and 0.427 mW cm^{-2}, respectively, which means that the discharge performance becomes better as the operating frequency increases. Meanwhile, compared with the peak power densities without sonication, percentage increase of the peak power density at different operating frequencies are 13.86%, 16.57%, 18.98% and 28.61%, respectively.

For the Mg-air one, as shown in Fig. 5.29(b, c), when the operating frequency is 122.1, 409.4, 584.7 and 608.4 kHz, the peak power density is 7.859, 8.076, 8.300 and 8.643 mW cm^{-2}, respectively. The percentage increase of peak power density of battery at different operating frequencies is 21.64%, 25.00%, 28.46% and 33.77%, respectively. Thus, using a higher ultrasonic operating frequency is beneficial to the enhancement of airborne ultrasound catalysis effect and improvement of discharge

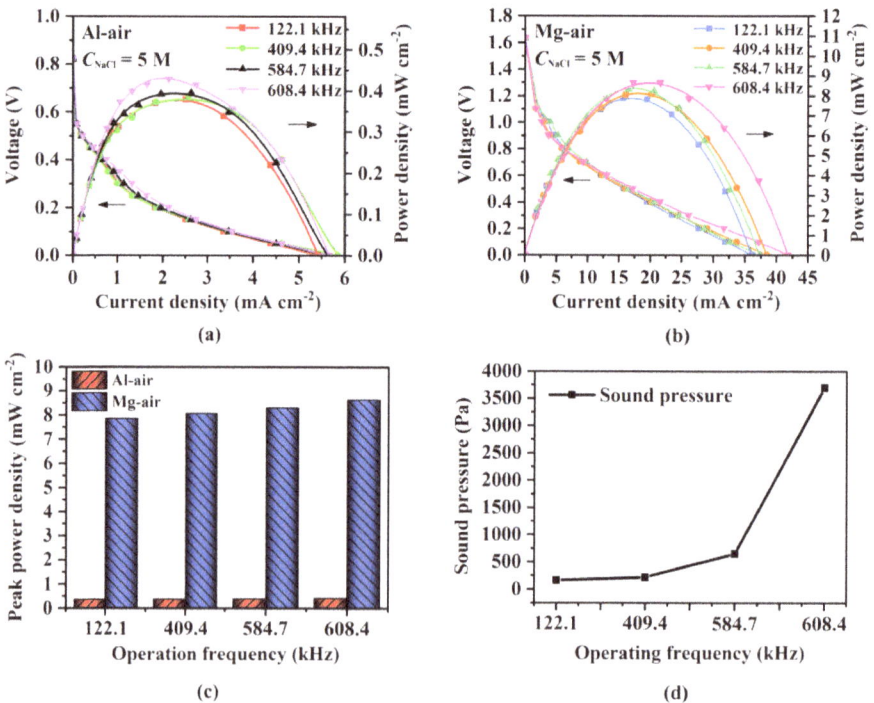

FIGURE 5.29   Discharge performance of the Al/Mg-air flow batteries catalyzed by the FAU under different operating frequencies. (a) Polarization and power density curves of the saltwater Al-air flow battery. (b) Polarization and power density curves of the saltwater Mg-air flow battery. (c) Peak power density of the Al/Mg-air flow battery. (d) Sound pressure at the three-phase interface on the cathode under different operating frequencies. Reproduced from Ref. 17 with permission from Elsevier.

performance. This is attributed to the increase of sound pressure (from 161.18 to 3695.8 Pa) when the operating frequency rises, as shown in Fig. 5.29(d). For the given ultrasonic transducer in this work, the peak power density of the Al/Mg-air flow batteries is maximum at the operating frequency of 608.4 kHz.

5.4.3.4 Effect of NaCl Electrolyte Concentration

We investigated and demonstrated the battery performance under different concentrations of NaCl electrolyte when the ultrasonic working conditions were kept at 12.5 V_{rms} and 608.4 kHz. As shown in Fig. 5.30(a), for the Al-air flow battery, it is seen that when the electrolyte concentration increases from 1 to 5 M, the peak power density increases from 0.272 to 0.427 mW cm^{-2}. Figure 5.30(b) shows the EIS plots of the Al-air flow battery, with the equivalent circuit used to fit the EIS plots included as the inset. The polarization resistance of the cathode dramatically reduces from 1451 $\Omega \cdot cm^2$ (under 1 M) to 94.11 $\Omega \cdot cm^2$ (under 5 M). This indicates that the increase of electrolyte concentration can effectively reduce the polarization resistance of the cathode and improve the discharge performance of the battery. There are similar results for the Mg-air one, as shown in Fig. 5.30(c).

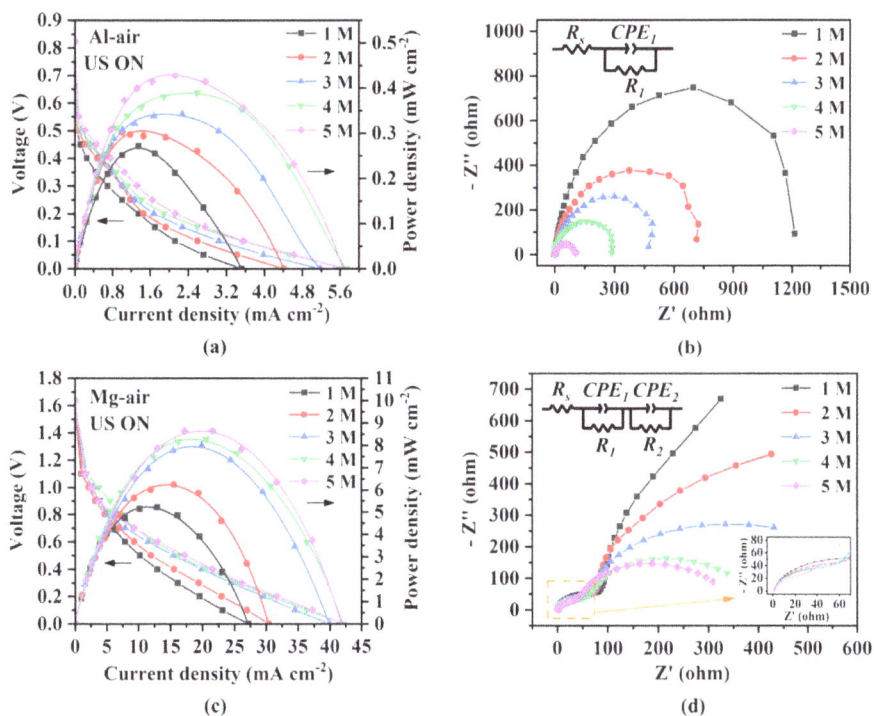

FIGURE 5.30 Discharge performance of the saltwater Al/Mg-air flow batteries catalyzed by the FAU under different NaCl electrolyte concentrations. (a) Polarization and power density curves of the Al-air flow battery. (b) EIS of the Al-air flow battery. (c) Polarization and power density curves of the Mg-air flow battery. (d) EIS of the Mg-air flow battery. Reproduced from Ref. 17 with permission from Elsevier.

For the Al-air one, the peak power density increases by 21.43%, 22.27%, 25.50%, 24.68% and 28.61%, respectively, compared to that of the battery not sonicated. While for the Mg-air one, it increases by 25.46%, 29.21%, 33.37%, 29.49% and 33.77%, respectively.

5.4.3.5 Effect of Electrode Reaction Area

To further investigate the effect of electrode reaction area on battery discharge performance, MAFBs with different electrode areas were fabricated in this work, and the experimental results are shown in Fig. 5.31. The experiments were carried out under 5M NaCl concentration, and the ultrasonic working conditions were kept as 12.5 V_{rms} and 608.4 kHz. To form the chambers with different electrode reaction areas, a series of square reaction channels were cut out in the middles of the PMMA plates, with the dimensions of 0.5 cm × 0.5 cm, 1.0 cm × 1.0 cm, 2.0 cm × 2.0 cm and 3.0 cm × 3.0 cm, respectively, as shown in Fig. 5.32. Figure 5.31(a) shows that for the Al-air flow battery, as the electrode reaction area increases from 0.25 to 9.0 cm², the peak power increases from 0.107 to 3.842 mW (about 36 times). Figure 5.31(b) shows a comparison of the peak power under different electrode reaction areas. It is seen that the enhancement percentage of the peak power is 24.71%, 29.26%, 27.78% and 28.61%, respectively. Herein, it must be pointed that increasing the electrode reaction area does not necessarily mean an increase of peak power density. As shown in Fig. 5.31(c), the peak power density is enhanced from 0.429 to 0.857 mW cm⁻² as the electrode reaction area

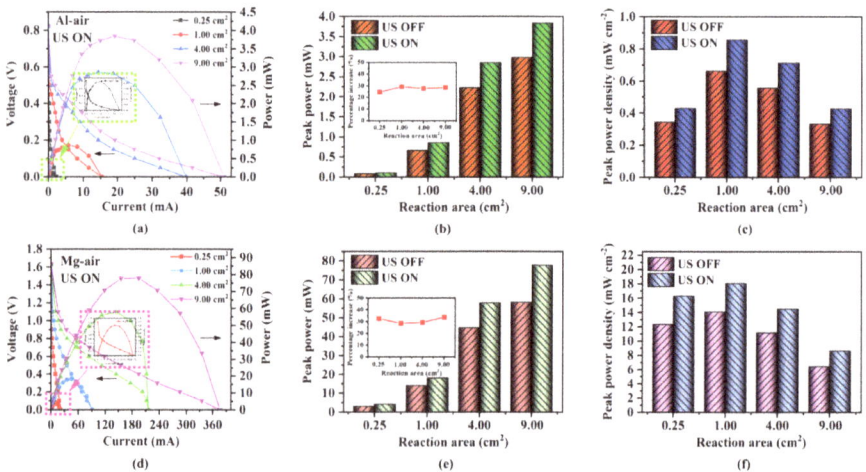

FIGURE 5.31 Discharge performance of the saltwater Al/Mg-air flow batteries catalyzed by the FAU under different reaction areas. (a) Polarization and power density curves of the Al-air flow battery. (b) Peak power of the Al-air flow battery under different electrode reaction areas with and without FAU. (c) Peak power density of the Al-air flow battery under different electrode reaction areas. (d) Polarization and power density curves of the Mg-air flow battery. (e) Peak power of the Mg-air flow battery under different electrode reaction areas with and without FAU. (f) Peak power density of the Mg-air flow battery under different electrode reaction areas. Reproduced from Ref. 17 with permission from Elsevier.

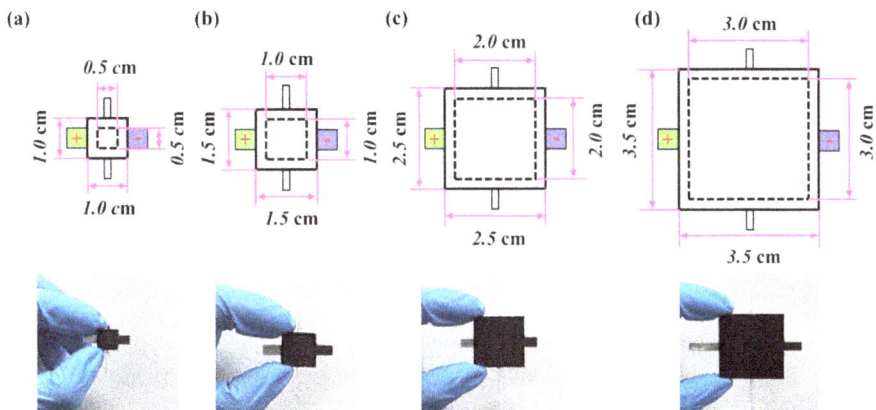

FIGURE 5.32 Structural size and photos of different electrode reaction areas for the Al/Mg-air flow batteries. (a) 0.5 cm × 0.5 cm. (b) 1.0 cm × 1.0 cm. (c) 2.0 cm × 2.0 cm. (d) 3.0 cm × 3.0 cm. Reproduced from Ref. 17 with permission from Elsevier.

increases from 0.25 to 1.00 cm^{-2} and decreases from 0.857 to 0.427 mW cm^{-2} as the electrode reaction area increase further. The former phenomenon is because of the decrease of negative effect of the flow resistance of the reaction chamber on the electrolyte refreshing. The latter one is because as the electrode reaction area increases, the volume of the reaction chamber also increases, which causes insufficient refreshing of the electrolyte in the reaction chamber at a given flow rate. The above explanation indicates that the electrode reaction area at which the peak power density reaches the maximum should not be affected by the anode material. The flow resistance of the electrolyte would be affected by the size of the reaction chamber, the reaction surface roughness of the cathode and anode, and the flow rate of electrolyte. For the Mg-air flow battery, similar results are shown in Fig. 5.31(d–f), resulting from a similar mechansim.

5.4.3.6 Effect of Using Pt/C as ORR Electrocatalyst

In the above studies, carbon paper was used as the cathode without any ORR electrocatalyst coating. Here, we further coated the Pt/C electrocatalyst on the cathode and investigated the discharge performance of the modified batteries under the FAU. The experiments were carried out at 12.5 V_{rms} ultrasonic operating voltage, 608.4 kHz ultrasonic operating frequency, 5 M NaCl concentration, and 1.0 cm^2 electrode reaction area. The overall Pt/C load on carbon paper was 2 mg cm^{-2}. As shown in Fig. 5.33(a), for the Al-air flow battery, the peak power density increases from 15.6 to 18.9 mW cm^{-2} (21.34%) with the sonication. For the Mg-air one, as shown in Fig. 5.33(b), the peak power density increases from 56.4 to 73.4 mW cm^{-2} (30.04%) with the sonication. Figure 5.33(c) summarizes the peak power densities of the MAFBs with and without the Pt/C electrocatalyst, which indicate that the FAU has a stronger effect on the discharge performance when the electrocatalyst is employed for the cathode.

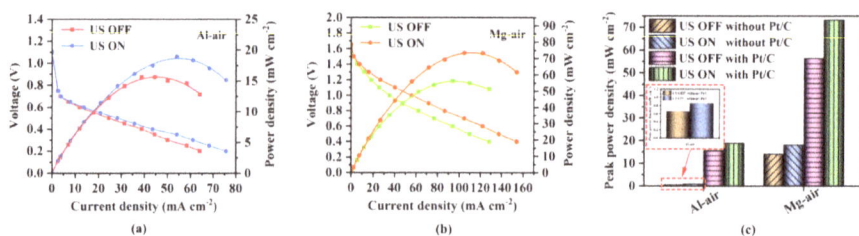

FIGURE 5.33 Discharge performance of the saltwater Al/Mg-air flow batteries using Pt/C catalysts (2 mg cm^{-2}) for the air cathodes. (a) Polarization and power density curves of the Al-air flow battery with catalysts loaded. (b) Polarization and power density curves of the Mg-air flow battery with catalysts loaded. (c) Comparison of peak power densities of the Al/Mg-air flow batteries under different conditions. Reproduced from Ref. 17 with permission from Elsevier.

5.4.4 CONCLUSIONS

In this work, we have proposed and demonstrated a physical catalysis method for promoting the discharge performance of saltwater Al/Mg-air flow batteries, which is realized by the FAU. Near the TPRI, the acoustic pressure and acoustic streaming are generated, which could enhance the oxygen diffusion and ORR rate. Our experimental results show that compared with the batteries not sonicated, the percentage increase of peak power density can reach up to 28.6% for the Al-air flow battery and 33.8% for the Mg-air one, respectively, when the ultrasonic frequency is 608.4 kHz. The achieved optimal peak power density is 18.938 and 73.377 mW cm^{-2} for the Al-air and Mg-air flow batteries sonicated by the FAU, respectively, when the electric catalyst loaded with Pt/C (2 mg cm^{-2}) is employed. Different from other existing ultrasound-assisted battery techniques, the FAU-enabled catalysis method for MFABs enhances the ORR directly. This ultrasonic catalysis method offers a simple but powerful tool to improve the discharge performance of MAFBs.

5.5 REMARKS

In this chapter, it is demonstrated that the ultrasonic catalysis effect, which is essentially a controlled driving of target gas molecules in the adjacent of a gas-solid interface, can effectively enhance the reaction at a gas-solid interface. As the ultrasonic catalysis effect results from the ultrasonic driving, theoretically it can be applied to the catalysis process of various reactions at the gas-solid interface, and can be employed in combination with the existing catalytic methods such as the use of catalytic materials and higher reaction temperature. Therefore, it is worth investigating the feasibility of applications of the gas-borne ultrasound based catalysis effect in more gas-solid reactions.

In terms of the principle of the gas-borne ultrasound based catalysis effect, although a preliminary study has been carried out, quantitative and deeper research is definitely necessary. In this aspect, there are two major works. One is to solve the thermo-acoustic field near the reaction surface by a numerical method such as the FEM. Another is to investigate the dynamic behavior of gas molecules near the reaction surface by molecular dynamic simulation.

REFERENCES

1. J. Hu, *Ultrasonic Micro/Nano Manipulations: Principles and Examples*, (World Scientific, New Jersey, London, Singapore, 2014), Chapters 3 and 5.
2. X. Qi, Q. Tang, P. Liu, I. V. Minin, O. V. Minin and J. Hu, "Controlled concentration and transportation of nanoparticles at the interface between a plain substrate and droplet," *Sens. Actuators B Chem.*, 274, pp. 381–392, 2018.
3. P. Liu, Z. Tian, K. Yang, T. D. Naquin, H. Hao, J. Huang, Q. Chen, H. Ma, P. Bachman, X. Zhang, J. Xu, T. J. Hu and Huang, "Acoustofluidic black holes for multifunctional in-droplet particle manipulation," *Sci. Adv.*, 8(13), 2022. DOI: 10.1126/sciadv.abm2592
4. Y. Wang and J. Hu, "Ultrasonic removal of coarse and fine droplets in air," *Sep. Purif. Technol.*, 153, pp. 156–161, 2015.
5. P. Liu, Z. Tian, N. Hao, H. Bachman, P. Zhang, J. Hu and T. J. Huang, "Acoustofluidic multi-well plates for enrichment of micro/nano particles and cells," *Lab Chip*, 20, pp. 3399–3409, 2020.
6. Q. Liu, K. Chen, J. Hu and T. Morita, "An ultrasonic tweezer with multiple manipulation functions based on the double-parabolic-reflector wave-guided high-power ultrasonic transducer," *IEEE Trans. Ultrason. Ferroelectr. Freq. Control*, 67(11), pp. 2471–2474, 2020.
7. Q. Liu, Q. Tang and J. Hu, "A new strategy to capture single biological micro particles at the interface between a water film and substrate by ultrasonic tweezers," *Ultrasonics*, 103, p. 106067, 2020.
8. Q. Liu, J. Hu, I. V. Minin and O. V. Minin, "High-performance ultrasonic tweezers for manipulations of motile and still single cells in a droplet," *Ultrasound Med. Biol.*, 45(11), pp. 3018–3027, 2019.
9. S. Su and J. Hu, "Ultrasound assisted low-concentration VOC sensing," *Sens. Actuators B Chem.*, 254, pp. 1234–1241, 2018.
10. S. Su, P. Liu, Q. Tang and J. Hu, "Physical principle of enhancing the sensitivity of a metal oxide gas sensor using bulk acoustic waves," *J. Appl. Phys.*, 124, 244902, 2018.
11. P. M. Morse and K. U. Ingard, *Theoretical Acoustics*, (McGraw–Hill, New York, 1968), pp. 241–249 and 285–286.
12. J. Lighthill, *Waves in Fluids*, (Cambridge University Press, Cambridge, 1978), p. 329 and 344–350.
13. Q. Tang and J. Hu, "Diversity of acoustic streaming in a rectangular acoustofluidic field," *Ultrasonics*, 58, pp. 27–34, 2015.
14. O. V. Abramov, *High–Intensity Ultrasonics*, (Gordon and Breach Science Publishers, Singapore, 1998), pp. 124–139.
15. A. P. Lee and B. J. Reedy, "Temperature modulation in semiconductor gas sensing," *Sens. Actuators B Chem.*, 60(1), pp. 35–42, 1999.
16. F. Xu and H. P. Ho, "Light-activated metal oxide gas sensors: a review," *Micromachines*, 8(11), p. 222, 2017.
17. H. Huang, P. Liu, Q. Ma, Z. Tang, M. Wang and J. Hu, "Airborne ultrasound catalyzed saltwater Al/Mg-air flow batteries," *Energy*, 270, 126991, 2023.

6 Ultrasonic Driving of Microfluid and Its Applications

In this chapter, principles of ultrasonic driving of microfluid are explained first in Section 6.1. Then applications of ultrasonic driving of microfluid in metal-air batteries (MABs) are demonstrated in Sections 6.2 and 6.3. After that, a cooling strategy for small solid heat sources based on micro acoustic streaming eddies is demonstrated in Section 6.4. This chapter demonstrates that the acoustic streaming is capable of improving the discharge performance of a metal-air battery and of cooling a hotspot on a hard surface, and the ultrasonic capillary effect can also effectively improve the performance of a metal-air battery.

6.1 PRINCIPLES OF ULTRASONIC DRIVING OF MICROFLUID

Ultrasound can drive microfluid in two ways [1]. One is to utilize the acoustic streaming, and another is to employ the so-called ultrasonic capillary effect.

The acoustic streaming is caused by the spatial gradient F_j of the Reynolds stress [2], which is

$$F_j = -\partial(\overline{\rho_0 u_i u_j})/\partial x_i, \qquad (6.1)$$

where u_i is the vibration velocities in the acoustic field, the bar signifies the mean value over one time period, ρ_0 is the fluid density in the undisturbed state, and repeated suffixes i and j represent the orthogonal directions such as the x and y in a two-dimensional (2D) acoustic field. Thus, the acoustic streaming is a second-order physical effect of sound waves. The acoustic streaming field may be computed by Navier-Stokes (N-S) equations. For the steady acoustic streaming field, it can be computed by the following simplified N-S equations [2]:

$$\rho_0(\bar{u}_i \, \partial \bar{u}_j / \partial x_i) = F_j - \partial \bar{p}_2 / \partial x_j + \eta \nabla^2 \bar{u}_j, \qquad (6.2)$$

where $\bar{u}_i$ is acoustic streaming velocity and $\bar{p}_2$ is the time average of the second order or mean pressure. The mean pressure $\bar{p}_2$ is calculated by [3]

$$\bar{p}_2 = \frac{1}{2\rho_0 c_0^2} \frac{B}{A} \langle p_1^2 \rangle, \qquad (6.3)$$

where < > represents the time average over one time period and $\dfrac{B}{A}$ is the nonlinear parameter of the acoustic medium, depending on the acoustic medium and

DOI: 10.1201/9781003404705-6

temperature (B/A = 5 for water at 25°C). The acoustic streaming also satisfies the continuity equation

$$\rho_0 \, \partial \overline{u}_i / \partial x_i = 0. \tag{6.4}$$

The acoustic streaming field exists in the form of vortices and has been applied in ultrasonic nano manipulations, ultrasonic cleaning, micro mixing, biological sample disruption, etc. [1, 4].

The ultrasonic capillary effect refers to the capillary action enhanced by ultrasound, which is supposed to result from the decrease of intermolecular force in the sonicated liquid [1, 4]. In the early research, the ultrasonic capillary effect was thought to be caused by the acoustic cavitation [4]. But later experiments and analyses indicate that the acoustic cavitation is not a necessary condition for the generation of ultrasonic capillary effect. This is because the related experiments show that the ultrasonic capillary effect can also happen even if the ultrasonic strength is lower than the cavitation threshold [1]. The ultrasonic capillary effect already has very good applications in various acoustofluidic devices.

The mechanism of ultrasonic capillary effect may be explained by the following model. In the negative half-cycles of sound pressure, the intermolecular forces in liquid decrease due to an increase of the distance among the liquid molecules. In the positive half-cycles of sound pressure, the intermolecular forces in liquid do not increase so much due to the repulsive action among the nuclei. Thus, the intermolecular force in sonicated liquid decreases by the sound pressure.

6.2 SALTWATER AL-AIR FLOW BATTERIES WITH ULTRASONICALLY DRIVEN ELECTROLYTE

As an application example of the ultrasonic capillary effect, the saltwater Al-air flow battery (AAFB) in which the electrolyte is circulated by the ultrasonic capillary action (rather than a mechanical pump) and the reaction chamber is agitated by ultrasonic vibration [5], is demonstrated in this section. A traveling ultrasonic wave in the electrolyte flow system causes the capillary flow and agitation. The percentage increase of the peak power density (PPD) (relative to that with static electrolyte) can be up to about 7.5 times of that with the electrolyte flow driven by a mechanical pump, under the same electrolyte flow rate and concentration (3.3 ml min^{-1} and 3 M NaCl). The optimal PPD, which can be achieved by optimizing the reaction chamber thickness, electrolyte concentration and ultrasonic vibration velocity is 43.88 mW cm^{-2}. In addition, analyses based on experimental results show that the energy gain of a series/parallel battery system formed by multiple identical cells can be larger than one, if the number of cells in the system is large enough. This example illustrates that the acoustofluidic method only improves the discharge performance of the saltwater AAFB effectively, but also greatly decreases the energy consumption, weight and volume of the electrolyte driving unit in the AAFB.

6.2.1 INTRODUCTION

With concerns for fossil energy consumption and environmental pollution, the development of renewable clean energy such as solar, wind and tidal power is becoming more and more important. However, these renewable clean energy sources are unstable, which brings great challenges to real-world applications. Therefore, it is necessary to develop stable and reliable energy conversion and storage systems. Among various energy storage systems, batteries have huge advantages due to their flexibility and stability. Since 1990s, lithium-ion batteries (LIBs) have been widely used in portable electronic devices and electric vehicles (EVs) due to their high energy density, low self-discharge rate and long cycle life. Nevertheless, LIBs have disadvantages such as high cost, safety problems, etc. Therefore, it is necessary to develop new battery energy storage technologies.

Metal-air batteries (MABs) are regarded as a type of clean and sustainable energy storage device of next-generation due to their high theoretical energy density, good safety, low cost and eco-friendliness. An MAB includes the metal anode, electrolyte and air cathode. Oxygen serving as the reactive material for the cathode comes from air in the working environment. Hence, the weight of an MAB can be greatly reduced, and its energy density can be very high. In recent years, various MABs with different metal anode materials have been developed, such as Li-air battery, Zn-air battery, Al-air battery and Mg-air battery, etc.

The theoretical energy density of the above MABs is much higher than that of LIBs. This also makes them capable of serving as energy sources for EVs. In these MAB systems, Li-air batteries (LABs) have the highest theoretical energy density (11.6 kW·h·kg^{-1}). However, the lithium dendrites not only cause serious safety and stability issues during recharging, but also lead to low coulombic efficiency of the battery. Although Al-air batteries (AABs) are usually electrochemically non-rechargeable, they are safe, lightweight, stable, cost-effective, environmentally friendly and easy to source the metal material (Al). Meanwhile, the theoretical energy density of AABs (8.1 kW·h·kg^{-1}) is competitive, compared to Zn-air batteries (1.3 kW·h·kg^{-1}) and Mg-air batteries (6.8 kW·h·kg^{-1}). Apart from these merits, the Al-air battery can be mechanically charged, and it takes only a few minutes to replace a new Al electrode. Therefore, AABs have received widespread attention and have been developed rapidly over the recent years.

Despite the above listed advantages, AABs are still facing some scientific and technical challenges, such as reduced energy efficiency of aluminum anode due to self-corrosion, short battery shelf life due to self-discharge and finite electrolyte, blocking of the cathode electrode holes due to the deposition of reaction by-products, and limited practical power density and energy efficiency due to the limited conductivity of electrolyte. To data, researchers have put great efforts to improve the performance of AABs, which includes the development of aluminum alloy as anode, gel electrolyte or corrosion inhibitor, and redox reaction catalysts with high catalytic activity for air cathode, etc. Among them, AAFBs are a competitive and emerging technique due to their capability of continuously refreshing the electrolyte, which can remove the reaction by-products and improve the energy efficiency. Thus, AAFBs have outstanding electrochemical performance such as excellent discharge performance and long cycle life.

The electrolyte flow methods in AAFBs can be divided into two categories, that is, external pump-assisted driving (such as peristaltic or syringe pumps) and paper-based capillary effect. The external pump-assisted electrolyte driving can not only drive the electrolyte flow fast and stably, but also significantly improve the discharge performance. However, the external pump-assisted electrolyte flow system is only suitable for large AAFBs system due to its bulky and complex structure. To achieve miniaturization and simplification of AAFBs, a paper-based microfluidic Al-air battery is developed and investigated, in which the electrolyte flow is driven via capillary action in the microchannel of the cellulose paper. Nevertheless, due to the limited energy storage and short battery-shelf life, the discharge performance of this microfluidics-based AAFB is quite weak.

In this work, the author's team has proposed a high-performance saltwater Al-air battery via ultrasonically driving electrolyte flow with decent battery discharge performance. Ultrasonic capillary action is utilized in the AAFB to circulate the electrolyte, while the ultrasonic wave transmitted into the reaction chamber is used to agitate the electrolyte to promote the oxidation-reduction reaction rate (ORR). Figure 6.1 shows a schematic of the acoustofluidic saltwater Al-air battery.

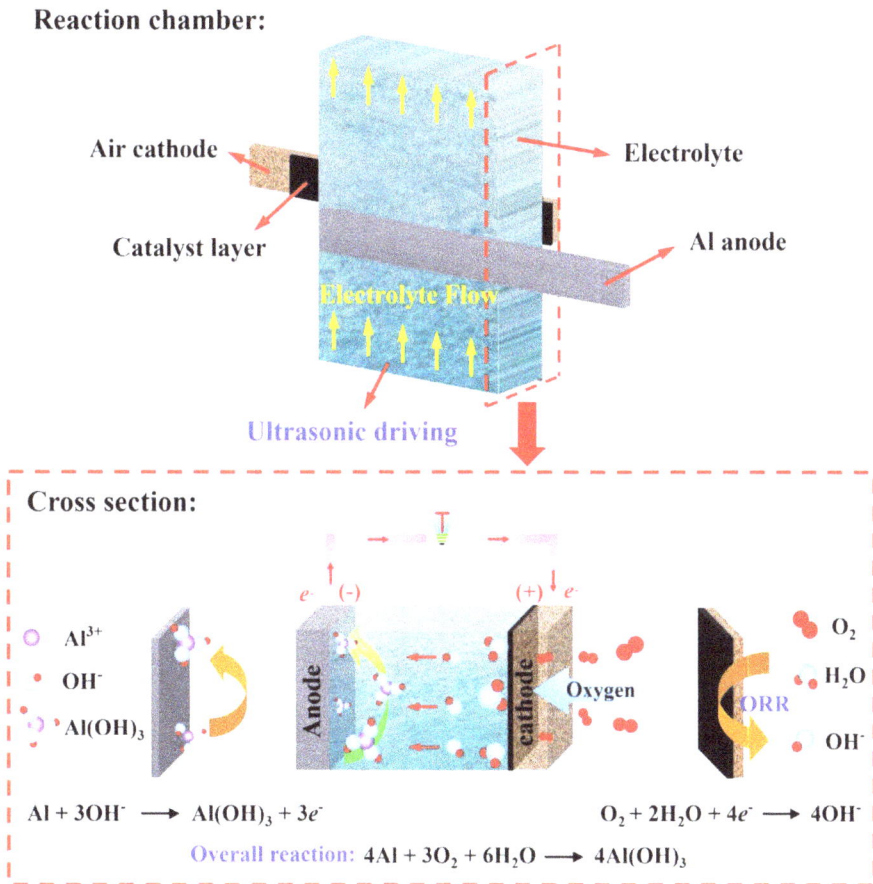

FIGURE 6.1 Schematic of the acoustofluidic saltwater Al-air battery. Reproduced from Ref. 5 with permission from Elsevier.

6.2.2 EXPERIMENTAL AND COMPUTATIONAL METHODS

6.2.2.1 Battery Design and Fabrication

Figure 6.2(a) shows the structure of the acoustofluidic saltwater Al-air battery proposed and investigated in this work, which is mainly composed of an ultrasonic transducer (HNC-8SH-3840N, Hainertec Co., Ltd.), a 3-mm-wall-thick polymethyl methacrylate (PMMA) electrolyte tank, a capillary glass tube and an Al-air battery cell. Resonance frequency of the ultrasonic transducer is 40 kHz.

As shown in Fig. 6.2(b), the battery has a sandwich structure, which is composed of two PMMA plates cut by a five-axis vertical machining center (DMU 60 monoBLOCK, DMG), and the middle one in the sandwich structure is a polyvinyl chloride (PVC) plate cut by a laser precision machining system (ProtoLaser U3, LPKF). To form an electrochemical reaction chamber, a rectangular channel is cut out in the middle plate, with dimensions of 18 mm × 5 mm. The shape of the middle plate is square (30 mm × 30 mm) with a thickness of 3 mm, unless otherwise

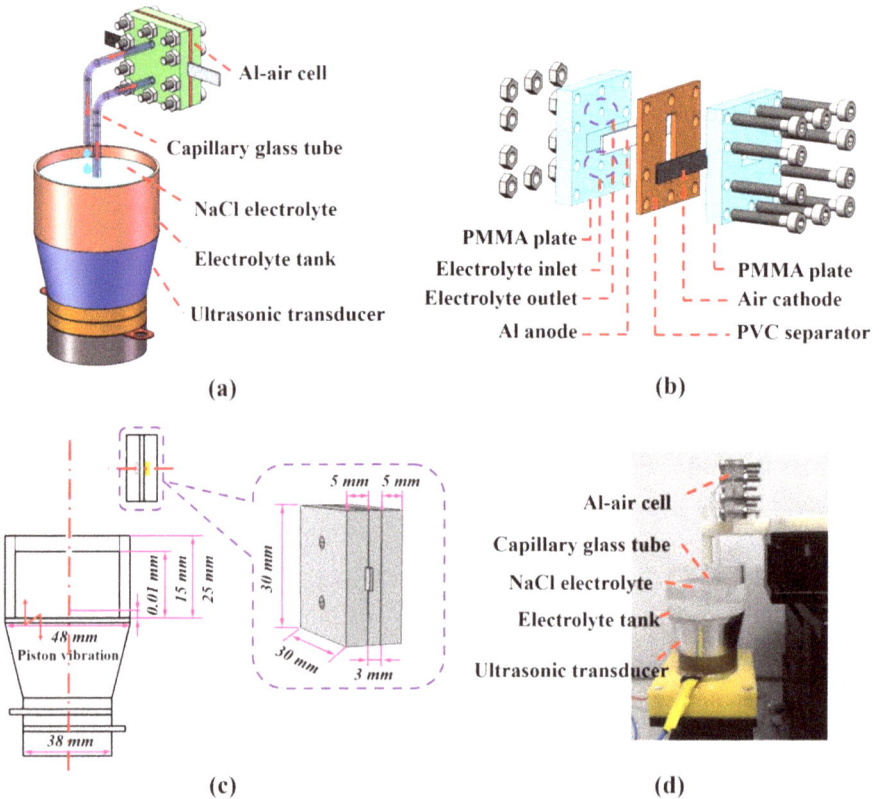

FIGURE 6.2   (a) Structure diagram of the acoustofluidic saltwater Al-air battery. (b) Exploded view of the battery structure. (c) Structure and size of the battery. (d) Photograph. Schematic of the acoustofluidic saltwater Al-air battery. Reproduced from Ref. 5 with permission from Elsevier.

specified. To dispose the two electrodes, a rectangular cavity is cut out in each PMMA plate. The PMMA plates have identical dimensions of 30 mm × 30 mm × 5 mm, and each electrode reaction area is 0.25 cm^2 (0.5 cm × 0.5 cm). An inlet and outlet are disposed in the anode PMMA plate for the flow of electrolyte. Inside the cathode PMMA plate, there is an air-breathing window (1 cm × 0.3 cm) for oxygen transport.

As shown in Fig. 6.2(c), the electrolyte tank has a cylinder shape, and its height, inner and outer diameters are 25, 42 and 48 mm, respectively. The electrolyte tank's bottom is bonded to the radiation surface of the ultrasonic transducer. The height of the NaCl solutions in the electrolyte tank is 15 mm, and the capillary glass tube is installed perpendicularly to the surface of the vibrating surface, in which the gap between the end of the capillary glass tube and the vibrating surface is 0.01 mm. The outer and inner diameters of the capillary glass tube are 3.0 mm and 1.5 mm, respectively. When the vibration velocity of the transducer is large enough, the electrolyte can be driven into the capillary glass tube by the ultrasonic capillary effect.

6.2.2.2   Material and Chemicals

Saltwater used as the electrolyte was prepared with deionized water produced by a deionized water preparation system (CJL-20-15, Ultrapure Water Co., Ltd.) and analytical pure sodium chloride (≥99.8%, Sinopharm Chemical Reagent Co., Ltd.). In the experiment, the aqueous NaCl solution electrolyte had several concentrations including 1, 2, 3, 4 and 5M. Pure aluminum with a purity of 99.99% was used as the battery anode. The cathode was a gas diffusion layer composed of a catalyst layer and carbon paper (HCP120 Shanghai Hesen Electric Co., Ltd.). The catalyst suspension was prepared by mixing catalyst powder (40 wt.% Pt/C, Johnson Matthey), Nafion solution (5 wt.%, 45 ± 3 mg cm^{-3}, DuPont) and ethanol-deionized water solvent (1:1 volume ratio of ethanol to deionized water). The mass ratio of catalyst powder, Nafion solution and ethanol-deionized water solvents was 1:10:40. After about 1 h sonication (UH600, Shanghai OuHor Machinery Equipment Co., Ltd.), the catalyst suspension was sprayed onto the carbon paper with an overall Pt/C load of 2 mg cm^{-2}. Then the carbon paper electrode was dried at 90°C for 6 h in a drying oven (DHG500-00, Supo Instrument Co., Ltd.). After the carbon electrode was dried, the copper foil used as the current collector was bonded onto a carbon paper with conductive silver paste to link to the external circuit.

6.2.2.3   Electrochemical Testing

The electrochemical measurement was carried out by an electrochemical workstation (CHI 760E, Shanghai Chenhua Instrument Co., Ltd.) at room temperature (298 ± 2 K). Before the measurement data were collected, the acoustofluidic saltwater Al-air battery was activated by carrying out multiple cyclic polarization tests until the peak discharge power density became stable, in order to remove the Al_2O_3 film on the anode. The polarization curves were obtained by the potentiostatic measurement method, in which the testing voltage value ranged from 0.8 to 0 V with a step of 0.1 V. Each measurement lasted for 60 s, and the stable value in the second-half 30 s was averaged to obtain the current value of the battery. During the measurement, the single electrode potential data were also obtained, using a reference electrode (Ag/AgCl, Shanghai Chenhua Instrument Co., Ltd.). The cathode was connected to the reference electrode through a Fluke digital multimeter to obtain the single electrode potential. In addition, in order to obtain the impedance of the battery, the electrochemical impedance spectroscopy (EIS) measurement method was used with a testing frequency range from 100 kHz to 0.1 Hz, with the A.C. (alternative current) amplitude of 5 mV at 0.3 V (voltage at peak power density). The power density (PD) was calculated by $PD = P/A$, where P is the battery power and A is the active electrode area. To investigate the discharge and open-circuit voltage stability, we conducted discharge experiments under different current densities (1, 10, 20, 40, 60, 80, 100 and 120 mA cm^{-2}), in which the discharge duration in each test was 600 s and the total running time of open-circuit voltage stability test was 4800 s. Every experiment was repeated five times.

6.2.2.4   Finite Element Analyses

To clarify the reason for the ultrasonic effect on the discharge process, the ultrasonic field in the battery was investigation by the finite element analysis (FEA) with COMSOL Multiphysics software R5.6. In the FEA, a three-dimensional (3D) FE model was established for the experimental battery. In the FEA, the working frequency and driving voltage used were 40.5 kHz and 36.8 V$_{0-p}$ (zero-peak value), respectively, and the maximum element size in the fluid domain was 1.2 mm, which was about 3.2% of the ultrasonic wavelength λ ($\lambda = c/f \approx 37.04$ mm, here $c = 1500$ m/s is the sound speed in saltwater and f is the working frequency).

6.2.3   RESULTS AND DISCUSSION

The discharge performance of the saltwater Al-air battery with the electrolyte flow driven by ultrasound was measured. For comparison, those of the same saltwater Al-air battery with static electrolyte and electrolyte flow driven by a mechanical pump (LM60A-YZ1515X-6B, Nanjing Runze Fluid Control Equipment Co., Ltd.) were also measured. The mechanical pump has an apparent size of 206 mm × 143 mm × 199 mm and a weight of 3.5 kg, whereas the ultrasonic transducer used in the work has a top diameter of 48 mm, bottom diameter of 38 mm, height of 45 mm, and weight of 0.3 kg, respectively. Thus, the ultrasonic transducer is much smaller and lighter than the mechanical pump. Unless otherwise specified, the experimental conditions were as follows: The working point of the ultrasonic system was at the resonance frequency of 40.5 kHz; the working current and voltage were 70 mA (rms) and 26 V (rms), respectively; the flow rate of ultrasonic capillary action and mechanical pumping was 3.3 ml min^{-1}. In addition, all comparison was made with identical electrolyte flow rate values.

6.2.3.1   Performance Improvement by the Ultrasonic Driving Method

The polarization and power density curves of the saltwater Al-air battery with ultrasonically driven electrolyte flow were measured, using the potentiostatic measurement method with 3M NaCl electrolytes and 3.3 ml min^{-1} electrolyte flow rate, and the result is shown in Fig. 6.3(a). It was found that compared with the static electrolyte, both short-circuit current density and PPD were increased by the ultrasonic-driven electrolyte flow. For the mechanical pumping method, the short-circuit current density slightly increases from 121.08 mA cm^{-2} (static electrolyte) to 127.88 mA cm^{-2}, which is 5.62% higher than that of the static electrolyte method. The PPD slightly increases from 18.68 mW cm^{-2}

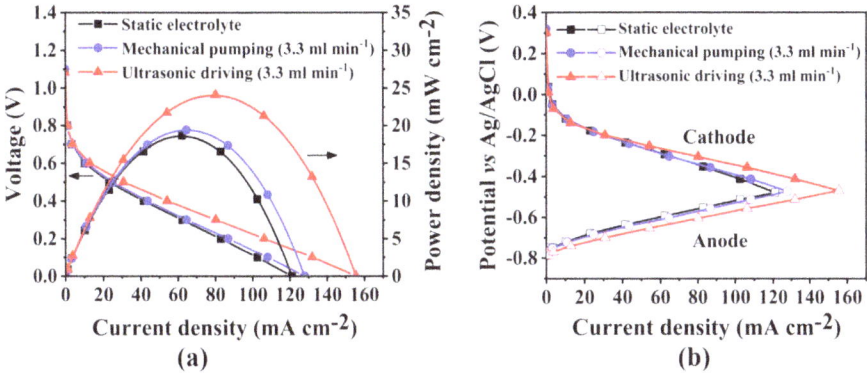

FIGURE 6.3   The electrochemical performance of the saltwater Al-air batteries under different driving methods of electrolyte flow under 3M NaCl and 3.3 ml min⁻¹ flow rate. (a) Polarization and power density curves. (b) Single electrode potential curves. Reproduced from Ref. 5 with permission from Elsevier.

(the static electrolyte) to 19.40 mW cm⁻² with an increase percentage of 3.85%. Nevertheless, the short-circuit current density and PPD of the acoustofluidic saltwater Al-air battery system dramatically increase to 155.60 mA cm⁻² and 24.08 mW cm⁻², with percentage increase of 28.51% and 28.91% (relative to the static electrolyte battery), respectively. Thus, the percentage increase of PPD of the acoustofluidic saltwater Al-air battery (relative to that with static electrolyte) can be up to 7.5 times of that with mechanical pumping, under the same electrolyte flow rate of 3.3 ml min⁻¹.

From Fig. 6.3(b), it is observed that both anode and cathode sides have less overpotential loss in the acoustofluidic saltwater Al-air battery than that with mechanical pumping. This performance results from the following two processes which are induced by ultrasound. The first one is the circulation of electrolytes inside and outside the reaction chamber, resulting from the ultrasonic capillary action. Another one is the agitation induced by the ultrasonic field in the reaction chamber, which can also weaken the concentration and electric double polarization.

To further understand the mechanisms of ultrasound-induced flow and discharge performance enhancement, distributions of sound pressure amplitude and phase in the capillary glass tube and reaction chamber were computed. Figure 6.4(a) shows a meshed FE model of the acoustofluidic saltwater Al-air battery, and Fig. 6.4(b, c) shows the computed distributions of sound pressure amplitude and its phase in the battery system, respectively.

Figure 6.4(b) confirms that there is ultrasonic field in the capillary tube and reaction chamber. Figure 6.4(c) shows that the sound pressure phase in the capillary glass tube decreases as the distance from the tube's inlet increases, which indicates that there is a traveling wave transmitting from the capillary tube's inlet to the reaction chamber. The traveling wave in the capillary glass tube causes the NaCl solution to flow upward in its length direction due to the decreased inter molecular force.

The EIS was also measured under different driving methods with the same electrolyte flow of 3.3 ml min⁻¹ and under 3M NaCl concentration, and the results are shown in Fig. 6.5. The equivalent circuit used to fit the EIS plots is shown in the

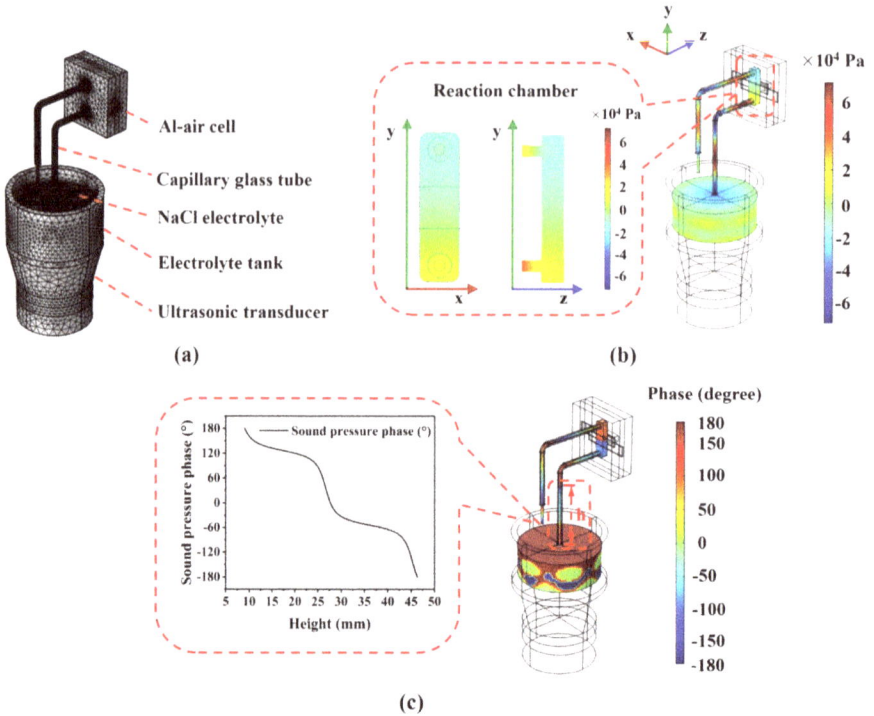

FIGURE 6.4 Results of FEM analyses. (a) A meshed 3D FEM computation model. (b) The distribution of sound pressure in capillary glass tube and battery. (c) The distribution of sound pressure phase in capillary glass tube and battery. Reproduced from Ref. 5 with permission from Elsevier.

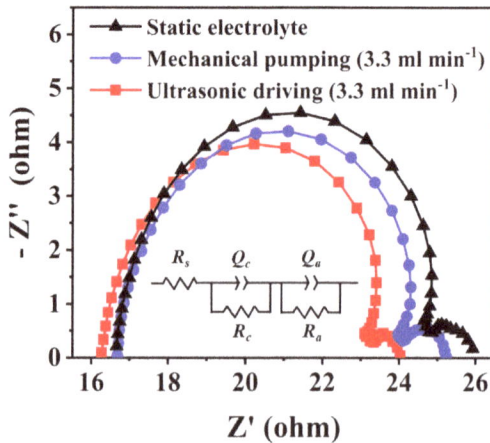

FIGURE 6.5 Electrochemical impedance spectroscopy under different driving methods of electrolyte flow at 3 M NaCl concentration and 3.3 ml min^{-1} flow rate. Reproduced from Ref. 5 with permission from Elsevier.

TABLE 6.1

Fitting Parameters of the Equivalent Circuit for the Measured EIS at 3 M NaCl Concentration and Flow Rate of 3.3 ml min⁻¹

Element	R_s ($\Omega \cdot cm^2$)	R_c ($\Omega \cdot cm^2$)	R_a ($\Omega \cdot cm^2$)	Q_1 (S·sn)	Q_2 (S·sn)
Static electrolyte	16.87	8.05	0.67	7.04×10^{-6}	0.0013
Mechanical pumping	16.94	7.31	0.65	5.43×10^{-6}	0.0010
Ultrasonic driving	16.56	6.80	0.43	4.81×10^{-6}	0.0012

inset, and the equivalent parameters are listed in Table 6.1. In the equivalent circuits, R_s represents the solution resistance, R_c and R_a represent the polarization resistance values of the cathode and anode, respectively, which include the charge transfer and diffuse layer resistances, and other resistances. Meanwhile, Q_c and Q_a are the constant phase elements of the cathode and anode, respectively. The EIS plot is composed of a high-frequency semicircle (the larger one) which is caused by the cathode interfacial electrochemical reactions, and a low-frequency semicircle (the smaller one) which is caused by the anode interfacial electrochemical reactions.

From Table 6.1, it is seen that the ultrasonic driving has a better effect on the decrease of polarization resistances than the mechanical pumping. This phenomenon confirms that the electrolyte flow resulting from the ultrasound is not the only cause of the ultrasound-induced performance enhancement. The difference of R_c (or R_a) between the ultrasonic driving and mechanical pumping means that there is other ultrasound-induced process(es) affecting the polarization resistances. We attribute this difference to the physical effects of ultrasound in the reaction chamber, such as the electrolyte vibration and ultrasound-induced decrease of inter-molecular force in the electrolyte.

6.2.3.2   Effect of NaCl Electrolyte Concentration

The battery performance under different NaCl electrolyte concentrations with the ultrasonic driving method was further investigated, and the results are shown in Fig. 6.6(a). It is seen that increasing the NaCl concentration from 1 to 4M, the PPD of the battery increases from 12.00 to 26.24 mW cm⁻², and the short-circuit current density increases from 98.40 to 159.88 mA cm⁻². This is because the conductivity of the electrolyte increases when the electrolyte concentration increases. However, further increasing the electrolyte concentration to 5M only slightly improves the performance of the Al-air battery. This is because a too high electrolyte concentration may hinder the transportation of the hydroxide ions (OH^-) toward the anode, which leads to a limited electrode reaction rate.

Figure 6.6(b) shows the measured single-electrode potential versus current density at different electrolyte concentrations. As the NaCl concentration increases, the conductivity of the electrolyte for OH^- becomes better, which enhances the oxidation reaction at the anodic surface and thus lowers the anodic potential. For the same reason, the cathodic potential increases as the NaCl concentration increases from 1 to 2 M. When the NaCl concentration increases from 2 to 5 M, the cathodic potential has little change. This is because the electrical double-layer polarization becomes stronger as

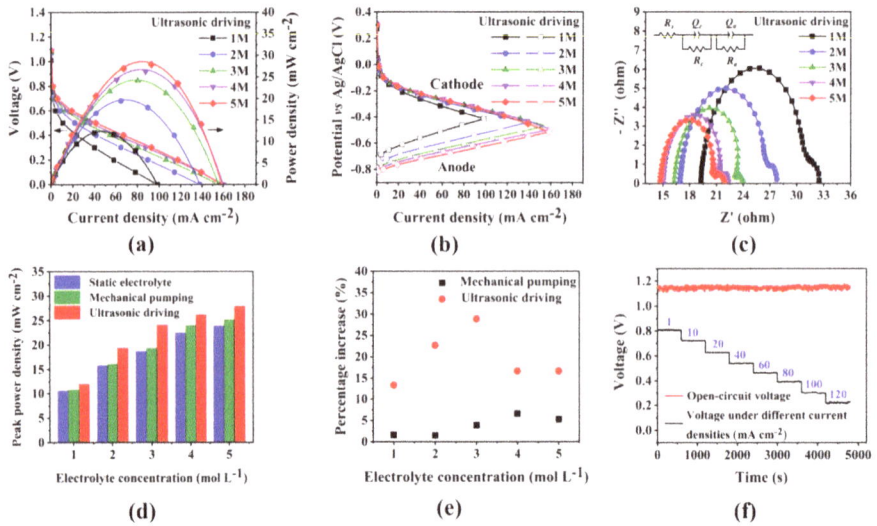

FIGURE 6.6 The electrochemical performance of the acoustofluidic saltwater Al-air battery under different NaCl concentrations and 3.3 ml min⁻¹ flow rate. (a) Polarization and power density curves. (b) Single electrode potential curves. (c) Electrochemical impedance spectroscopy (EIS). (d) The maximum power density versus electrolyte concentration under different driving methods of electrolyte flow. (e) Percentage increase of the peak power density relative the static electrolyte battery. (f) Open-circuit voltage curve and long-term discharge under different current densities with 5M NaCl electrolyte concentration. Reproduced from Ref. 5 with permission from Elsevier.

the NaCl concentration increases. The change of single-electrode potentials with the current density in Fig. 6.6(b) is consistent with the voltage change in Fig. 6.6(a). The measured EIS in Fig. 6.6(c) suggests that the EIS of the acoustofluidic saltwater Al-air battery is affected by the electrolyte concentration, just like the traditional MAFBs.

Figure 6.6(d) summarizes the PPD versus NaCl electrolyte concentrations under different electrolyte driving methods, which indicates that the ultrasonic method brings in the best PPD of the three methods at the electrolyte concentrations. Figure 6.6(e) shows the percentage increase of the PPD of the flow batteries with the ultrasonic and mechanical pump methods (relative to the battery with static electrolyte). Under the mechanical pumping method, it only increases by 1.60%, 1.52%, 3.85%, 6.58% and 5.32% for the saltwater concentrations, respectively. While under the ultrasonic driving method, it drastically increases by 13.40%, 22.73%, 28.91%, 16.73% and 16.70%, respectively.

Furthermore, the stability of discharge performance was tested with the multi-step current discharge method at 5M NaCl electrolyte solution under the ultrasonic driving method, and the result is shown in Fig. 6.6(f). The open-circuit voltage was measured in a duration of 4800 s, and the voltage at each current density was measured separately in a duration of 600 s. It shows that the open-circuit voltage is kept at 1.15 V with a standard deviation of 0.01 V, which indicates that the open-circuit voltage is quite stable. When the current density is less than 120 mA cm⁻², the voltage has little fluctuation, which also indicates that the acoustofluidic Al-air battery system has a high discharge stability.

6.2.3.3 Effect of Vibration Velocity

Vibration velocity of the radiation surface of the ultrasonic transducer was used to represent the strength of ultrasonic vibration of the battery system. The polarization and power density curves of the battery at different vibration velocities were measured with a 5 M NaCl electrolyte, and the result is shown in Fig. 6.7(a). It is observed that when the vibration velocity increases from 38.68 to 55.35 mm s^{-1}, the PPD increases from 24.80 to 27.88 mw cm^{-2}.

Figure 6.7(b) shows the measured flow rate and computed sound pressure (in the middle of the reaction chamber) versus the vibration velocity. It is seen that the ultrasonic capillary action starts to appear when the vibration velocity or sound pressure excels a critical value and then increases as the vibration velocity increases. The former phenomenon is due to the flow resistance of the inner wall of the capillary glass tube, and the latter one is because the increase of sound pressure causes the increase of electrolyte flow rate in the battery system, enhancing the ultrasonic agitation effect in the reaction chamber. In the experiments, the maximum flow rate was 3.3 ml min^{-1}, owing to the performance limitation of the experimental transducer-electrolyte tank subsystem. Also, it is worth noting that the battery exhibits the best discharge performance (27.88 mW cm^{-2}) at the vibration velocity of 55.35 mm s^{-1} (3.3 ml min^{-1}), which is the upper limit of the vibration velocity that the experimental transducer can provide.

A comparison of the energy consumption and PPD enhancement with the ultrasonic driving and mechanical pumping methods is listed in Table 6.2.

Here, the energy consumption refers to the input electric power of the ultrasonic transducer and mechanical pump, and the PPD enhancement is defined as

$$PPD \ enhancement = \left(PPD_{u(m)} - PPD_0\right)/PPD_0, \qquad (6.5)$$

where $PPD_{u(m)}$ is the peak power density of the saltwater Al-air battery under ultrasonic driving method or mechanical pumping method, and PPD_0 is the peak power density of the saltwater Al-air battery with static electrolyte. It is seen that to increase

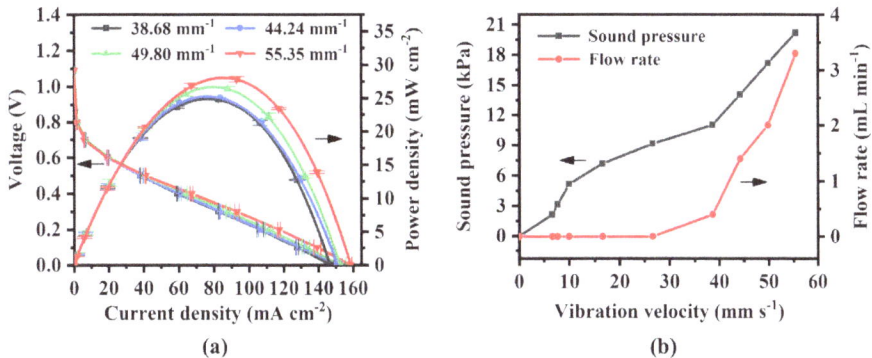

FIGURE 6.7 (a) Polarization and power density curves of the saltwater Al-air batteries under different ultrasonic vibration velocities. (b) Sound pressure in the middle of the reaction chamber and flow rate of ultrasonic capillary action versus ultrasonic vibration velocity. Reproduced from Ref. 5 with permission from Elsevier.

TABLE 6.2

A Comparison of the Energy Consumption and Peak Power Density (PPD) Enhancement with the Ultrasonic Driving and Mechanical Pumping Methods

	Ultrasonic Driving		Mechanical Pumping	
	Energy Consumption	PPD Enhancement	Energy Consumption	PPD Enhancement
Flow Rate (ml min⁻¹)	(W)	(%)	(W)	(%)
0.4	0.75	3.81	3.12	0.96
1.4	1.06	4.98	3.60	2.97
1.9	1.41	10.51	6.00	3.47
3.3	1.82	16.70	6.48	5.32
Weight and volume of electrolyte driving unit	0.3 kg; 59393 mm³		3.5 kg; 5862142 mm³	

the electrolyte flow rate from 0.4 to 3.3 ml min⁻¹, the energy consumption of the ultrasonic driving and mechanical pumping systems must increase from 0.75 to 1.82 W and from 3.12 to 6.48 W, respectively, whereas the PPD enhancement increases from 3.81% to 16.70% and from 0.96% to 5.32%, respectively. Therefore, at the same electrolyte flow rate, the acoustofluidic saltwater AAFB has a better PPD enhancement and lower energy consumption than the conventional AAFB. Moreover, the volume and weight of the ultrasonic transducer are 59393 mm³ and 0.3 kg, respectively, which are about 1.01% and 8.57% of those of the mechanical pump, respectively.

6.2.3.4   Effect of the Reaction Chamber Thickness

The reaction chamber thickness (distance between the two electrodes) can also affect the performance of the acoustofluidic saltwater Al-air battery, as indicated by the experimental results in Fig. 6.8(a). In the experiments, PVC plates with different thicknesses of 0.1 mm, 0.5 mm, 1.0 mm, 2.0 mm and 3.0 mm were used as the electrode spacing

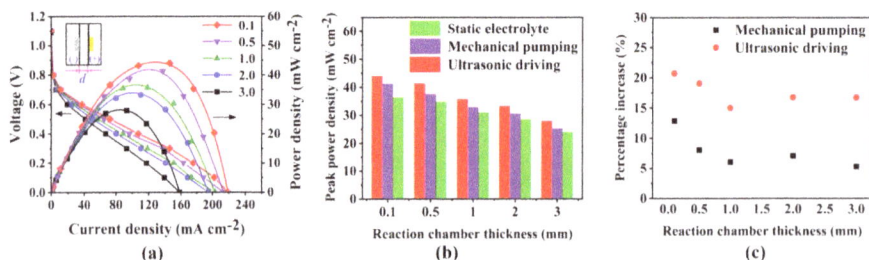

FIGURE 6.8   (a) Polarization and power density curves of the acoustofluidic saltwater Al-air battery for various reaction chamber thicknesses at 5M NaCl concentration and 3.3 ml min⁻¹ flow rate. (b) The peak power density of the battery versus reaction chamber thickness for different driving methods of electrolyte flow. (c) The percentage increase of the peak power density under The acoustofluidic and mechanical pumping methods with a reaction chamber thickness of 0.1 mm. Reproduced from Ref. 5 with permission from Elsevier.

separator. It shows that both PPD and short-circuit current density increase with the decrease of the thickness. As the thickness decreases from 3.0 to 0.1 mm, the PPD increases from 27.88 to 43.88 mW cm^{-2}, that is, an increase of 57.39%. This is because as the reaction chamber thickness decreases, the electrolyte resistance between the two electrodes decreases and sound pressure in the reaction chamber increases, both of which facilitate the mass transportation. Figure 6.8(b) summarizes the PPD at different reaction chamber thicknesses under the ultrasonic driving, mechanical pumping and static electrolyte methods with a constant NaCl electrolyte concentration of 5M. At the thickness of 0.1 mm, the PPD under different driving methods of electrolyte decreases in the following order: Ultrasonically driving > mechanical pumping > static electrolyte. The percentage increase of the PPD (relative to the static electrolyte) was calculated for the ultrasonic driving and mechanical pumping methods, and the results are shown in Fig. 6.8(c). With the mechanical pumping method, it is increased by 12.87%, 8.07%, 6.10%, 7.09% and 5.32% for the thickness values, respectively, whereas with the ultrasonic driving method, it is increased by 20.68%, 19.03%, 15.00%, 16.74% and 16.70%, respectively. As a conclusion, a smaller reaction chamber thickness is beneficial to increasing the PPD. According to the experiment results, the PPD (=43.88 mW cm^{-2}) is the maximum when the reaction chamber thickness is 0.1 mm.

6.2.3.5   Acoustofluidic Saltwater Al-Air Battery Stacks

A single acoustofluidic Al-air cell not only has a limited output voltage (V) and peak power (mW), which is not sufficient for most of electronic devices in practical application, but also limits the utilization rate of ultrasonic energy. To investigate the performance of a battery pack formed by multiple acoustofluidic saltwater Al-air cells, an eight-series-cell battery pack was fabricated, as shown in Fig. 6.9(a). It is a series of eight identical acoustofluidic saltwater Al-air cells, and the capillary rubber tubes for each cell are also in series.

The experiments were carried out under the conditions of 5 M NaCl, 3.3 ml min^{-1} flow rate and 0.1 mm reaction chamber thickness. From Fig. 6.9(b, c), it is seen that the peak power and open circuit voltage of battery packs are almost proportional to the cell number. The peak power of the eight-series-cell battery pack is 89.61 mW, that is, about eight times of that of the single cell (10.97 mW). From Fig. 6.9(d), it is known that the open-circuit voltage of the battery pack is approximately proportional to the cell number, that is, it is 8.9 V at a cell number of 8 (open-circuit voltage of a single cell is 1.1 V). Figure 6.9(e) shows a comparison of the peak powers of the eight-series-cell battery pack under the ultrasonic driving and mechanical pumping, with the same working conditions. It is seen that the peak power of the battery pack under the ultrasonic driving is higher than that under the mechanical pumping by 1.2–2.1 times.

Although the ultrasound can increase the peak output power of the Al-air saltwater battery, it consumes electrical energy. Thus, whether the ultrasound-induced output power increase can be more than the consumed electric power by the ultrasonic transducer was investigated. To investigate this problem quantitatively, energy gain (EG) of an Al-air saltwater battery system is defined as follows:

$$EG = (P_u - P_0)/P_{in}, \qquad (6.6)$$

where P_u is the peak output power of the saltwater Al-air battery under the ultrasonic driving method, P_0 is the peak output power of the saltwater Al-air battery with

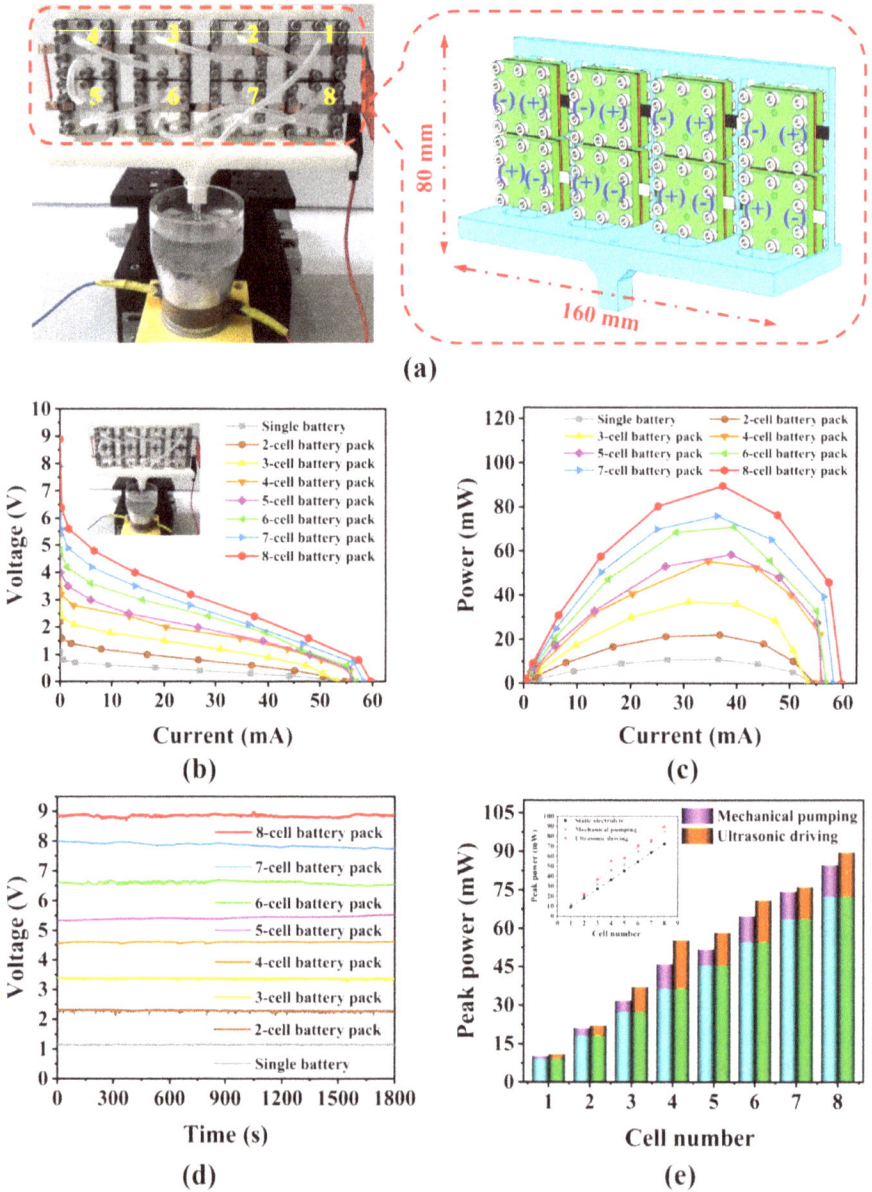

FIGURE 6.9 (a) An eight-series-cell battery pack formed by eight acoustofluidic saltwater Al-air cells connected in series. (b) Polarization (*j-v*) and (c) power curves for eight-series-cell battery packs of acoustofluidic saltwater Al-air batteries with 5 M NaCl, 3.3 ml min⁻¹ and 0.1 mm reaction chamber thickness. (d) Open-circuit voltage of the cell packs with different cell numbers. (e) The peak power with different cell numbers. Reproduced from Ref. 5 with permission from Elsevier.

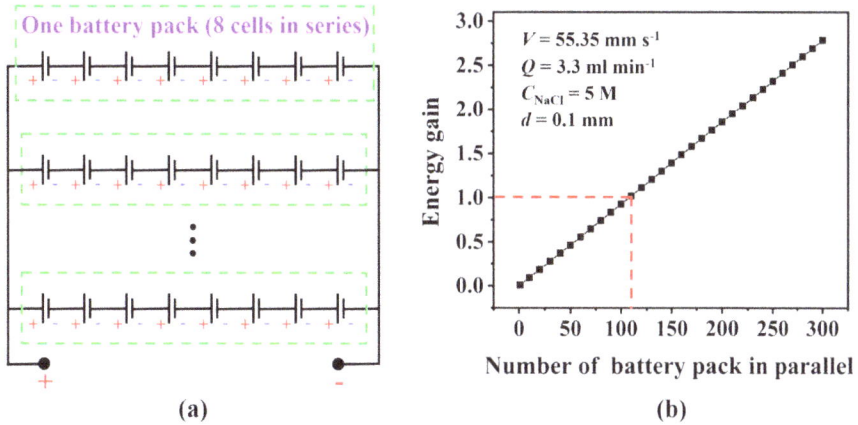

FIGURE 6.10   (a) Electric circuit of the Al-air saltwater battery system to achieve an energy gain larger than one. (b) Energy gain versus number of the identical eight-series-cell battery packs in parallel. Reproduced from Ref. 5 with permission from Elsevier.

static electrolyte, and P_{in} is the input power (power consumption) of the ultrasound transducer.

According to the experimental results of the eight-series-cell battery pack, the energy gain of the battery system shown in Fig. 6.10(a), which is formed by the parallel of multiple identical eight-series-cell battery packs, was simulated. In each battery pack, the electrolyte flow system is in series. Thus, the electrolyte flow system of each battery pack only has one inlet tube which is inserted into the electrolyte tank, and one outlet tube which is suspended above the electrolyte tank. The simulated energy gain versus the number of battery packs N in parallel is shown in Fig. 6.10(b). It shows that the energy gain becomes larger one when N is more than 108, which indicates the ultrasound-induced increase of peak output power can be more than the electric power consumed by the ultrasonic transducer if the number of battery packs in parallel is more than 108. As the surface area of the electrolyte in the tank and the capillary glass tube is 1385 mm² and 7 mm², respectively, the total cross-sectional area of the inlet capillary glass tubes of the 108 battery packs only shares 54.6% of the electrolyte surface area.

If the number of battery cells in each series branch is more than eight, then the line of energy gain in Fig. 6.10(b) will have a larger slope, and the number of battery packs in parallel N will be less, to keep $EG > 1$. Moreover, the diameter of the electrolyte tank may be designed to be larger than that of the radiation surface of the ultrasonic transducer. In this case, more inlet capillary glass tubes can be inserted into the electrolyte tank, which means a much larger energy gain at an identical number of the battery packs.

In addition, one may use a piezoelectric disk working in the thickness vibration mode to excite the electrolyte tank, which can decrease the power consumption and increase the energy gain. With this ultrasound excitation method, the volume of the ultrasonic vibration excitation system may be reduced greatly, and the volume power density of the system will be increased greatly.

6.2.4 CONCLUSIONS

In this work, an acoustofluidic saltwater Al-air battery, in which ultrasonic vibration is employed to drive the electrolyte flow and generate ultrasonic field in the reaction chamber, to enhance the battery performance, has been proposed and investigated. The percentage increase of PPD relative to that with static electrolyte is up to 7.5 times of that with the mechanical pumping method, under 3.3 ml min^{-1} electrolyte flow rate and 3M NaCl solution. The optimal PPD of the acoustofluidic saltwater Al-air battery, which is achieved by optimizing the reaction chamber thickness, electrolyte concentration and ultrasonic vibration velocity, is 43.88 mW cm^{-2}. This work provides a new and effective method to increase the peak power of a given saltwater AAFB while keeping a good stability of its discharge performance. The energy consumption, weight and volume of electrolyte driving unit of the acoustofluidic saltwater AAFB are much less than those of the conventional AAFB. The energy gain of a series/parallel battery system formed by multiple identical cells can be larger than one, if the number of cells in the system is large enough. The acoustofluidic saltwater Al-air battery system proposed in the work may have potential applications in the energy storage systems that have strict requirements in eco-friendliness, cost-effectiveness, safety and discharge process stability.

6.3 AN ULTRASONICALLY EXCITED BUTTON ZINC-AIR BATTERY

6.3.1 INTRODUCTION

In recent years, zinc-air batteries have attracted renewed interest as a possible energy storage solution due to their high energy density (1086 W·h kg^{-1} in theory), environmental friendliness, abundance of Zn, low cost and excellent safety. Therefore, a series of researches on high-performance zinc-air batteries have been carried out over the last few years. Most of these researches have been focused on the development of high-performance electrocatalysts, such as the adoption of precious metals and alloys, hetero-atoms doped carbonaceous materials, perovskite materials and transition metal carbides and oxides. Efforts also have been made in the optimization of zinc species allocation in the zinc electrode, which includes utilization of high-surface-area porous structures and 3D conductive host materials, etc. In addition to the above methods, the electrolyte, separator and other components of zinc-air batteries have also been studied to improve the performance.

Up to the present, with the development of zinc-air battery performance, primary zinc-air batteries have been commercially implemented for some applications such as hearing-aid and outdoor lighting. However, despite the primary commercialization and promising applications, development of zinc-air batteries has been impeded by some limitations associated with the metal and air electrodes, and the current of existing zinc-air batteries is far from satisfactory. The urgent issues that need to be solved in the development of high-performance zinc-air batteries include slow oxygen reduction reaction (ORR), non-uniform zinc dissolution and deposition, extended cycle life required, electrolyte reaction with CO_2 to form carbonates, etc.

In this work, a method to enhance the output power and increase the rating capacity of a commercialized button zinc-air battery is proposed and demonstrated [6]. In the structural design of the device, a piezoelectric ring is used as a vibration source, which is bonded onto the top of the battery to excite an ultrasonic field inside the battery. The experimental result shows that the output power of the battery increases from 22.2 to 32.6 mW, and the rating capacity increases from 330.56 to 396.89 mA·h when the vibration velocity is 52.8 mm/s at a working frequency of 161.2 kHz. Also, the impedance of the battery is measured to analyze the working principle. The analyses show that the ultrasonic effects in electrolyte solution, such as acoustic microstreaming vortices and viscosity decrease, contribute to the discharge performance improvement through enhancing the uniformity of OH⁻ distribution and decreasing the resistance of mass transfer.

6.3.2 RESULTS AND DISCUSSION

The front and back views of the button zinc-air battery are shown in Fig. 6.11(a, b), respectively. A piezoelectric ring of 6 mm (inner diameter) × 12 mm (outer diameter) × 8 mm (thickness) is bonded onto the top of the outer surface of air electrode of the battery (A675/PR44, Fujian Nanping Nanfu Battery Co., Ltd.). There are six air inlets in the cathode shell. One is at the center of the top surface, and the rest are distributed on a circumference with a diameter of 5 mm. The inner diameter of the piezoelectric

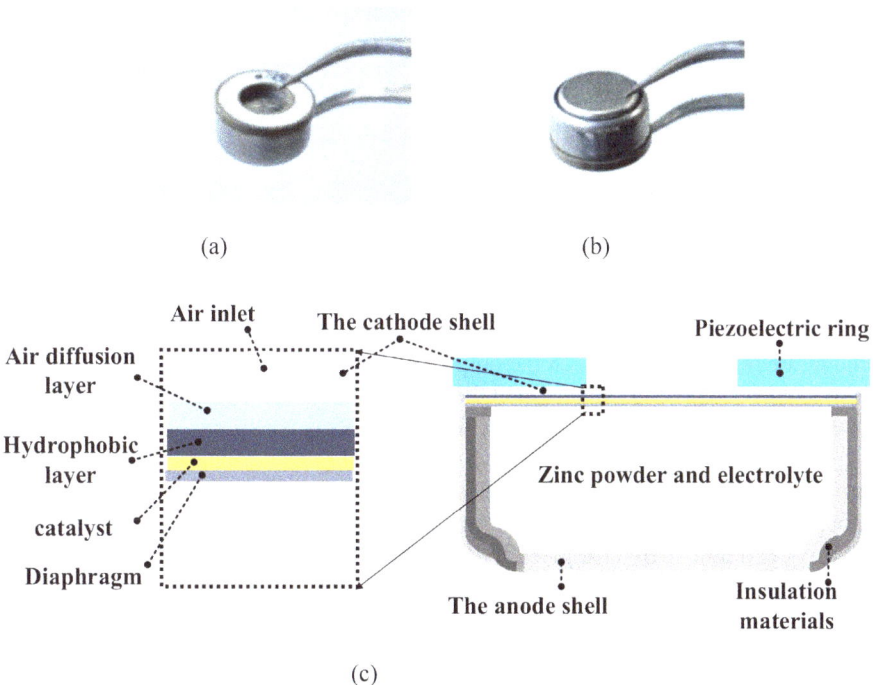

FIGURE 6.11 Photos and schematic of the button zinc-air battery with a piezoelectric ring. (a) Front view. (b) Back view. (c) Structural schematic.

ring is chosen to be 6 mm, to ensure the air inlets are not covered by the piezoelectric ring. During the operation procedure, the piezoelectric ring acts as a vibration source to mechanically excite the battery, by applying an AC (alternative current) voltage from a power amplifier (HFVP-83A, Nanjing Foneng Science and Technology Industrial Co., Ltd.), which receives a sinusoidal signal generated by a function generator (ARG 3022B, Tektronix). The abovementioned function generator and power amplifier are used only for the convenience of our experiments, and the electrical driving system for the piezoelectric ring can be miniaturized in practical applications. The driving voltage and current of the piezoelectric ring, and the phase difference between them are measured and monitored by an oscilloscope (DPO2014, Tektronix). Vibration distribution of the piezoelectric ring are measured by a 3D laser Doppler vibrometer (PSV-500, POLYTEC).

Figure 6.11(c) shows the detailed structure of the button zinc-air battery with a piezoelectric ring. Teflon® (polytetrafluoroethylene) air diffusion layer, Teflon® hydrophobic layer, wire mesh framework with catalytic materials layer (nickel-plated wire mesh) and diaphragm (polyamide) are stacked beneath the end surface of the cathode shell. The inner cavity of the battery, which is formed by the cathode and anode shells, is full of electrolytes mixed with zinc powders. Insulating material is filled in between the anode and cathode shells to prevent a short-circuit. The electrode reactions are expressed as follows:

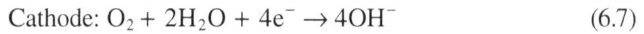

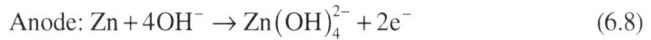

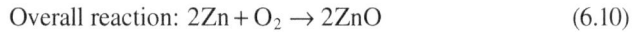

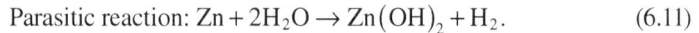

$$\text{Cathode: } O_2 + 2H_2O + 4e^- \rightarrow 4OH^- \tag{6.7}$$

$$\text{Anode: } Zn + 4OH^- \rightarrow Zn(OH)_4^{2-} + 2e^- \tag{6.8}$$

$$Zn(OH)_4^{2-} \rightarrow ZnO + H_2O + 2OH^- \tag{6.9}$$

$$\text{Overall reaction: } 2Zn + O_2 \rightarrow 2ZnO \tag{6.10}$$

$$\text{Parasitic reaction: } Zn + 2H_2O \rightarrow Zn(OH)_2 + H_2. \tag{6.11}$$

Figure 6.12 shows the measured characteristics of the button zinc-air battery ultrasonically excited under different vibration velocities at 161.2 kHz. Figure 6.12(a) shows the relationship between the output voltage and current of the battery, and Fig. 6.12(b) shows the relationship between the output power and current. An adjustable resistor is used as the load of the battery, and the output voltage can be measured by a digital multimeter (FLUKE 8845A) under different excitation conditions. The abbreviation v_{p-p} in the figure is the averaged peak-peak value of the out-of-plane vibration velocity on the upper surface of the piezoelectric ring, which is measured by the 3D laser Doppler vibrometer. For comparison, the results when there is no ultrasonic excitation ($v_{p-p} = 0$ mm/s) are also given in the figure. It is seen that for a given discharge current, the ultrasonic excitation can increase the output voltage and power. Also, the effect becomes more obvious with the increase of the discharge current. When the vibration velocity v_{p-p} is 52.8 mm/s, the maximum output power of the battery is 32.6 mW, which is 48% higher than that without ultrasonic excitation (22.2 mW).

In order to elucidate the mechanism of discharge performance improvement by ultrasonic vibration, the AC impedance of the battery is measured by an electrochemical

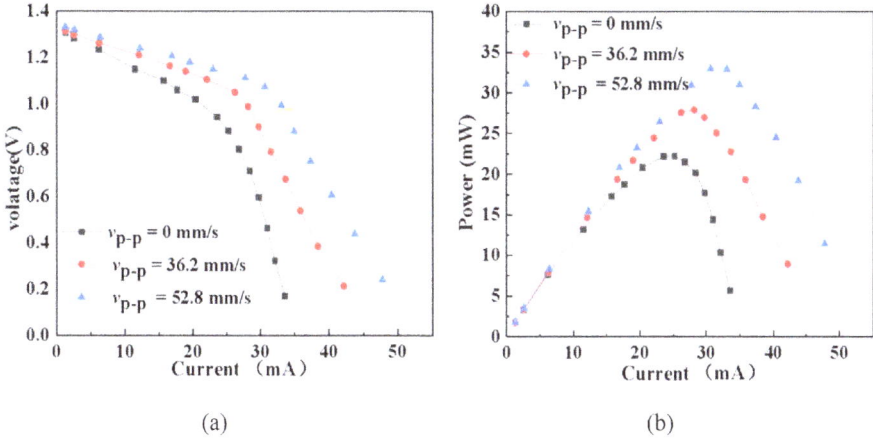

(a)　　　　　　　　　　　　　　(b)

FIGURE 6.12　Characteristics of the button zinc-air battery ultrasonically excited under different vibration velocities at 161.2 kHz. (a) Voltage versus current. (b) Power versus current. The abbreviation v_{p-p} is the averaged peak-peak value of the out-of-plane vibration velocity on the upper surface of the piezoelectric ring.

workstation (CHI760E, Shanghai Chen Hua Co., Ltd.) for the cases with different vibration velocities and without sonication, and the results are shown in Fig. 6.13.

Figure 6.13(a) shows the Nyquist plots under different vibration velocities, indicating that both the resistance and capacitive reactance decrease with the increase of ultrasonic vibration velocity. The inset in Fig. 6.13(a) shows that the ohmic resistance R_s, which is the intercept with the x-axis (at high frequency) and consists of the electrolyte solution resistance and contact resistance, has a slight decrease with the increase of ultrasonic vibration velocity. Figure 6.13(b) shows the Bode plots under different vibration velocities. It is well known that the impedance in the low-frequency range in a Bode plot is mainly caused by the concentration polarization and that in the middle-frequency range is mainly caused by the electric double-layer effect. Therefore, it indicates that the

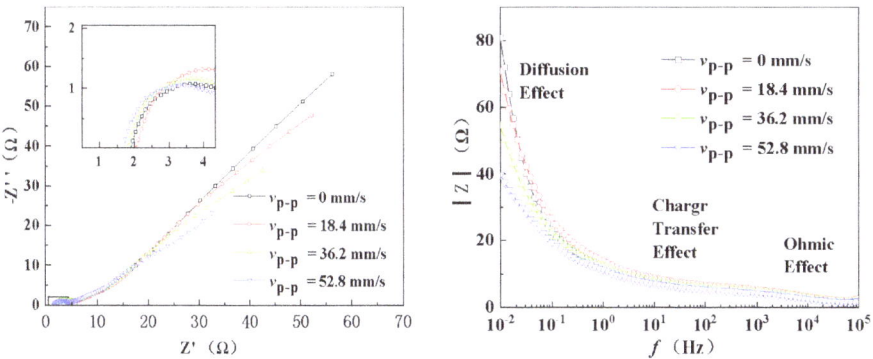

FIGURE 6.13　Measured AC impedance of the ultrasonically excited zinc-air battery under different vibration velocities at 161.2 kHz. (a) Nyquist plots; (b) Bode plots.

ultrasonic excitation affects both of the concentration polarization and electric double layer polarization, and the former's effect is larger.

It is known that the diffusion layer thickness δ in the forced convection is related to the diffusivity D, electrolyte kinetic viscosity μ_f and flow velocity u_0 of electrolyte [7]. The diffusion layer thickness δ decreases as the electrolyte flow velocity u_0 increases, and the electrolyte kinetic viscosity μ_f decreases [8]. Considering the propagation of ultrasonic field into the inner cavity of the battery, the weakening of both concentration polarization and electric double-layer effect is attributed to the acoustic microstreaming vortices and viscosity decrease of the electrolyte solution induced by the ultrasound. The acoustic microstreaming vortices can flush the ions in the electrolyte solution away from the electric double layer and weaken the electric double layer effect. The slight decrease of the ohmic resistance due to ultrasonic excitation is attributed to the viscosity decrease induced by ultrasound.

To confirm the existence of the ultrasonic field and acoustic streaming vortices in the battery, the vibration of the piezoelectric ring was measured, and the acoustofluidic field in the inner cavity of the battery was computed by the commercial FEM software COMSOL Multiphysics (version 5.4). Unless otherwise specified, all material parameters used in the simulation process are listed in Table 6.3. Vibration distributions on the top surface of the piezoelectric ring were measured at 161 kHz and 18 V_{p-p} by the 3D laser Doppler vibrometer, and the measured in-plane and out-of-plane vibration distributions are shown in Fig. 6.14(a, b). It is seen that the piezoelectric ring mainly vibrates in the radial direction at the working frequency, which can excite the

TABLE 6.3

Material Property Parameters Used in the FEM Simulation

Material Parameters	Value
KOH (at temperature $T = 26°C$)	
Dynamic viscosity (Pa·s)	2.4
Density (kg/m³)	1305
Diffusion coefficient (m²/s)	3.79×10^{-9}
Velocity of sound (m/s)	1988
Lead Zirconate Titanate (PZT-4)	
Density (kg/m³)	7500
Piezoelectric coefficient e_{33} (C/m²)	15.1
Relative dielectric constant ε_{r33}	663.2
Structural steel	
Density (kg/m³)	7850
Young's modulus (Pa)	200×10^9
Poisson's ratio	0.3
Nylon	
Density (kg/m³)	1150
Young's modulus (Pa)	2×10^9
Poisson's ratio	0.4

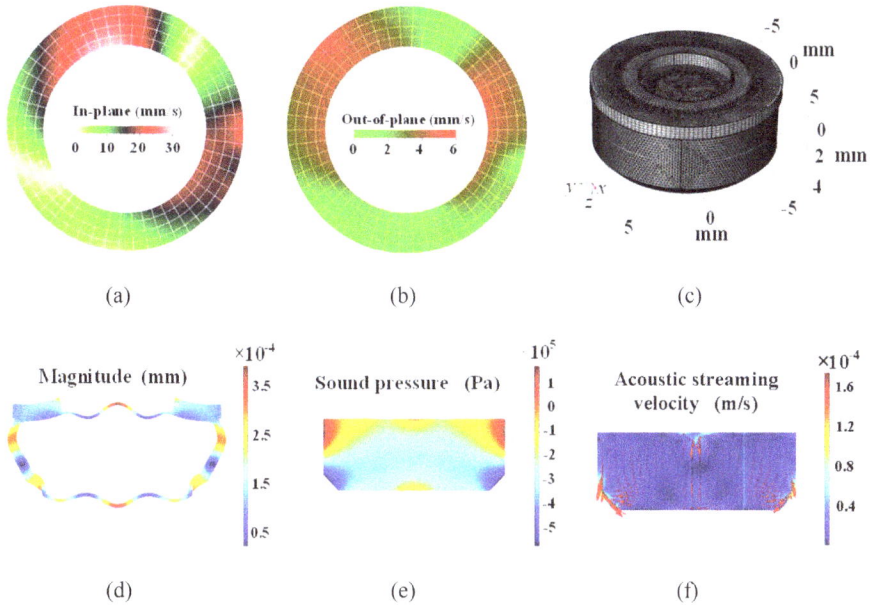

FIGURE 6.14 Measured vibration distribution at 161 kHz and 18 V$_{p-p}$ and computed acoustofluidic field in the battery. (a) The in-plane vibration of the piezoelectric ring's top surface. (b) The out-of-plane vibration of the piezoelectric ring's top surface. (c) A 3D grid meshed model of the battery. (d) Vibration displacement pattern of the battery. (e) Sound pressure distribution in the central plane of the inner cavity. (f) Velocity distribution of acoustic streaming in the central plane of the inner cavity.

anode shell to vibrate flexurally. The vibration distribution of the battery and ultrasonic field in the electrolyte can be clarified in the following simulation results.

Figure 6.14(c) shows a 3D meshed model of the button zinc-air battery, and Fig. 6.14(d) shows the computed vibration displacement of the whole shell, in which the color represents the total vibration displacement magnitude. It is seen that the shell of the battery vibrates flexurally with the ultrasonic excitation of the piezoelectric ring.

Figure 6.14(e) shows the sound pressure distribution in the central plane of the inner cavity of the battery. During the negative half periods of sound pressure, the distance among the molecules in the electrolyte increases, causing the cohesive forces among the molecules to become small. During the positive half periods of sound pressure, the molecules are compressed. However, the cohesive forces cannot increase too much because the molecules repel each other when they are too close. Thus, during one vibration period, the averaged cohesive forces among the liquid molecules become weaker under sonication [1]. As a result, the equivalent viscosity of the electrolyte decreases, which contributes to the weakening of the concentration polarization and double layer polarization. Figure 6.14(f) shows the computed acoustic streaming field in the central plane of the inner cavity. Mixing function of the acoustic streaming vortices improves the uniformity of the ion distribution, reducing the concentration polarization and double layer polarization. Therefore, the internal resistance of the button zinc-air battery reduces, and the output power increases.

More characteristics of the ultrasonically excited button zinc-air battery have been investigated. The influence of acoustic streaming on the battery under constant current discharge was measured by the battery test system BT2018A (Hubei Lanbo New Energy Equipment Co., Ltd.), and the results are shown in Fig. 6.15(a). In the experiments, the vibration velocity was 52.8 mm/s, the working frequency was 161.2 kHz, and the discharge current was 20.3 mA. The capacity C is calculated by the following equation:

$$C = I \cdot t, \qquad (6.12)$$

where I and t are the discharge current and time, respectively. Figure 6.15(a) shows that the working time, discharge voltage and capacity of the battery with the ultrasonic assistance are 19 h, 0.85–1.15 V and 396.89 mA·h, respectively, and those without the ultrasonic assistance are 16.3 h, 0.75–1.0 V and 330.56 mA·h, respectively. Thus,

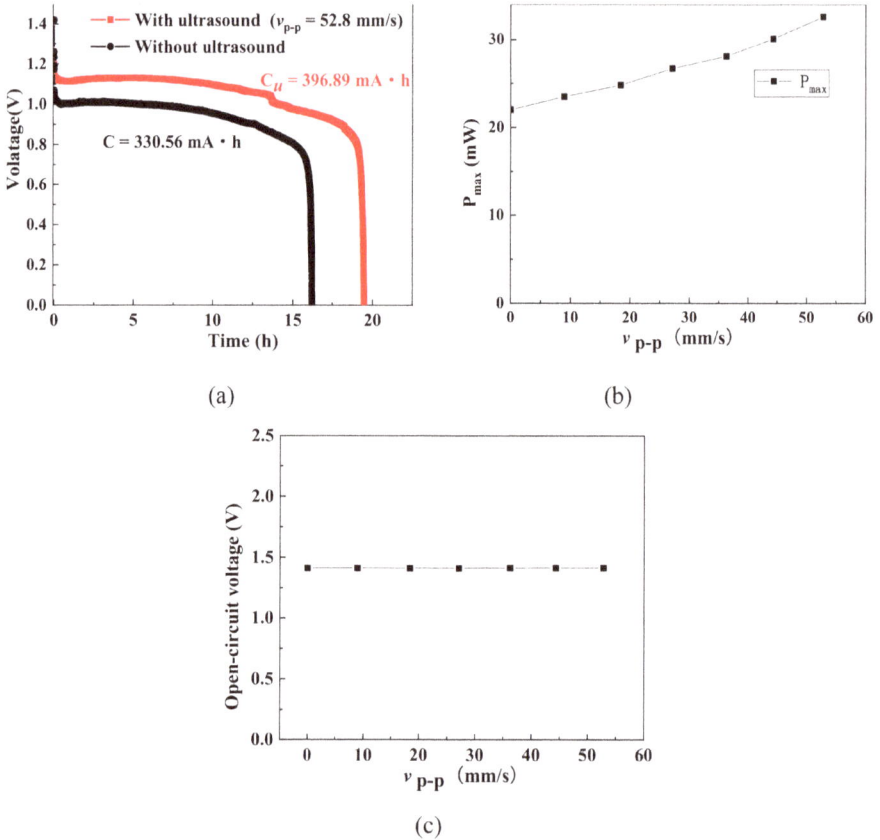

(a)

(b)

(c)

FIGURE 6.15   Measured characteristics of the ultrasonically excited button zinc-air battery. (a) Discharge profile under a constant discharging current (20.3 mA) with and without sonication. (b) Maximum power versus ultrasonic vibration velocity. (c) Open-circuit voltage versus ultrasonic vibration velocity.

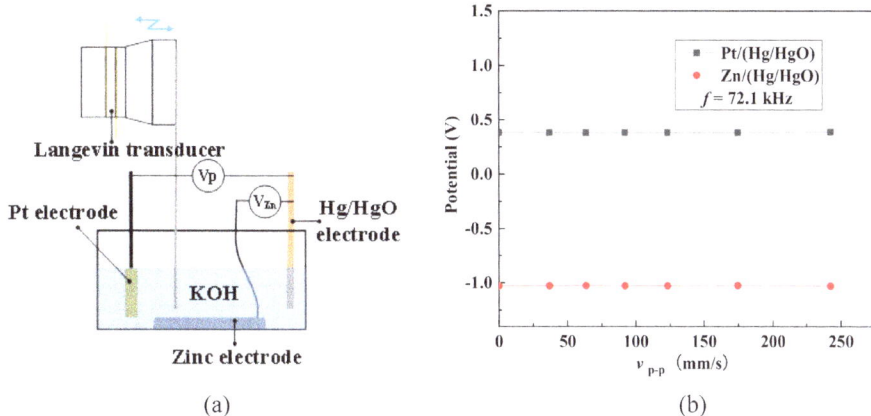

(a)   (b)

FIGURE 6.16 Independency of single-electrode potentials of the zinc-air battery on the electrolyte's ultrasonic vibration. (a) Schematic diagram of testing device. (b) Measured Pt/(Hg/HgO) and zinc/(Hg/HgO) potentials.

with the assistance of acoustofluidic field, the discharge time is extended for about 3 h and the capacity is increased by 20%.

Figure 6.15(b) shows the measured maximum output power of the battery versus vibration velocity at the upper surface of the piezoelectric ring ($r = 9$ mm). It is seen that the maximum output power increases with the increase of ultrasonic vibration velocity. When the vibration velocity is 52.8 mm/s, the maximum output power is 32.6 mW, which is 48% higher than the maximum output power of the battery without sonication (22.2 mW). Figure 6.15(c) shows the measured open-circuit voltage of the button zinc-air battery under different vibration velocities at 161.2 kHz. It can be seen that the open-circuit voltage remains unchanged at around 1.41 V, which means that the ultrasonic vibration has little influence on the open-circuit voltage.

In order to analyze and explain the abovementioned influence in Fig. 6.15(c), a zinc-air galvanic cell shown in Fig. 6.16(a) is designed to investigate the ultrasonic effect on the single-electrode potential. The zinc, platinum, Hg/HgO electrodes and 30 wt% KOH are used in the experimental setup. A copper plate is bonded to the radiation surface of a Langevin transducer to transmit vibration into the electrolyte to generate the acoustic streaming vortices. Figure 6.16(b) shows the measured potentials of the zinc and platinum electrodes under different vibration velocities of the Langevin transducer's radiation surface. It is seen that the increase of the vibration velocity has little influence on the anode and cathode potentials, which indicates that the ultrasonic vibration does not affect the single-electrode potential. This is because in the open circuit condition, the redox reaction of the battery does not occur and the ultrasonic vibration has little influence on the temperature and ion concentration on the electrodes.

6.3.3 CONCLUSION

A method to enhance the output power of a zinc-air battery by ultrasonic vibration is proposed and demonstrated. A piezoelectric ring bonded onto a commercialized button zinc-air battery is used to produce an ultrasonic field inside the battery. There is a 48% increase in the output power of the battery when the vibration velocity is 52.8 mm/s at 161.2 kHz, and the ultrasonic vibration can increase the rating capacity by about 20%. Based on the measured AC impedance and computed acoustofluidic field, the discharge performance improvement is attributed to the acoustic microstreaming vortices and viscosity decrease of the electrolyte solution, caused by ultrasonic excitation. This principle can also be applied to a battery pack to enhance the discharge performance and rating capacity by using a suitable ultrasonic vibration system.

6.4   A FOCUSED ULTRASOUND-BASED COOLING STRATEGY FOR SMALL SOLID HEAT SOURCES

6.4.1   Introduction

Controlling the temperature rise of functional devices, especially electronic devices, is significantly important to prevent performance deterioration of the whole system. In order to ensure the reliability of electronic systems, effective cooling for the key devices or hotspots has become increasingly important. Conventional cooling methods for heat sources include heat sinks, axial fans, microfluidic systems, piezoelectric fans and synthetic jets, and so on.

With the rapid increase of integration level and miniaturization degree, the size of the key devices or hotspots in electronic systems is decreasing. For effective cooling of these small-size heat sources, one may bond the heat source onto a heat sink, in order to increase the heat dissipation area. Novel and effectively cooling methods for small heat sources may not only reduce the structural complexity of cooling systems but also make the cooling systems more energy effective.

The acoustic streaming, which is the vortex flow resulting from the spatial gradient of the Reynolds stress in an ultrasonic field and has been employed for ultrasonic manipulation and mixing, can also result in a convective heat transfer. Before this work, acoustic streaming fields generated by the standing wave, traveling wave and near field have been used in cooling. In ref. [9], the acoustic streaming generated by the resonant acoustic field between a heat sink and a 30 kHz ultrasonic transducer with a flat radiation face was employed to enhance the convective heat transfer of a heat sink. At a heat sink temperature of 85°C and ultrasonic power of 0.8 W, the additional heat dissipation caused by the acoustic streaming was 5 W. In Ref. [10], the acoustic streaming generated by a 28 kHz standing wave field in between the cooled device and flexural beam was utilized to enhance the convective heat transfer. At a vibration amplitude of 25 μm, the device at 98°C was cooled down to 58°C in 5 min. Although the limit of cooling capability of acoustics streaming is still unknown, it usually has unique merits such as compact structure, no acoustic noise, no rotary part and good durability. Moreover, without a rotary part, it has merits such as no dust accumulation, no wearing, silent operation and ease of miniaturization.

Focused ultrasound concentrates acoustic energy onto a specific area. Due to the higher Reynolds stress gradient in this area, the acoustic streaming velocity generated by focused ultrasound is much higher than other kinds of ultrasound fields. Conventionally the focused ultrasound was used for ablation of tumors at its focal region. In this work [11], we proposed a focused ultrasound-based cooling method in which a small solid heat source in air located at the focal region could be effectively cooled by the acoustic streaming, and investigated the cooling characteristics. After confirming the cooling effect and analyzing the cooling principle, the effects of the ultrasound strength, working frequency and acoustic lens' eigen mode were investigated and clarified, and the cooling effect was compared with that by a commercial axial fan. Using the focused ultrasound rather than other types of ultrasonic field can significantly raise the utilization rate of acoustic energy, and thus further improve the ultrasonic cooling effect.

6.4.2 MATERIAL AND METHODS

6.4.2.1 Experimental Setup

Figure 6.17 shows a schematic of the testing system to investigate the effects and characteristics of the cooling method proposed in this work. It mainly consists of the cooling device, solid heat source, electrical driving system, single chip computer (SCM, STM32F103RCT6, China) and upper computer. The electrical driving system includes a power amplifier (NF HAS 4052, Japan), signal generator (Tektronic AFG 3022B, USA), oscilloscope (Tektronic DPO 2014, USA), current sensor (HIOKI 3272-50, Japan) and a DC (direct current) power supply (GWINSTEK GPS-3303C, China). The cooling device is located above the heat source and fixed onto the platform of an XYZ stage (LD125-LM-2, Shengling Precise Machinery Co., Ltd., China), in order to tune the relative location of the heat source and cooling device.

Figure 6.18 is a schematic of the cooling device, which consists of a commercial Langevin transducer (HNC-4AH-2560, Hainertec Co., Ltd., Suzhou China) and an acoustic lens made of aluminum. The acoustic lens has a cylindrical appearance of 30 mm diameter × 18 mm length and a hemispherical radiation face of 14 mm radius. Its radiation face was polished to be concave, and the circular upper-end face was bonded onto the radiation face of the Langevin transducer by epoxy resin adhesive. When a sinusoidal voltage of a certain frequency is applied to the transducer, the out-of-plane vibration can be generated at the radiation surface of the ultrasonic transducer, due to the converse piezoelectric effect, exciting the acoustic lens to vibrate.

FIGURE 6.17 Schematic diagram of the testing system. Reproduced from Ref. 11 with permission from Elsevier.

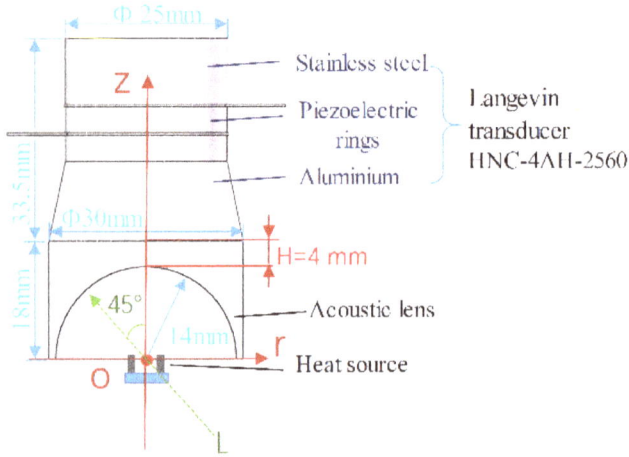

FIGURE 6.18 Schematic diagram of the cooling device. Reproduced from Ref. 11 with permission from Elsevier.

The concave shape of the acoustic lens makes acoustic energy converge onto the focal region. As the acoustic attenuation increases with the working frequency, a low-frequency ultrasonic transducer is suitable for the cooling device.

Prior to assembling the testing system, vibration characteristics of the cooling device were measured by a Laser Doppler Vibrometer (PSV-500, POLYTEC). It shows that the cooling device has two resonance frequencies which are 49.4 kHz and 61.6 kHz, respectively. Our FEM analyses indicate that the resonance at 49.4 kHz and at 61.6 kHz is caused by the half-wavelength resonance of the Langevin transducer and the flexural resonance of the acoustic lens, respectively. In the experiments, the Langevin transducer was driven by the AC (alternative current) electrical driving system. The cooling device was kept in resonance by adjusting the working frequency, and the ultrasonic vibration velocity was controlled by the driving current of the transducer.

The shape and size of the heat source used in the testing system are shown in Fig. 6.19(a). It is an ellipsoid approximately, with axial length values of 0.9 mm × 0.6 mm × 0.6 mm, and connected by lead wires to the electrodes fixed on a support platform. The structure of the heat source is shown in Fig. 6.19(b). It consists of a coil made of high-purity platinum wire with a diameter of 26 μm and metal oxide cover. When a DC current passes through the platinum wire, the temperature of the whole heat source rises rapidly, and then reaches a stable value. As the temperature of the heat source has a linear relationship with its resistance, the heat source temperature can be obtained by measuring its resistance. Figure 6.19(c) shows the temperature measurement circuit, which consists of a DC voltage source U, heat source resistor R_C and divider resistor R_L, all of which are in series. By measuring the output voltage of the divider resistor V_L, the heat source resistance R_C can be calculated by

$$R_c = R_L (U - V_L)/V_L, \qquad (6.13)$$

where $R_L = 10\ \Omega$ in this work. From the resistance of the heat source, the heat source temperature can be deduced with the resistance-temperature relationship T-Rc of the heat source (Fig. 6.19(d)), which was measured beforehand.

(a)   (b)

(c)   (d)

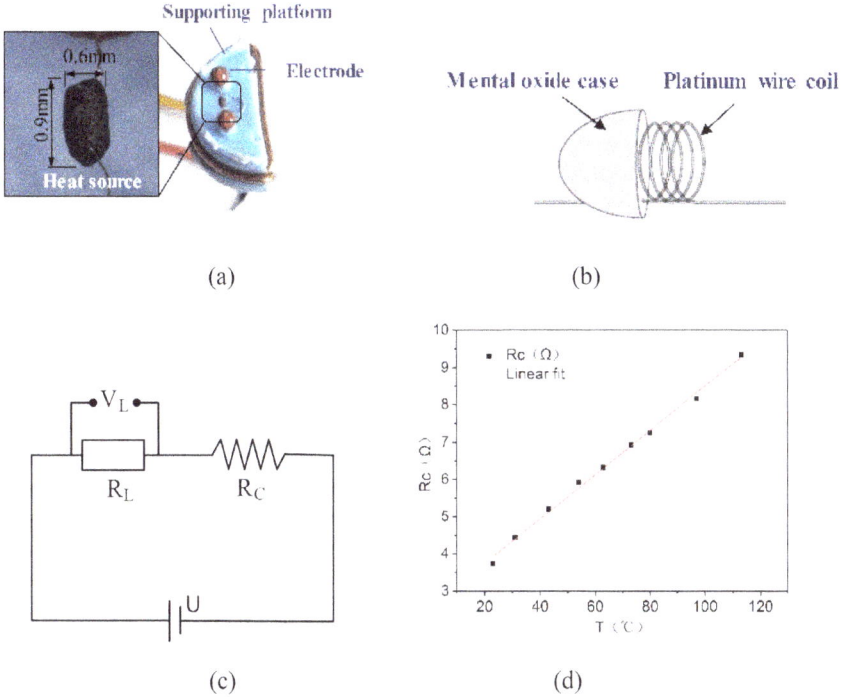

FIGURE 6.19 A small heat source for cooling experiment. (a) Photography of the heat source and its support. (b) Schematic diagram of the heat source. (c) Temperature measurement circuit. (d) Measured T-Rc correlation. Reproduced from Ref. 11 with permission from Elsevier.

In the testing system, the heat source is powered by an adjustable DC supply, and the output voltage is read out by an SCM, and recorded by a computer for further data processing. The temperature in T-Rc was measured by an infrared thermal imager (compact pro, SeekThermal), and Rc in T-Rc was calculated by Eq. (6.13) with the measured V_L.

The heat source is cooled down by three heat transfer processes, that is, the convection between the heat source surface and environment, radiation from the heat source surface and conduction through the lead wires. An estimation based on the experimental results in this work shows that the heat radiated from the surface Q_r is around 0.6%–0.7% of the total Joule heat Q, and that conducted through the Pt lead wire Q_l is around 4.6%–7.0% of Q. Thus, Q_r and Q_l were ignored for the convenience of analyses, and the convection heat transfer coefficient h was obtained from

$$h = Q/A/(T - T_a),\qquad(6.14)$$

where A is the surface area of the heat source, T is the surface temperature of the heat source and T_a is the ambient temperature. The Nusselt number Nu is

$$Nu = hd/k = Qd/A/k/(T - T_a),\qquad(6.15)$$

where k is the thermal conductivity of the air at the film temperature (the average of T and T_a), and d is the characteristic length of the heat source (the averaged value of the length of the long and short axes). In terms of the experimental conditions, unless otherwise specified, the environmental temperature T_C was 21°C ± 1°C, and the relative humidity was 65% ± 5%.

6.4.2.2   Adjustment of the Cooling Position

To find out the optimal relative position of the cooling device to the heat source, cooling effect for the heat source at different relative locations in the ultrasonic field was measured when the working frequency was 61.6 kHz and 49.4 kHz, respectively. The temperature drop of the heat source was measured by shifting the heat source along the r-axis at $z = 0$, along the z-axis at $r = 0$ and along the L-axis, as shown in Fig. 6.18. It was found that the cooling effect was the best when the center of the heat source was at the focal point ($r = 0$ and z = 0), and the cooling effect at 61.6 kHz was much better than at 49.4 kHz (about 15°C temperature drop at 61.6 kHz and 3°C at 49.4 kHz when input power of the cooling device was 1.5 W and initial temperature of the heat source is 82°C). Thus, in the subsequent experiments, the heat source was located at the focal region of the cooling device, and the working frequency of the cooling device was 61.6 kHz, unless otherwise specified.

6.4.3   COOLING EFFECT AND PRINCIPLE ANALYSIS

Figure 6.20 shows the measured temperature versus heat flux of the heat source at different power inputs of the cooling device (P_{in}) at 61.6 kHz. At the same heat flux, the stronger the ultrasound, the lower the heat source temperature due to a better convective heat transfer. For the surface temperature around 100°C, the heat flux is increased by 150% when the power input to the cooling device is 9.64 W. Also, it

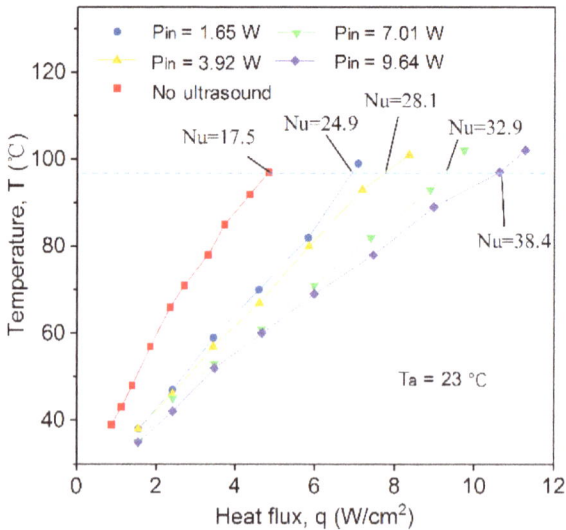

FIGURE 6.20   Measured temperature versus heat flux of the heat source at different power input of the cooling device. P_{in} denotes the power input of the cooling device. Reproduced from Ref. 11 with permission from Elsevier.

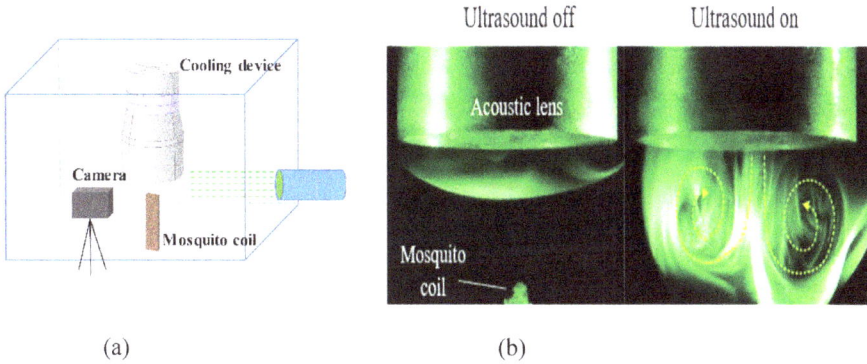

(a)   (b)

FIGURE 6.21   Flow field below the cooling device working at 61.6 kHz. (a) Flow visualization by smoke. (b) Flow pattern with ultrasound off and on. Reproduced from Ref. 11 with permission from Elsevier.

is seen that Nu increases as the power input P_{in} of the cooling device increases, and Nu is doubled when the power input is 9.64W (at the temperature of 97°C). Thus, experimental results in Fig. 6.20 indicate that the focused ultrasound can enhance the convective heat transfer.

To explain the cooling phenomenon described above, a smoke-based flow visualization method was used to observe the flow field below the acoustic lens, and the experimental setup is shown in Fig. 6.21(a). A 500 mW green laser beam generated by a laser (M-D532-50-G, Guangzhou Mingtuo Opto-Electronic Technology Co., Ltd.) was used to illuminate the observed region. A burning mosquito coil continually supplied the smoke streaming below the cooling device. A large transparent chamber was built to contain the flow field and eliminate the influence of air flow in the room. A camera (Mate10, Huawei Technologies Co., Ltd., 30fps) was located outside the chamber and used to record the smoke flow pattern. The smoke flow pattern when the cooling device worked at 61.6 kHz was observed, and the photos are given in Fig. 6.21(b). The smoke floated under the acoustic lens smoothly when there was no ultrasound. When the cooling device was turned on, the smoke flowed downwards along the central axis (z-axis) of the cooling device and then the vortices appeared on both sides of the central axis. The yellow dotted lines and arrows roughly show the smoke flow pattern and directions. The asymmetry of the two vortices was caused by a slight deviation of the smoke source position from the center axis and the asymmetry of the smolder.

The observed smoke flow pattern is supposed to be caused by the acoustic streaming eddies in the focused ultrasonic field. To confirm this, the acoustic streaming field was computed by the FEM. As the structure of the cooling system was axisymmetric, a 2D axial symmetry FEM model was built, which included the cooling device, ultrasonic field and mosquito coil, as shown in Fig. 6.22(a). The acoustics/solid-coupling and laminar flow modules of COMSOL Multiphysics software (R5.4) were used. Parameters used in the computation are listed in Table 6.4. The mosquito coil was regarded as a hard cylinder, and its radius and length were 2 mm and 15 mm, respectively. The acoustic-structure boundary condition was applied between the solid

(a)

(b)

(c)

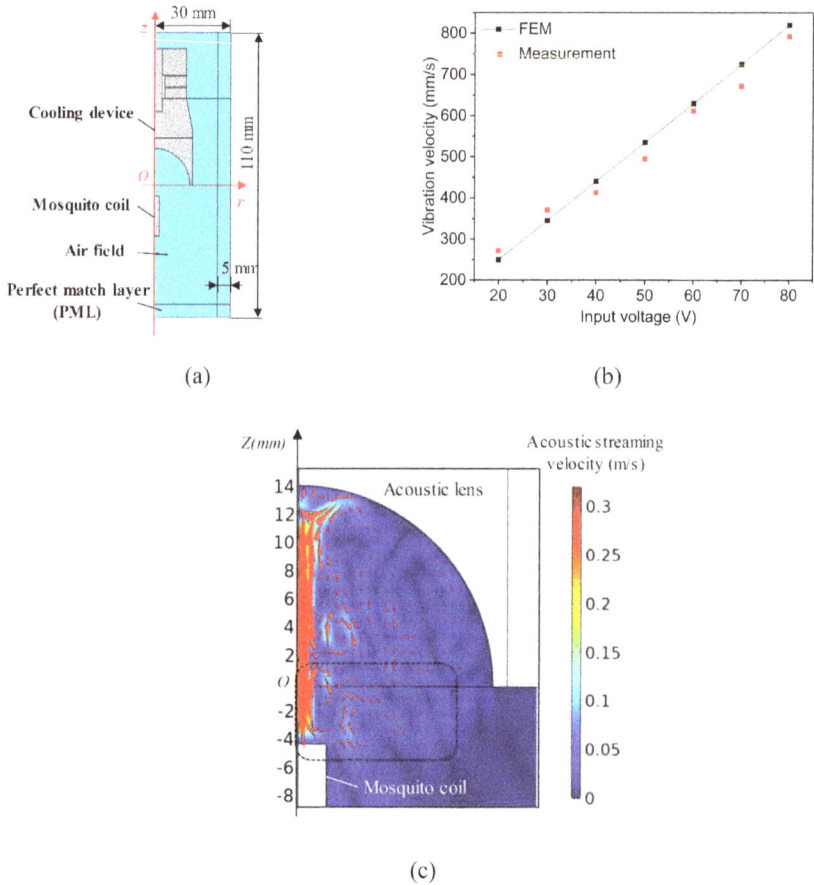

FIGURE 6.22 Computed acoustic field below the cooling device. (a) FEM model and air acoustic field dimension. (b) Comparison between the computed and measured vibration velocity at different input voltage when the ultrasonic transducer is at resonance. (c) Acoustic streaming field pattern. Reproduced from Ref. 11 with permission from Elsevier.

material and air acoustic field, and the perfectly matched layer (PML) was built around the acoustic field to make the simulated region finite with the function to eliminate the wave reflections.

The triangular mesh element was employed for the solid vibrator and ultrasonic field, and the mapping grid was built for the PML. The maximum mesh size of the ultrasonic field was one-tenth of the wave length to meet the calculation requirements. To verify the above computational method, the z-directional vibration velocity at the rim of the acoustic lens was calculated by the FEM, and compared with that measured by the laser Doppler vibrometer. The result is shown in Fig. 6.22(b), which shows that the computational result agrees quite well with the measured one.

For the working frequency of 61.6 kHz and transducer input voltage of 5 V, the computed acoustic streaming field is shown in Fig. 6.22(c). It is confirmed that the focused ultrasound can generate small acoustic streaming eddies in the focal region,

TABLE 6.4

Property Constants of the Materials Adopted for Simulation

Property Constants	Value
Air (at temperature $T = 20°C$)	
Dynamic viscosity [kg/(ms)]	1.810^{-5}
Ratio of specific heats	1.4
Density (kg/m³)	1.2
Heat capacity at constant pressure [J/(kg·K)]	1005.4
Thermal conductivity [W/(m·K)]	2.57710^{-2}
Speed of sound (m/s)	343.2
Copper	
Density (kg/m³)	8960
Young's modulus (Pa)	11010^9
Poisson's ratio	0.35
Aluminum	
Density (kg/m³)	2700
Young's modulus (Pa)	7010^9
Poisson's ratio	0.33
Lead Zirconate Titanate (PZT-4)	
Density (kg/m³)	7500
Piezoelectric coefficient e_{33} (C/m²)	15.1
Relative dielectric constant ε_{r33}	663.2
Structural steel	
Density (kg/m³)	7850
Young's modulus (Pa)	20010^9
Poisson's ratio	0.3

which qualitatively explains the enhancement of convective heat transfer between the heat source surface and the ambient. Thus, the flow that enhances the convection process in the focused ultrasound method is in the form of eddies.

The acoustic streaming field when the heat source was located at the focal region was computed for the working frequency of 61.6 and 49.4 kHz, and the flow patterns are shown in Fig. 6.23. In the computation, the input voltage of the transducer was 5 V, and the heat source was ignored in the model for its small size. The support platform was regarded as a disc with a radius of 5 mm and a height of 1 mm. From Fig. 6.23, it is seen that in the region near the z-axis above the support platform, the acoustic streaming exists at both working frequencies. The acoustic streaming velocity around the focal point at 61.6 kHz is 0.83 m/s, which is much larger than that (0.43 m/s) at 49.4 kHz. This can well explain why the measured cooling effect at 61.6 kHz is much better than that at 49.4 kHz.

The computed acoustic streaming velocity near the heat source along the z direction is shown in Fig. 6.24, as well as the measured Nu, for the 61.6 kHz ultrasonic field, in which the upper surface of the heat source is at the focal point $z = 0$. It is seen

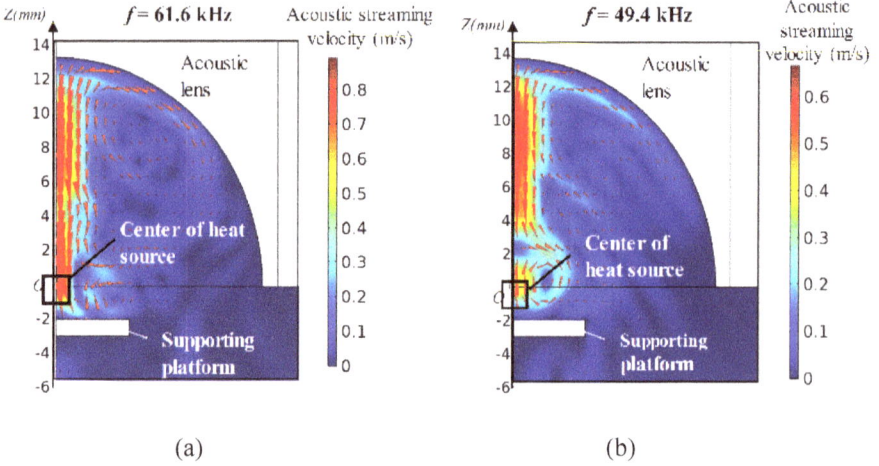

(a)   (b)

FIGURE 6.23 Computed acoustic streaming field with the heat source. (a) f = 61.6 kHz. (b) f = 49.4 kHz. Reproduced from Ref. 11 with permission from Elsevier.

that the acoustic streaming velocity (red dotted line) and Nu reach the maximum at the same location $z = 0$, which confirms the conclusion that the cooling effect is caused by the acoustic streaming.

A fundamental question about the working principle is why the acoustic stream-ing in conventional applications of focused ultrasound (like heating and killing of tumors) does not cool the treated tissues, whereas it cools solid heat source in air. The biological tissues treated by focused ultrasound are acoustically soft, and they have quite strong capability of absorbing the acoustic energy. As most of the absorbed acoustic energy is converted into heat, the cooling effect of acoustic streaming is

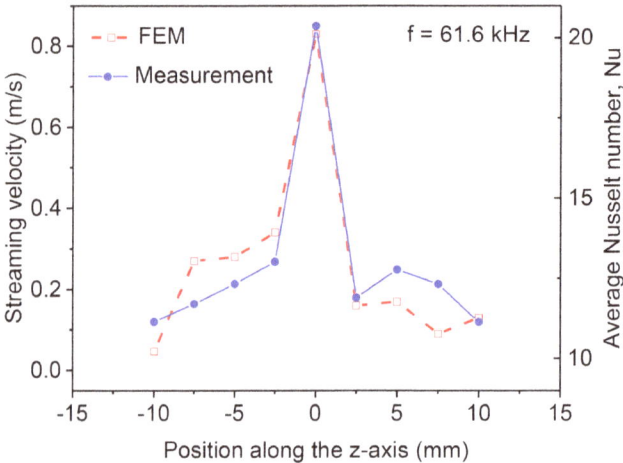

FIGURE 6.24 Comparison between the spatial dependence of computed streaming veloc-ity and measured Nusselt number. Reproduced from Ref. 11 with permission from Elsevier.

covered up by the heating effect. For nonflexible solid heat sources made of acoustically hard materials such as silicon, as they absorb little acoustic energy, the acoustic energy applied to them is hardly converted into heat.

6.4.4 CHARACTERISTICS AND DISCUSSION

To investigate the influence of working frequency on the cooling effect, Nu of the heat source versus power input P_{in} of cooling devices shown in Fig. 6.25(a), was measured, and the result is shown in Fig. 6.25(b). In the experiments, the initial temperature of the heat source was 81°C, which could be controlled by the voltage U in the circuit shown in Fig. 6.19(c). From Fig. 6.25(b), it is seen that the working frequency and ultrasonic strength do affect the cooling effect, and a higher working frequency and stronger ultrasonic strength are beneficial to improving the cooling effect. This is because a higher working frequency increases the Reynolds stress gradient which is the driving force of the acoustic streaming, due to its shorter wavelength, and thus

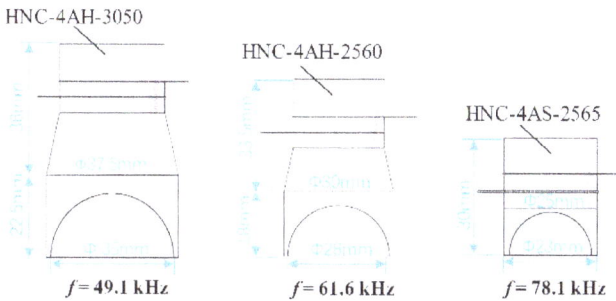

(a)

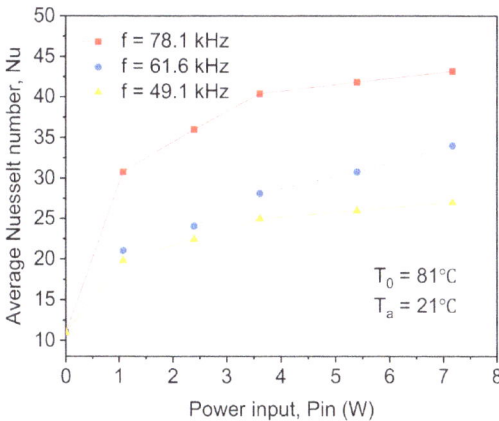

(b)

FIGURE 6.25 The influence of the working frequency of the cooling device on the cooling effect. (a) Diagram of cooling devices with different working frequencies. (b) Nusselt number versus power input of the cooling device. Reproduced from Ref. 11 with permission from Elsevier.

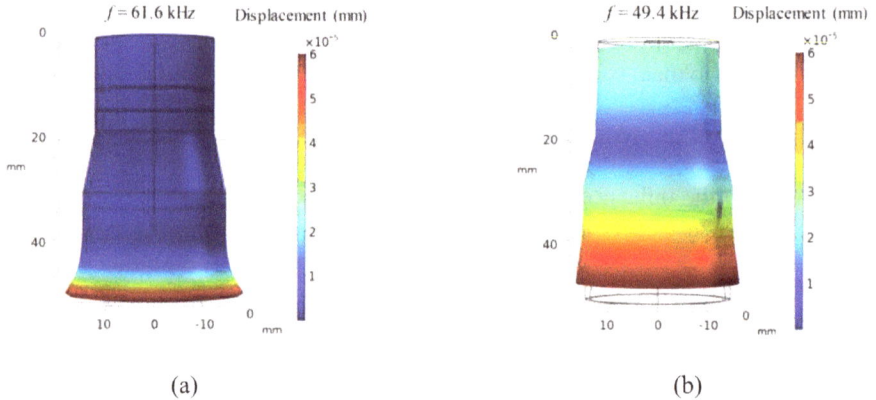

(a)   (b)

FIGURE 6.26 Computed vibration distribution patterns of the cooling device. (a) $f = 61.6$ kHz. (b) $f = 49.4$ kHz. Reproduced from Ref. 11 with permission from Elsevier.

increases the acoustic streaming velocity. But it must be pointed out that a too high working frequency is not suggested, as it is difficult for ultrasound at high frequency like MHz to propagate in the air. The conclusion that the cooling capability may be strengthened further by optimizing the working frequency may be used in the miniaturization of the cooling device.

To investigate the effect of vibration distribution of the cooling device, the vibration distribution of the cooling device shown in Fig. 6.18, at the resonance frequency of 61.6 kHz and 49.4 kHz, was computed by the FEM, respectively, with the same driving voltage (5 V_{p-p}), and the results are shown in Fig. 6.26.

From Fig. 6.26, it is seen that although they have similar maximum vibration displacement, they have different vibration distribution patterns. At 61.6 kHz, the vibration displacement concentrates on the rim of the acoustic lens, and this region does the bending vibration, as shown in Fig. 6.26(a). At 49.4 kHz, the whole device does the axial extension vibration, as shown in Fig. 6.26(b). As the measured cooling effect at 61.6 kHz is much better than at 49.4 kHz, it is deduced that the flexural vibration at the rim of the acoustic lens contributes more to the cooling effect, and a flexural vibration at the rim is more helpful to the cooling than a longitudinal one. When the cooling device works at the flexural vibration mode (at 61.6 kHz), the vibration displacement concentrates on the rim of the acoustic lens, which has direct contribution on the acoustic streaming. Whereas when the device vibrates longitudinally, the vibration displacement distributes in the whole device, most of which has no direct effect on the acoustic streaming.

Another cooling device with Langevin transducer HNC-4AH-3050 (shown in Fig. 6.25(a)) was used to investigate the dependency of cooling effect on the vibration distribution pattern (at different resonance frequencies). Figure 6.27(a) shows the measured temperature drop when the cooling device vibrates at resonance frequencies of 41.1, 49.1, 59.1 and 85 kHz, with the same input power of 1.6 W. The initial temperature of the heat source was 80°C in the experiments. It can be seen that the temperature drop varies from 12 to 30°C, and reaches the maximum at 49.1 kHz.

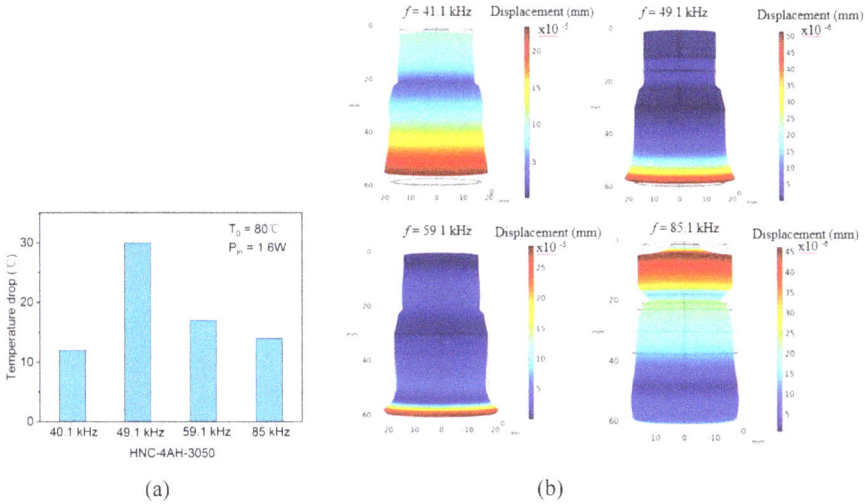

(a)   (b)

FIGURE 6.27   The temperature drop and vibration distribution patterns of the cooling device (HNC-4AH-3050) at different resonance frequencies. (a) Measured temperature drop. (b) Computed vibration distribution patterns. Reproduced from Ref. 11 with permission from Elsevier.

Then the vibration distribution patterns at these four frequencies were computed by the FEM, and the results are shown in Fig. 6.27(b). The conclusion obtained from Fig. 6.26 can well explain that the cooling device working at 49.1 kHz, at which the rim of the acoustic lens vibrates flexurally as shown in Fig. 6.27(b), has a better cooling performance than that at 41.1 kHz at which the whole device vibrates longitudinally as shown in Fig. 6.27(a). The weak vibration in the acoustic lens working at 85.1 kHz well explains its poor cooling effect. Moreover, it is seen that although the rim of the acoustic lens does the flexural vibration at both 49.1 and 59.2 kHz, the cooling effect has large difference at these two frequencies. This phenomenon is attributed to the height of the rim region with bending vibration. The larger the height of the rim region with bending vibration, the better the cooling effect.

The temperature change process of the heat source, caused by a commercial axial fan (QG 5010-12V, Hong Xing Shu Electronics Co., Ltd, China), was measured, and the result is shown in Fig. 6.28. For comparison, the result caused by the focused ultrasound is also given in the figure. The dimensions of the fan are 50 mm (height) × 50 mm (width) × 10 mm (thickness), and its rating power, rotor speed and air volume are 1.2 W, 5200 ± 10% RPM (revolutions per minute) and 0.17 m^3/min, respectively. In the experiments, the power input of both the axial fan and ultrasonic transducer was 1 W. The heat flux of the Pt heat source was 2.1 W/cm^2 and 4.8 W/cm^2, respectively, obtained by applying different input power to the heat source. The response time of the cooling process t_1 or t_2 (1 represents the focused ultrasound and 2 represents the axial fan) is defined as the time taken by the cooled sample to reach 90% change of the whole temperature drop from the initial to the steady state. It is seen that for the Pt wire heat source, the method proposed in this work has a similar cooling effect but faster cooling response than that based on the axial fan. The response time is

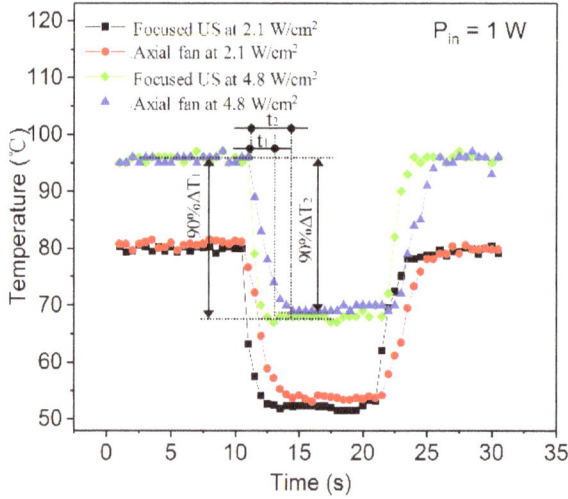

FIGURE 6.28 Cooling effect comparison between focused ultrasound and axial fan. Measured temperature change processes of the Pt wire heat source are used for the comparison. Reproduced from Ref. 11 with permission from Elsevier.

decreased by nearly 45% with focused ultrasound. This phenomenon is attributed to different nature of the flow fields near the cooled surface, generated by the axial fan and focused ultrasound. A faster cooling response can decrease the duration in which the temperature rise is harmful to the working device.

In the axial fan method, there is no cooling effect until the flow from the fan reaches the heat source with a speed not exceeding several tens per second. Whereas in the ultrasonic cooling method, the acoustic streaming starts to appear when the acoustic wave reaches the heat source. As the sound speed is 340 m/s in air at room temperature, which is much larger than the speed of flow generated by the fan, the cooling effect starts early in the ultrasonic method.

Apart from the heat source shown in Fig. 6.19, two disc-shaped metal ceramics heaters (MCHs) with the dimensions of 5 mm (diameter) × 0.6 mm (thickness) and 9 mm (diameter) × 1.5 mm (thickness), respectively, were used to test the cooling effect of focused ultrasound on larger devices. The ceramic heaters also have a linear relationship between their temperature and resistance. The measurement system in Fig. 6.17 was used for the cooling effect test. The surface temperature at different heat flux without and with sonication (input power of the cooling device = 6.7 W) was measured, and the results are shown in Fig. 6.29. It is shown that with the sonication, the heat flux increases more than 100% for both heaters (5.8 to 13.1 W/cm^2 for the 5 mm diameter heat source, and 3.7 to 7.5 W/cm^2 for the 9 mm diameter one) at a surface temperature of 500°C, and the Nusselt number is doubled compared to the natural convection. This indicates that the cooling device is still effective for the heat source with larger size.

The experiments and FEM analyses in this work have shown that the acoustic streaming generated by the focused ultrasound contributes to the cooling effect. Thus, an optimization of topological structure, working frequency, vibration distribution pattern and radiation performance (acoustic impedance matching) for the ultrasonic transducer may further strengthen the cooling capability and increase the heat flux

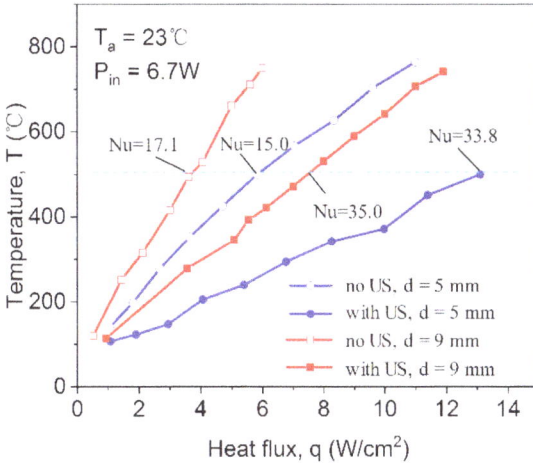

FIGURE 6.29   Measured surface temperature of a disc-shaped heat source versus the heat flux with and without sonication. Reproduced from Ref. 11 with permission from Elsevier.

which the device can stand. Due to the utilization of focused ultrasound, the method proposed and investigated by this work is suitable for the cooling of hot spots in electronic circuit systems such as the Central Processing Units and switching components.

6.4.5 CONCLUSIONS

In this work, a cooling method for small solid heat sources, which is based on the focused ultrasound in air, has been demonstrated and investigated. By the forced convective heat transfer resulting from the focused ultrasound, the heat flux is increased by 150% and the Nusselt number is doubled approximately (compared to the natural convection), for an ellipsoidal heat source made of Pt wire with a surface area of 6.44 mm^2 and initial temperature of 100°C. The cooling principle and characteristics were analyzed by the experiments and FEM. It is clarified that the cooling effect is affected by the vibration velocity, and working frequency and vibration mode of the cooling device. The vibration mode is a main design factor. Employing the flexural vibration mode of the rim of the acoustic lens is critical to achieve a good cooling effect. In the same vibration mode, increasing the vibration velocity and working frequency properly can further increase the cooling effect. Compared with the commercial axial fan, this method has similar cooling effect but 45% increase of the cooling response speed. The cooling effect for heat source of a 9 mm diameter metal ceramic heater was also investigated by experiment. The method reported in this paper provides an effective way for the targeted cooling of a device or hotspot.

6.5 REMARKS

In this chapter, it is demonstrated that the acoustic streaming is capable of improving the discharge performance of a metal-air battery and cooling a hotspot on a hard surface. It is also showed that the ultrasonic capillary effect can effectively improve the performance of a metal-air flow battery. For the ultrasound-assisted battery and cooling systems described in this chapter, more studies on energy-effective and structure-compact ultrasonic transducers are needed in order to make the ultrasonic units more energy-effective, lighter and smaller.

REFERENCES

1. J. Hu, *Ultrasonic Micro/Nano Manipulations: Principles and Examples*, (World Scientific, New Jersey, London, Singapore, 2014), Chapters 3 and 5.
2. J. Lighthill, *Waves in Fluids*, (Cambridge University Press, Cambridge, 1978), p. 329 and 344–350.
3. Q. Tang and J. Hu, "Diversity of acoustic streaming in a rectangular acoustofluidic field," *Ultrasonics*, 58, pp. 27–34, 2015.
4. O. V. Abramov, *High-Intensity Ultrasonics*, (Gordon and Breach Science Publishers, Singapore, 1998), pp. 124–139.
5. H. Huang, P. Liu, Q. Ma, Z. Tang, M. Wang and J. Hu, "Enabling a high-performance saltwater Al-air battery via ultrasonically driven electrolyte flow," *Ultrason. Sonochem.*, 88, p. 106104, 2022.
6. Z. Luo, Q. Tang and J. Hu, "Effect of ultrasonic excitation on discharge performance of a zinc-air button battery," *Micromachines*, 12(7), p. 792, 2021.
7. C. H. Hamann, A. Hamnett and W. Vielstich, *Electrochemistry*, (Weinheim, Wiley-VCH, 2007).
8. P. Gao, Y. Zhu and C. Yu, *Basic Course of Electrochemistry*, (Chemical Industry Press, Beijing, 2019), pp. 138–140.
9. T. Sattel, G. Mitic, M. Honsberg-Riedl, T. Vontz and R. Mock, "Inaudible cooling: A novel approach to thermal management for power electronics based on acoustic streaming," 30th *Annu. Semicond. Therm. Meas. Manage. Symp.*, 114–117, 09–13 March 2014. 10.1109/SEMI-THERM.2014.6892226.
10. P. I. Ro and B. Loh, "Feasibility of using ultrasonic flexural waves as a cooling mechanism," *IEEE Trans. Ind. Electron.*, 48(1), pp. 143–150, 2001.
11. Y. Hu, Z. Luo, Y. Zhou and J. Hu, "A focused ultrasound based cooling strategy for small solid heat sources," *Sens. Actuators A Phys.*, 331, p. 112932, 2021.

7 FEM Modeling and Computation for the Devices

The ultrasonic devices for nano/micro fabrication, handling and driving have irregular shape and so is their ultrasonic field. For this reason, it is not likely to obtain the analytical solutions of their vibration and ultrasonic field. Numerical computation is a feasible method to analyze the ultrasonic vibration and acoustofluidic field of the ultrasonic devices. The finite element method (FEM) modeling and computation play an important role in design and optimization of the ultrasonic devices. Although this issue has already been mentioned in the preceding chapters, more examples of the FEM modeling and computation are given in this chapter, in order that the authors can have a better understand on how to establish the FEM models of the devices and how to apply the FEM results in the device design and optimization. All of the FEM models in this chapter are based on practical ultrasonic devices, and verified by experimental results or analytical solution.

7.1 ACOUSTIC RADIATION FORCE ON PARTICLES WITH ARBITRARY SHAPE IN ARBITRARY ACOUSTIC FIELD

7.1.1 INTRODUCTION

In the design and optimization of ultrasonic tweezers, one often encounters the problem of computation of the acoustic radiation force on particles with arbitrary shape in arbitrary acoustic field. This section describes a solution to this problem, which was proposed and developed by the authors' team in 2009 [1, 2]. To demonstrate the computational principle and its physical essence more clearly, the particles trapped by acoustic radiation force are assumed to be rigid, and not in vibration with the acoustic field.

7.1.2 DEVICE ANALYSIS METHOD AND EXPERIMENTAL SETUP

The FEM model was built for the structure shown in Fig. 7.1 [1, 2], which was used to trap particles in air. In this structure, two identical aluminum strips are clamped to a Langevin transducer (FBL28452HS, Fuji Ceramics), as shown in Fig. 7.1(a). The aluminum strips have the shape and size shown in Fig. 7.1(b, c). The upper part of them is a rectangular aluminum plate, and the lower part is a V-shaped aluminum strip, tapered along the length. The upper part has a size of 40 mm × 45 mm × 1.5 mm with a 10 mm diameter hole at its center. The lower part with a length of 99 mm, width of 22.5 mm and thickness of 1.5 mm, is tapered off from the upper end to the lower end.

DOI: 10.1201/9781003404705-7

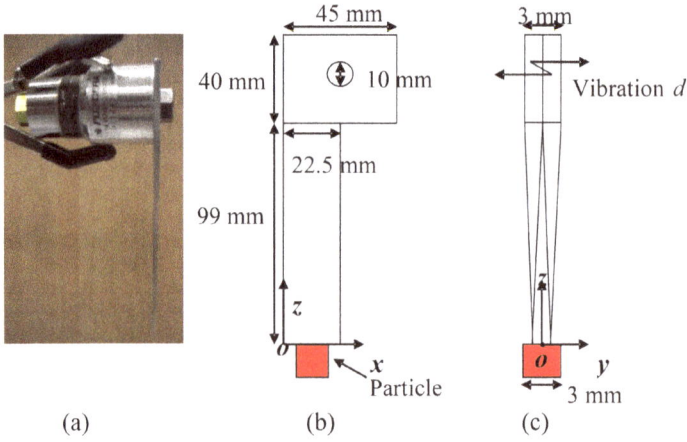

FIGURE 7.1 Structure and size of the ultrasonic transducer with two V-shaped aluminum strips in air. (a) Photo of structure of the ultrasonic transducer. (b) Shape and size of the aluminum strip. (c) Air gap formed by the two V-shaped strips. Reproduced from Ref. 1 with permission from IEEE.

The thickness of strip at the tip is around 200 μm. In this way, a triangular air gap is formed between the two V-shaped strips, which has a thickness of 1.3 mm at the tip. The Langevin transducer has a resonance frequency of 25.3 kHz. The two-dimensional (2D) finite element analysis (FEA) by the acoustic module of COMSOL Multiphysics shows that a flexural vibration is excited in the aluminum strips when an AC (alternative current) voltage with a working frequency close to the resonance frequency of the ultrasonic transducer is applied. This flexural vibration will generate a sound field, and the sound field near the lower end of the gap can generate an acoustic radiation force on the particles to suck the particles to the lower end of the strips.

The acoustic radiation force F on a rigid immovable object in a sound field in ideal fluid is given by the following integration over the surface of the object:

$$F = \left\langle \iint_S (K - U)\boldsymbol{n}\, dS \right\rangle, \tag{7.1}$$

where the notation <> denotes time average over one period, K is the kinetic energy density, U is the potential energy density, and $\boldsymbol{n}$ is the outward normal unit vector of the surface. The kinetic and potential energy densities K and U can be calculated by [3]

$$K = \frac{\rho_0 v^2}{2}, \tag{7.2}$$

$$U = \frac{p^2}{2\rho_0 c_0^2}, \tag{7.3}$$

where ρ_0 and c_0 are the density of and sound speed in the fluid when there is no sonication, v is the velocity, and p is the sound pressure.

The three-dimensional (3D) FEA is conducted by the harmonic analysis function of the acoustic module in COMSOL Multiphysics software, to solve the sound field surrounding a rigid particle under the two vibrating sharp edges in air, as shown in Fig. 7.1(c). It is found that when the two identical metal strips are vibrating in the $\pm y$ directions, the sound pressure in the air gap is symmetric about its central plane (the xz-plane). So, we split the whole structure (two metal strips, a particle and the surrounding sound field) into two parts about the xz-plane for the 3D FEA.

The boundary conditions for the sound field are: vibration velocity = 0 is used for the xz-plane; the rest of the sound field boundaries are radiation boundary or perfect matching layer. The excitation conditions for the transducer are: The excitation frequency f is 25.3 kHz, which is the resonance frequency of the transducer; the amplitude of the y-directional vibration displacement (0-peak) of the upper part of metal plates d is 10 μm; the loss factor of the vibration in aluminum is 0.02, which is defined as the ratio of the amount of energy dissipated as heat to that of total stored energy (COMSOL Multiphysics: Acoustic Module User's Guide).

Figure 7.2(a) shows the mesh of the half structure (a single strip, half of a 3 mm × 3 mm × 3 mm cube particle and the surrounding sound field) in the 3D FEA, where the maximum mesh size at the particle surface boundaries is around 0.66% of the wavelength. Figure 7.2(b, c) shows the amplitudes of the x-directional velocity, y-directional velocity and sound pressure on the top surface of the half particle ($z = 0$). The x-directional and z-directional vibration velocities on the top surface of the half particle are much smaller than y-directional vibration velocity. Thus, they are not listed in the figure. It is seen that the x-directional velocity is very small compared with the y-directional velocity; the y-directional velocity is the maximum at around $y = 0.75$ mm, which is near the sharp edge of the vibrating strip. From the 3D FEM results shown in Fig. 7.2(b, c), it is known that $\iint_S \langle K \rangle dS$ and $\iint_S \langle U \rangle dS$ on the top surface of the cube particle is 6.5×10^{-4} and 1.2×10^{-5} N, respectively. Also, it is found that $\iint_S \langle K \rangle dS$ and $\iint_S \langle U \rangle dS$ on the side and bottom surfaces of the particle are less than 1% of that on the top surface; thus, they are negligible. Thus, on the top surface of particle, $\iint_S \langle K \rangle dS \gg \iint_S \langle U \rangle dS$. So, F has the same direction as $\mathbf{n}$ (see Eq. (7.1)), and the particle may be sucked onto the sharp edge of the strips. Therefore, for the structure used in our experiment, the acoustic radiation force acting on particle is determined by the force on the top surface of particle, pointing upwards, and the trapping force mainly occurs near the two vibrating sharp edges of the strips.

Particles used in the experiment are made of clay. The sound speed of clay material is measured by a sound speed meter (SV-DH-7A, Guotai Electronics, China). In the method, two cylindrical clay samples with different lengths (L_1 and L_2, respectively) are used, and the time for sound to travel through each sample in its length direction is measured (t_1 and t_2, respectively). The sound speed in clay cylinder is calculated by the length difference ($\Delta L = L_1 - L_2$) and the time difference ($\Delta t = t_1 - t_2$), and the measured sound speed is 2716 m/s. The density of clay material is calculated by using the measured mass and volume of clay particles, and the measured density is 1554 kg/m³. There are six types of particles with different shapes but the same mass and volume. Each particle has a mass of 40 mg. The shapes of the six particles are

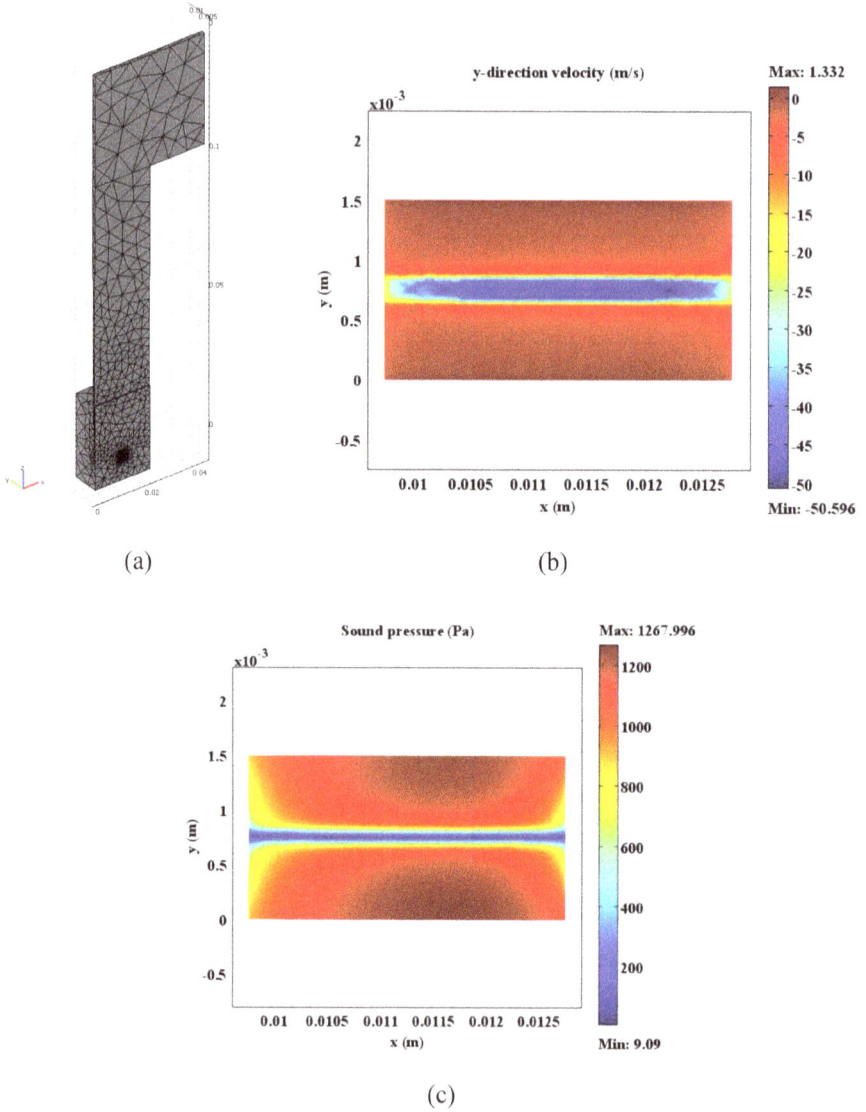

(a)

(b)

(c)

FIGURE 7.2   FEA results of the half structure (a single strip, and half of a 3 mm long cube particle and surrounding sound field) in air when a vibration displacement amplitude (0-peak) of upper part of metal plate is 10 μm. (a) Meshed model. (b) The y-directional velocity on top surface of the half particle. (c) The sound pressure on the top surface of the half particle.

sphere, cube, cylinder, cone, rectangular cuboid and hollow cylinder, respectively. The size of each particle is shown in Fig. 7.3.

In the experiment, the two vibrating sharp edges of strips are moved to the top surface of particle, and then the transducer is lifted up. It is found that all of the six particles can be trapped in air by the transducer operating at a proper vibration. For the

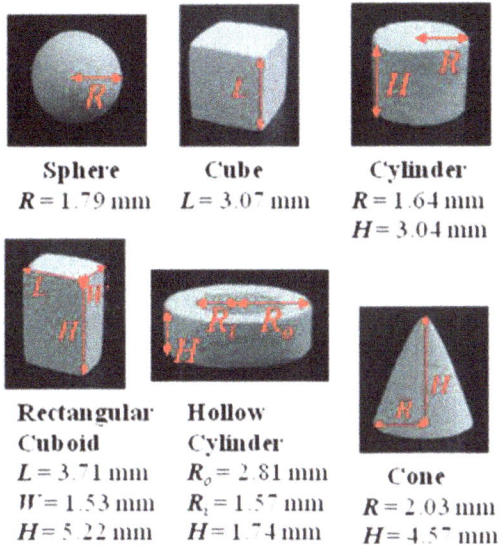

Sphere   Cube   Cylinder
$R = 1.79$ mm $L = 3.07$ mm $R = 1.64$ mm
                             $H = 3.04$ mm

Rectangular   Hollow
Cuboid   Cylinder
$L = 3.71$ mm $R_o = 2.81$ mm   Cone
$W = 1.53$ mm $R_i = 1.57$ mm $R = 2.03$ mm
$H = 5.22$ mm $H = 1.74$ mm $H = 4.57$ mm

FIGURE 7.3 Photos and dimensions of six different shaped clay particles with the same mass and volume. Reproduced from Ref. 1 with permission from IEEE.

comparison of acoustic trapping capability, the acoustic trapping capability coefficient C_{tr} is defined as

$$C_{tr} = \frac{1}{d_{mim}}, \tag{7.4}$$

where d_{min} is the minimum vibration displacement at the tip of the metal strip to trap one particle. The acoustic trapping capability for particles is stronger if its C_{tr} is larger. Figure 7.4 shows the trapped rectangular cuboid-shaped particle under the two sharp edges of the strips in air. For the convenience of analysis and discussion, the contact line between the sharp edge of a strip and the top surface of a particle below is defined as the action line.

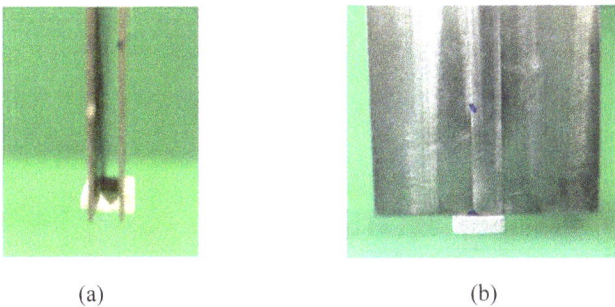

(a)   (b)

FIGURE 7.4 A trapped rectangular cuboid-shaped particle in air. Reproduced from Ref. 1 with permission from IEEE.

7.1.3 RESULTS AND DISCUSSION

The effect of orientation of rectangular cuboid particle on the acoustic trapping capability in air has been investigated both experimentally and theoretically. The rectangular cuboid particle trapped under the two sharp edges of metal strips has six possible trapping orientations, from *a* to *f*, as shown in Fig. 7.5(a). Figure 7.5(b) shows the experimental and calculated acoustic trapping capability coefficients for the rectangular cuboid particle at the different trapping orientations. In the experiment, the particle can be trapped at five orientations (*a*, *b*, *c*, *d* and *e*), and experimental and calculated results agree well. The action line's length of the rectangular cuboid particle from orientation *a* to *f* is 5.22, 5.22, 3.71, 3.71, 1.53 and 1.53 mm, respectively. It is found that the acoustic trapping capability coefficient C_{tr} increases mostly with the increase of the action line's length. Also, when the action line's length is identical, the particle with larger top surface area (shaded area in Fig. 7.5(b)) has stronger acoustic trapping capability. For example, the action line's length of cases *a* and *b* is identical, but case *a* has stronger trapping capability than case *b*. This is because the larger the top surface area, the larger the action area of the upward acoustic radiation force.

Figure 7.6 shows the experimental and calculated effects of the orientation of cylinder particle on the acoustic trapping capability in air. The cylinder particle has three possible trapping orientations, as shown in Fig. 7.6(a). From Fig. 7.6(b), it is seen that cylinder can be trapped at all the orientations. It is also seen that for orientation *a* and *b*, the experimental and calculated results agree well, and for orientation *c*, the difference between the experimental and calculated results is relatively large. The action line's length of the cylinder particle for orientation *c* is 0 mm. This causes a large calculation error for orientation *c*.

The acoustic trapping capability for particles in air with different shapes shown in Fig. 7.3 has been investigated experimentally. For each particle shape, there are

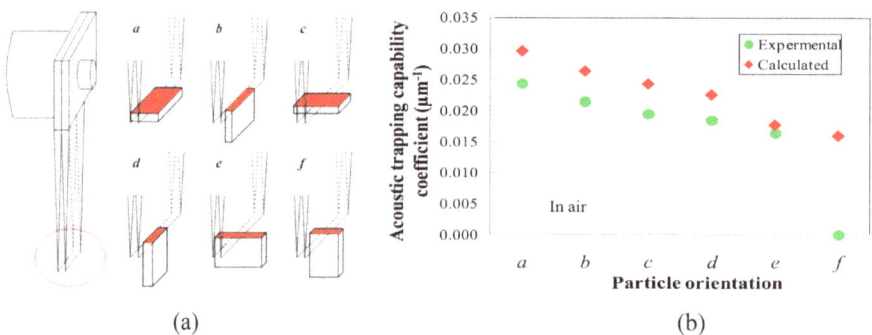

(a)   (b)

FIGURE 7.5   The effect of orientation of rectangular cuboid clay particle on the acoustic trapping capability in air. (a) Six possible trapping orientations of the same rectangular cuboid-shaped particle. (b) The experimental and calculated acoustic trapping capability coefficients for the rectangular cuboid particle at different trapping orientations. Reproduced from Ref. 1 with permission from IEEE.

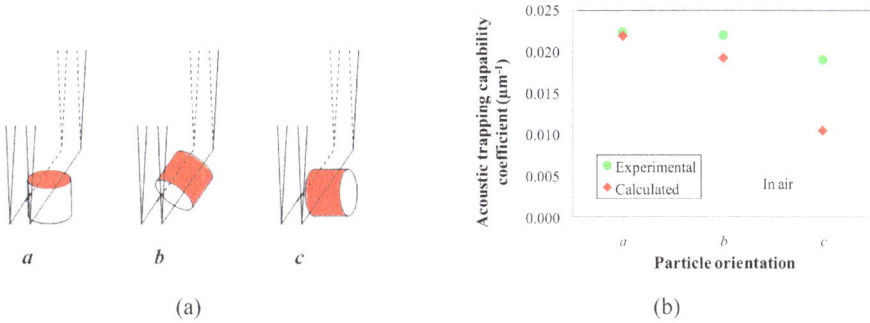

FIGURE 7.6 The effect of orientation of cylinder clay particle on the acoustic trapping capability in air. (a) Three possible trapping orientations of the same cylindrical-shaped particle. (b) The experimental and calculated acoustic trapping capability coefficients for the cylinder particle at different trapping orientations. Reproduced from Ref. 1 with permission from IEEE.

several possible trapping orientations, and every orientation has its own trapping capability. Figure 7.7(a) lists the orientations for the best trapping capability of each particle shape. Figure 7.7(b) shows the experimental and calculated best acoustic trapping capabilities for the particles. It is found that the experimental best acoustic trapping capability for each type of particle becomes worse in the sequence from rectangular cuboid, cylinder, cone, cube, sphere to hollow cylinder. The error between the experimental and calculated results is small except for sphere and hollow cylinder. The theoretical calculations are not applied to sphere and hollow cylinder for the reason given in the discussion of Fig. 7.6.

The effect of orientation of a rectangular cuboid particle on the acoustic radiation force acting on the particle in water and in air is investigated theoretically, and the results are shown in Fig. 7.8.

In the calculation, the vibration excitation conditions in water are the same as that in air ($d = 10$ μm and $f = 25.3$ kHz). The trapping orientations of the rectangular cuboid particle as shown in Fig. 7.5(a) are used. From Fig. 7.8, it is seen that the acoustic radiation force acting on the rectangular cuboid particle decreases from orientation *a* to *f* both in air and in water, and the acoustic radiation force in water is larger than that in air. The latter phenomenon is because of a stronger sound field in water and buoyancy force on the particle in water. In the above analysis, the effect of acoustic stream on a trapped particle is neglected.

Figure 7.9 shows the calculated acoustic radiation forces acting on the particles shown in Fig. 7.3 both in water and in air, which have different shapes but the identical volume and density. In the calculation, the vibration excitation conditions in water are the same as that in air ($d = 10$ μm and $f = 25.3$ kHz). The trapping orientation of each particle is shown in Fig. 7.7(a). From Fig. 7.9, it is seen that the acoustic radiation force in water decreases in the sequence of rectangular cuboid, cylinder, cone, cube, sphere and hollow cylinder, which has the same order as that in air.

(a)

(b)

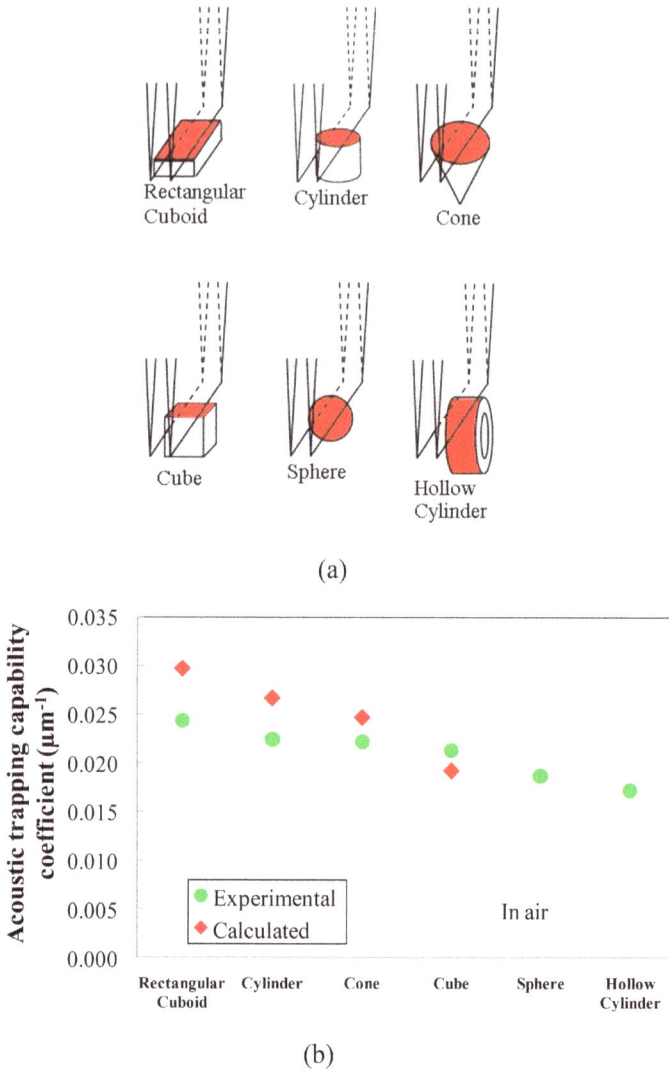

FIGURE 7.7 The effect of particle shape on the acoustic trapping capability in air. (a) Six particles with different shapes but the same volume and density, at the orientation at which the experimental trapping capability is the strongest for each particle. (b) The experimental and calculated acoustic trapping capability coefficients for the particles at the orientations are shown in (a). Reproduced from Ref. 1 with permission from IEEE.

Figure 7.10 shows the calculated acoustic radiation force on the rectangular cuboid particle at the interface of water and air. In the calculation, the vibration excitation conditions are the same as those in air and in water ($d = 10$ μm and $f = 25.3$ kHz); the trapping orientations of the rectangular cuboid particle shown in Fig. 7.5(a) are used, and half of the particle is in water and another half in air. From Fig. 7.10, it is seen that from orientation a to f, the acoustic radiation force on particle at the interface of water and air decreases correspondingly. Compared with Fig. 7.8, it is found that the

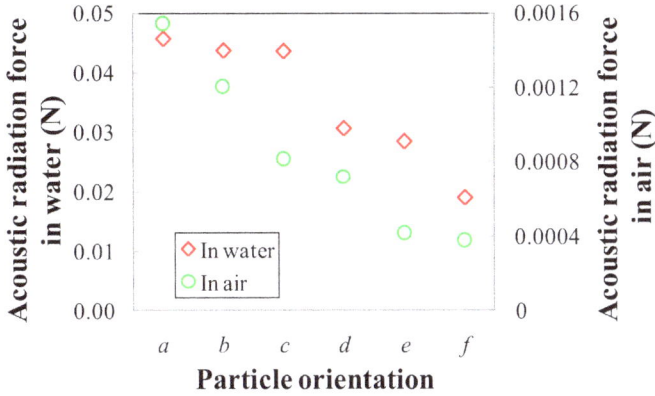

FIGURE 7.8 The effect of orientation of the rectangular cuboid particle on the calculated acoustic radiation force acting on particle in water and in air. Reproduced from Ref. 1 with permission from IEEE.

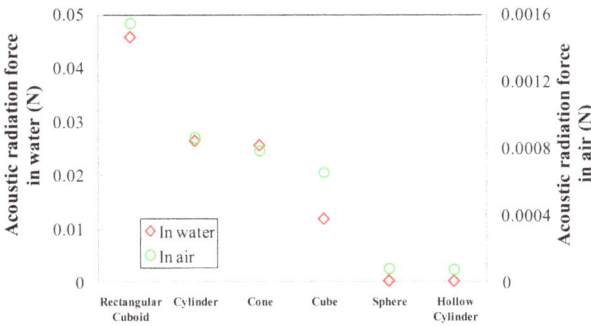

FIGURE 7.9 The effect of particle shape on the calculated acoustic radiation force acting on particle in water and in air. Reproduced from Ref. 1 with permission from IEEE.

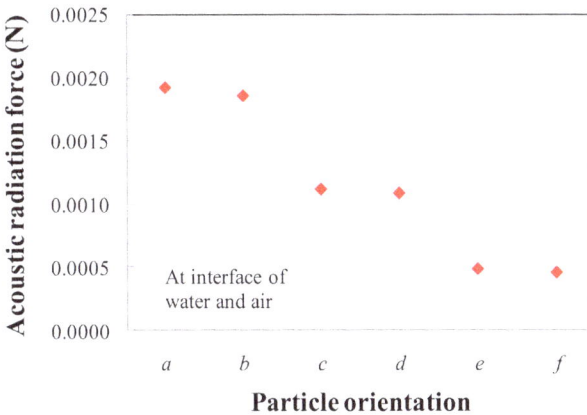

FIGURE 7.10 The effect of orientation of the rectangular cuboid particle on the calculated acoustic radiation force when the particle is at the interface of water and air. Reproduced from Ref. 1 with permission from IEEE.

acoustic radiation force on a particle at the interface between water and air is larger than that entirely in air, and less than that entirely in water.

7.1.4 CONCLUSIONS

This example demonstrates a FEM computation method of the acoustic radiation force on particles with arbitrary shape in arbitrary acoustic field. It combines the FEM analysis and the theory of acoustic radiation force. By comparing the experimental and calculated results, it is seen that the computational results can explain the experimental results well. With this numerical method, the following conclusions are obtained. When the density, volume and mass of a particle are constant, the acoustic trapping capability depends on the orientation and shape of the particle. This dependence is caused by the difference in the action line's length. The acoustic trapping capability increases with the increase of the action line's length at large, and it is the best at the orientation where the action line is the longest for a given particle. Acoustic trapping capability becomes worse in the sequence of rectangular cuboid, cylinder, cone, cube, sphere and hollow cylinder. Also, it is found that the acoustic radiation force on a certain particle in water is much larger than that in air; the acoustic radiation force on particle at the interface of water and air is larger than that entirely in air, and less than that entirely in water.

7.2 DIVERSITY OF ACOUSTIC STREAMING IN A RECTANGULAR ACOUSTOFLUIDIC FIELD

7.2.1 INTRODUCTION

Diversity of acoustic streaming field in a 2D rectangular chamber with a traveling wave and using water as the acoustic medium is numerically investigated by the FEM [4]. Commercial FEM software COMSOL Multiphysics is used to implement the computation tasks, which makes our method very easy to use. The computational method is partially verified by an established analytical solution. It is found that the working frequency, the vibration excitation source length and the distance and phase difference between the two separated symmetric vibration excitation sources can cause the diversity in the acoustic streaming pattern. It is also found that a small object in the acoustic field may result in additional eddies, and affects the eddy size in the original acoustic streaming field. In addition, the computational results show that with an increase of the acoustic medium's temperature, the speed of the main acoustic streaming decreases first and then increases, and the angular velocity of eddies in the corners increases monotonously, which can be well explained by the change of acoustic dissipation factor and shearing viscosity of the acoustic medium with the temperature.

7.2.2 CALCULATION METHOD AND CONDITIONS

The computational process consists of three steps. In the first step, the sound field is solved by the acoustic module of COMSOL Multiphysics software. In the second

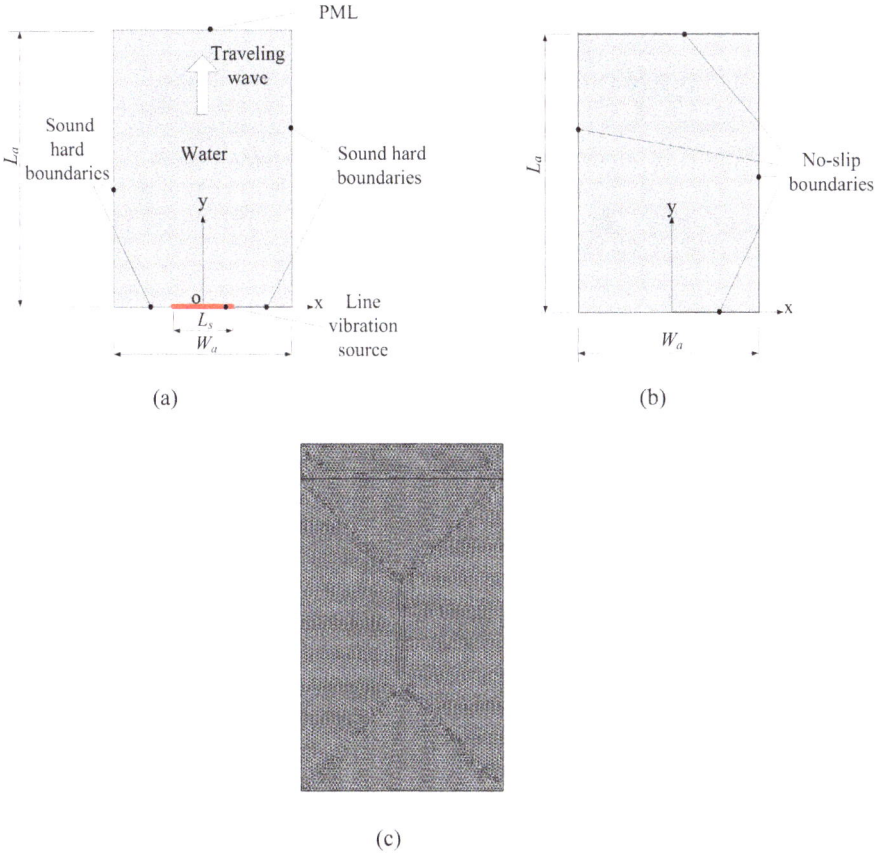

(a)   (b)

(c)

FIGURE 7.11 Model and mesh of the acoustofluidic field. (a) Computation model for the sound field. (b) Boundary conditions of the acoustic streaming field. (c) Mesh of the acoustofluidic field. Reproduced from Ref. 4 with permission from Elsevier.

step, the vibration velocity and sound pressure are used to calculate spatial gradients of Reynolds stress and mean pressure by the post-processing functions of the software, which generate the acoustic streaming. In the last step, the steady acoustic streaming is solved by the fluidic dynamics module, with a proper boundary condition setting.

FEM computational model of the acoustofluidic field is shown in Fig. 7.11. The acoustofluidic field is rectangular and 2D, in which water is used as the acoustic medium, and the vibration source (red line) is at $y = 0$. To generate a traveling wave in the field, the boundary opposite to the vibration source allows the sound wave to pass without any reflection and absorption, that is, a perfect matching layer (PML) is used to generate the traveling wave. The other boundaries of the sound field are acoustically hard ($\frac{\partial p}{\partial n} = 0$, where n denotes the unit normal vector of the boundaries). For the acoustic streaming field, all the boundaries are no-slip, as shown in Fig. 7.11(b). This is because our work is mainly interested in the overall pattern of acoustic streaming

TABLE 7.1

Parameters of the Acoustofluidic Field

Length of Line Vibration Source L_s (m)	Width of Acoustofluidic Field W_a (m)	Length of Acoustofluidic Field L_a (m)
0.02	0.06	0.09
Frequency of line vibration source f (kHz)	Amplitude of line vibration source v (m/s)	Volume-to-shear viscosity ratio in water η'/η
20	$4\pi \times 10^{-3}$	2.1
Ratio of specific heats γ in water	Isothermal compressibility β_T (Pa^{-1}) in water at 298 K	Isobaric thermal expansion coefficient α_p (K^{-1}) in water at 298 K
1	4.5×10^{-10}	2.5×10^{-4}

in the chamber, rather than the details of acoustic streaming near the boundaries. Figure 7.11(c) shows a meshed model for the acoustofluidic field, with the maximum element size of 0.001 m, which is about 1/75 of the wavelength of the sound field at 20 kHz ($\lambda = \dfrac{c}{f} \approx \dfrac{1500}{20}$ mm = 75mm). Unless otherwise specified, the dimensions and acoustic medium (water) properties shown in Table 7.1 are used.

The following wave equation is used to solve the sound field:

$$\rho \frac{\partial^2 p}{\partial t^2} = \rho_0 c^2 \nabla p + b \nabla \frac{\partial p}{\partial t}, \tag{7.5}$$

where p is the sound pressure, ρ_0 is the fluid density in the undisturbed state, and c is the sound speed. The acoustic dissipation factor b is calculated by

$$b = \frac{4}{3}\eta + \eta' + \kappa_t \left(\frac{1}{C_v} - \frac{1}{C_p} \right), \tag{7.6}$$

where η and η' are the shear and bulk coefficient of viscosity, respectively, κ_t is the thermal conductivity of the acoustic medium, and C_p and C_v are the heat capacities of the acoustic medium at constant pressure and volume, respectively.

The thermodynamic relation among pressure p, temperature T and density ρ in a sound field can be written as

$$d\rho/\rho = \gamma \beta_T dp - \alpha_p dT, \tag{7.7}$$

where γ is the specific heat ratio, β_T is the isothermal compressibility and α_p is the isobaric thermal expansion coefficient. The values of γ, β_T and α_p of water at 298 K are listed in Table 7.1. Thus, Eq. (7.7) can be simplified as $d\rho/\rho = 4.5 \times 10^{-10}$ $dp - 2.5 \times 10^{-4}$ dT. From Ref. [8], the pressure change is $dp = \rho_0 c \omega A$, in which $\rho_0 c$ is the specific acoustic impedance of the acoustic medium, and ω and A are the angular frequency and amplitude of the vibration source, respectively. The density and the sound speed of water are about 1000 kg/m^3 and 1500 m/s, respectively. In the ultrasonic manipulation based on acoustic streaming, the maximum order of magnitude of ω and A is 10^7 rad/s and 10^{-6} m, respectively. Thus, the order of magnitude of dp cannot be larger than 10^7 Pa, and that of term $4.5 \times 10^{-10}\, dp$ cannot be larger than 10^{-3}. In the acoustic streaming-based manipulations, the temperature change in water at the manipulation point is usually less than 10 K. This means that the order of magnitude of term $2.5 \times 10^{-4}\, dT$ cannot be larger than 10^{-3}. So, the order of magnitude of $d\rho/\rho$ cannot be larger than 10^{-3} in the water, which means that the water density can be treated as constant for the ultrasonic manipulations based on acoustic streaming. For the acoustofluidic model in this work, the order of magnitude of parameters ω, A and dT is less than the above-stated maximum ones. Therefore, the water density can also be treated as constant in this work.

The steady acoustic streaming satisfies the following equation:

$$\rho_0(\bar{u}_i\, \partial \bar{u}_j / \partial x_i) = F_j - \partial \bar{p}_2 / \partial x_j + \eta \nabla^2 \bar{u}_j, \tag{7.8}$$

where $\bar{u}_i$ is the acoustic streaming velocity, repeated suffix i and j represent x and y in the 2D model, respectively, ρ_0 is the fluid density in the undisturbed state, F_j is the gradient of Reynolds stress, which acts on the fluid as a driving force of the acoustic streaming, and $\bar{p}_2$ is the time average of the second order pressure or mean pressure. F_j is calculated by

$$F_j = -\partial(\overline{\rho_0 u_i u_j}) / \partial x_i, \tag{7.9}$$

where u_i is the vibration velocities in the sound wave, and the bar signifies the average value over one period.

The total pressure P in the acoustic medium can be expanded into a Taylor series as follows [5]:

$$P = P_0 + \left(\frac{\partial p}{\partial \rho}\right)_{s,\rho=\rho_0} (\rho - \rho_0) + \left(\frac{\partial^2 p}{\partial \rho^2}\right)_{s,\rho=\rho_0} \frac{(\rho-\rho_0)^2}{2} + \cdots, \tag{7.10}$$

in which s means that the terms are evaluated in the isentropic state, and P_0 is the fluid pressure in the undisturbed state. The (first order) sound pressure p_1 is

$$p_1 = \left(\frac{\partial p}{\partial \rho}\right)_{s,\rho=\rho_0} (\rho - \rho_0) = A\left(\frac{\rho-\rho_0}{\rho_0}\right), \tag{7.11}$$

where

$$A = \rho_0 \left(\frac{\partial p}{\partial \rho}\right)_{s,\rho=\rho_0} = \rho_0 c_0^2, \tag{7.12}$$

in which c_0 is the sound speed in the undisturbed state. The second order pressure p_2 is

$$p_2 = \left(\frac{\partial^2 p}{\partial \rho^2}\right)_{s,\rho=\rho_0} \frac{(\rho-\rho_0)^2}{2} = \frac{B}{2}\left(\frac{\rho-\rho_0}{\rho_0}\right)^2, \tag{7.13}$$

where

$$B = \rho_0^2 \left(\frac{\partial^2 p}{\partial \rho^2} \right)_{s,\rho=\rho_0}.$$ (7.14)

From Eqs. (7.11) to (7.13), there is

$$p_2 = \frac{1}{2\rho_0 c_0^2} \frac{B}{A} p_1^2.$$ (7.15)

Therefore, $\bar{p}_2$ is calculated by

$$\bar{p}_2 = \frac{1}{2\rho_0 c_0^2} \frac{B}{A} \langle p_1^2 \rangle,$$ (7.16)

where $<>$ represents the time average over one period, and $\frac{B}{A}$ is the nonlinear parameter of the acoustic medium, depending on the medium and temperature ($B/A = 5$ for water at 25°C). The acoustic streaming also satisfies the continuity equation

$$\rho_0 \, \partial \bar{u}_i / \partial x_i = 0.$$ (7.17)

7.2.3   RESULTS AND DISCUSSION

Unless otherwise specified, the acoustic streaming field is computed when the acoustic medium's temperature T is 298 K, the operating frequency f is 20 kHz, and the amplitude of source vibration A is 100 nm. In the calculation of the sound field and acoustic streaming, the relative tolerance of the solvers in the COMSOL Multiphysics software is 0.0001 and 0.001, respectively. Figure 7.12(a) shows us the distributions of the y-directional velocity of acoustic streaming at $y = 0.04$ m, calculated by our finite element analysis method and by the analytical solution [6], respectively. It is seen that the two acoustic streaming distributions are quite close at $y = 0.04$ m, which partially verifies our calculation method. Figure 7.12(b) shows the computed sound pressure in the sound field, and Fig. 7.12(c) shows the computed acoustic streaming field. In Fig. 7.12(c), the color denotes the magnitude of the acoustic streaming velocity and the arrow denotes the direction and magnitude of the acoustic streaming velocity. In addition to the two main eddies EL (eddy on the left side) and ER (eddy on the right side) in the center of the field, which have been predicted by the traditional analytical solutions, a pair of additional circular eddies, denoted as ECL (eddy in the corner on the left side) and ECR (eddy in the corner on the right side), are found in the two corners of the field.

The effect of the working frequency on the pattern of acoustic streaming field is computed, and Fig. 7.13 shows patterns of the acoustic streaming field at 10 kHz, 40 kHz, 80 kHz and 216 kHz, respectively, with the same velocity amplitude of the vibration source (1.26 cm/s). It is seen that as the working frequency increases, the number of acoustic streaming eddies increases, and the pattern becomes more complex. In the acoustic streaming field at 216 kHz, there are six eddies which have an

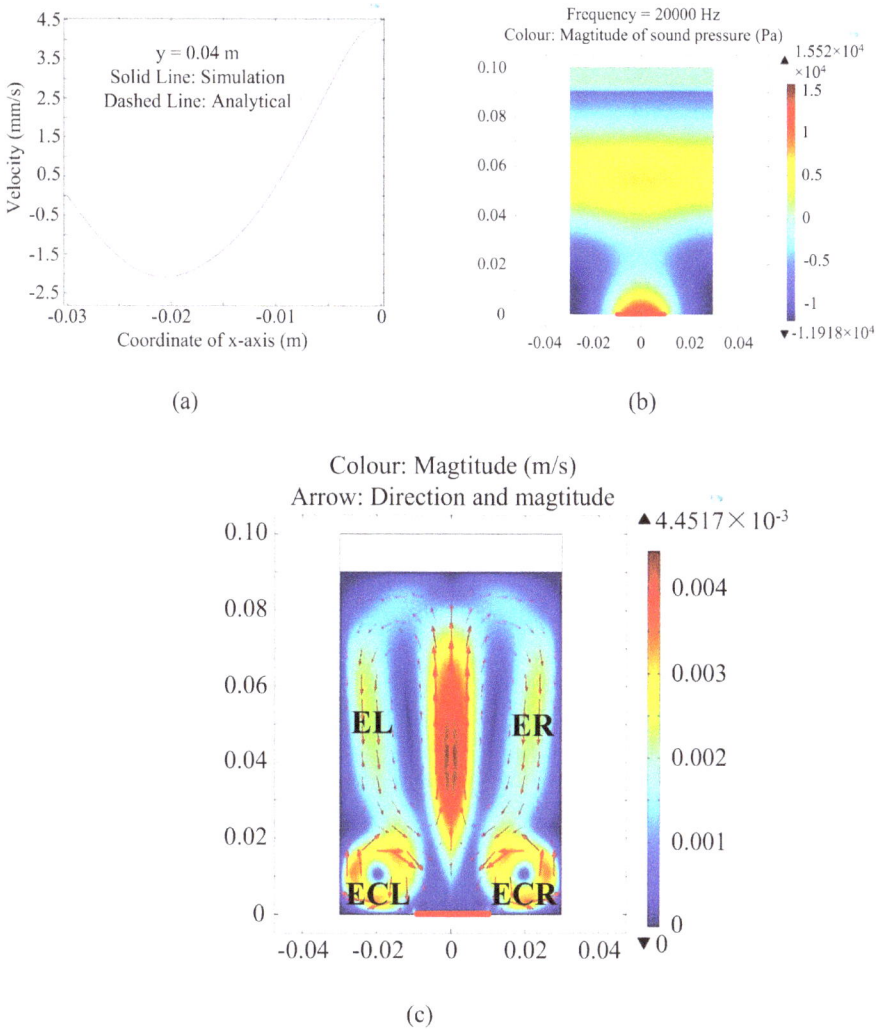

(a)

(b)

(c)

FIGURE 7.12   Sound pressure and acoustic streaming field at $f = 20$ kHz, $A = 0.1$ μm and $T = 298$ K. (a) Distributions of the y-directional velocity at $y = 0.04$ m. (b) Pattern of sound pressure field. (c) Pattern of acoustic streaming field. Reproduced from Ref. 4 with permission from Elsevier.

orderly arrangement. Thus, it provides a method for manipulating small particles at several points simultaneously.

The acoustic streaming field was computed for different vibration source lengths L_s, and the result is shown in Fig. 7.14. Figure 7.14(a) indicates that as the vibration source length increases, eddies ECL and ECR in the two corners will disappear, and the length of the main eddies will change. The acoustic streaming velocity has the maximum velocity at the boundary between eddies EL and ER. Figure 7.14(b) shows the maximum velocity of the acoustic streaming versus the vibration source

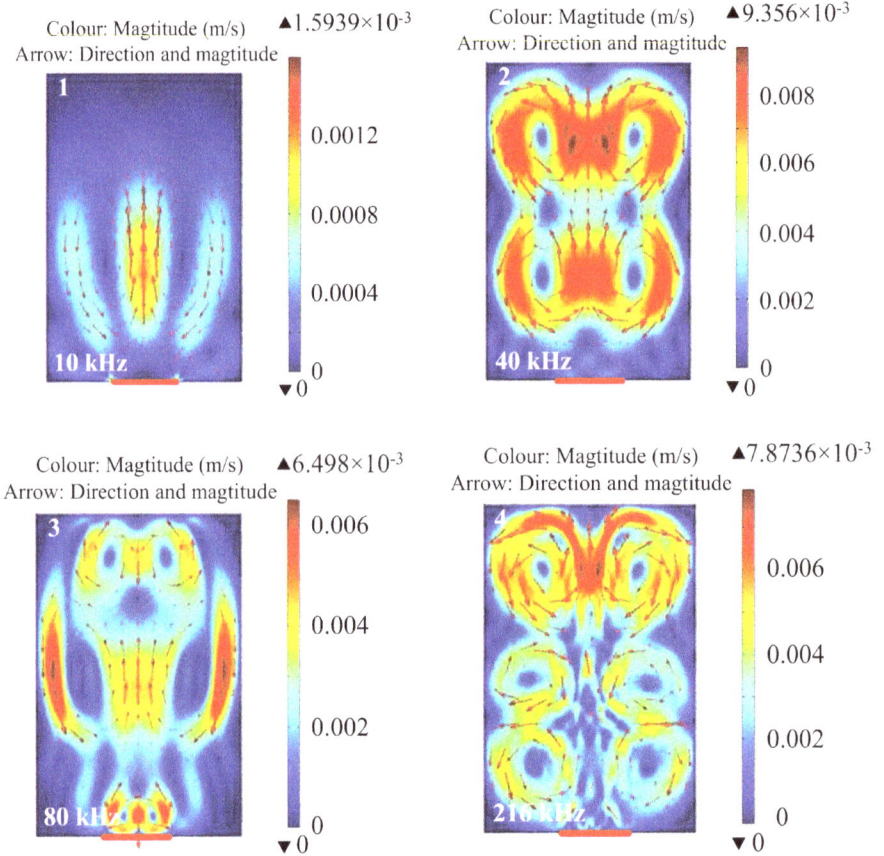

FIGURE 7.13   Patterns of acoustic streaming field at different frequencies with constant velocity amplitude. Reproduced from Ref. 4 with permission from Elsevier.

length. As the vibration source length increases, the maximum velocity increases and then decreases, and reaches the maximum when L_s/W_a is 2/3. The maximum velocity increase with L_s is because more acoustic energy is used to support the main eddies. The maximum velocity decrease with L_s is because a too long radiation line restrains the acoustic streaming from returning to the radiation boundary along the two sides.

The acoustic streaming field excited by two separated vibration line sources was computed. In the computation, the vibration line sources are spatially symmetric about the y-axis, with the identical vibration amplitude and frequency, and the total length of them is constant (= 0.02 m). Figure 7.15(a, b) is the computed acoustic streaming field versus separation distance between the two vibration sources when the vibration sources are in and out of phase, respectively. Figure 7.15(a) shows that the main eddy length decreases as the separation between the in-phase vibration sources increases, and increasing the separation can eliminate the two eddies in the corners. From a comparison between images $a1$ (in Fig. 7.15(a)) and $b1$ (in Fig. 7.15(b)), it is

(a)

(b)

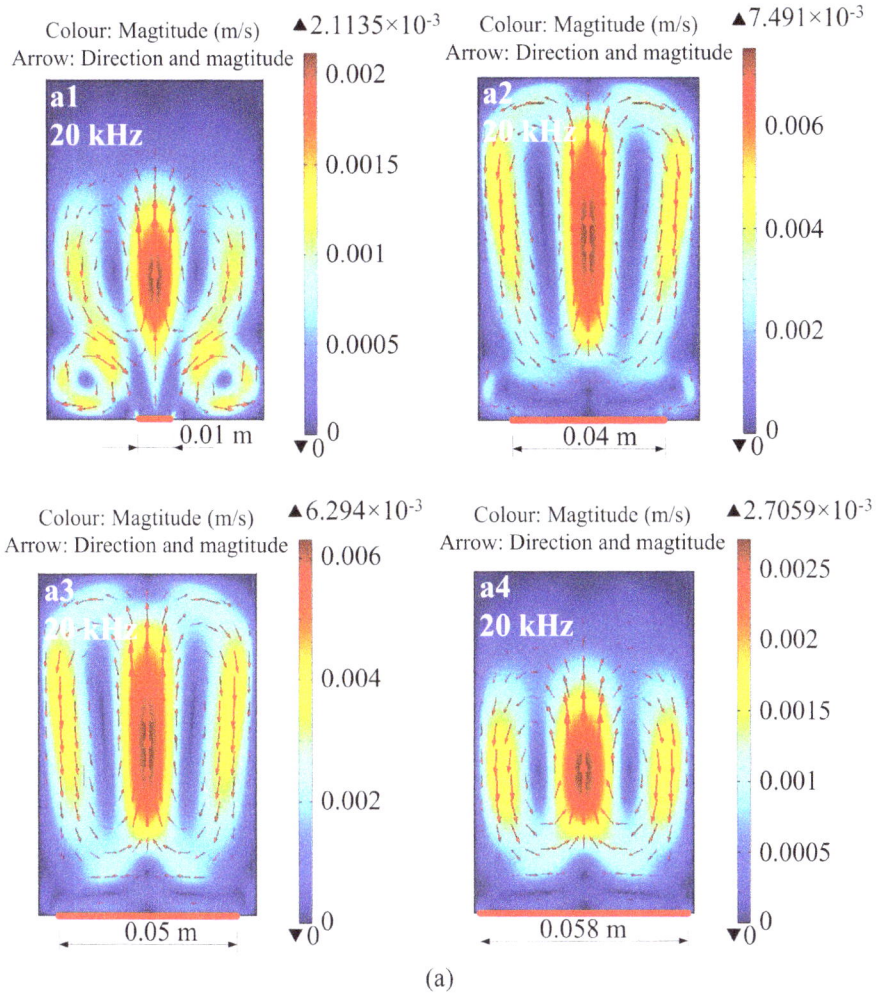

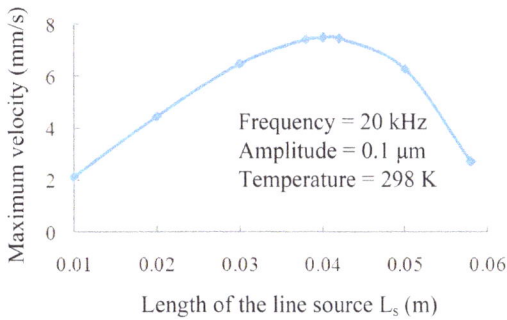

FIGURE 7.14   (a) Patterns of acoustic streaming field at different lengths of the line vibration source. (b) Maximum velocity of the main eddies versus the vibration source length. Reproduced from Ref. 4 with permission from Elsevier.

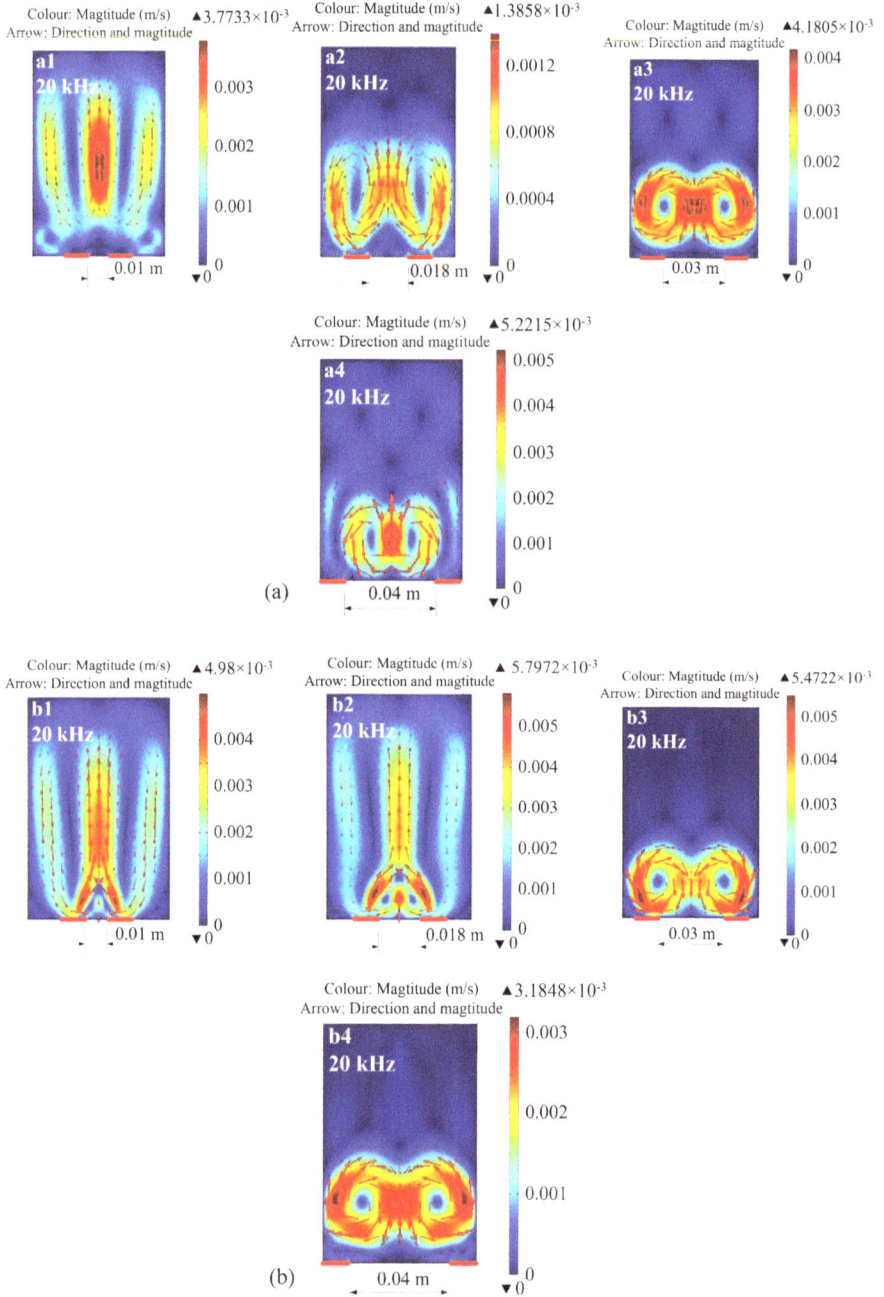

FIGURE 7.15 Patterns of acoustic streaming field generated by two separated vibration sources with constant total length and symmetric spatial distribution about the central axis. (a) Vibration sources in phase. (b) Vibration sources out-of-phase by π. Reproduced from Ref. 4 with permission from Elsevier.

(a)

(b)

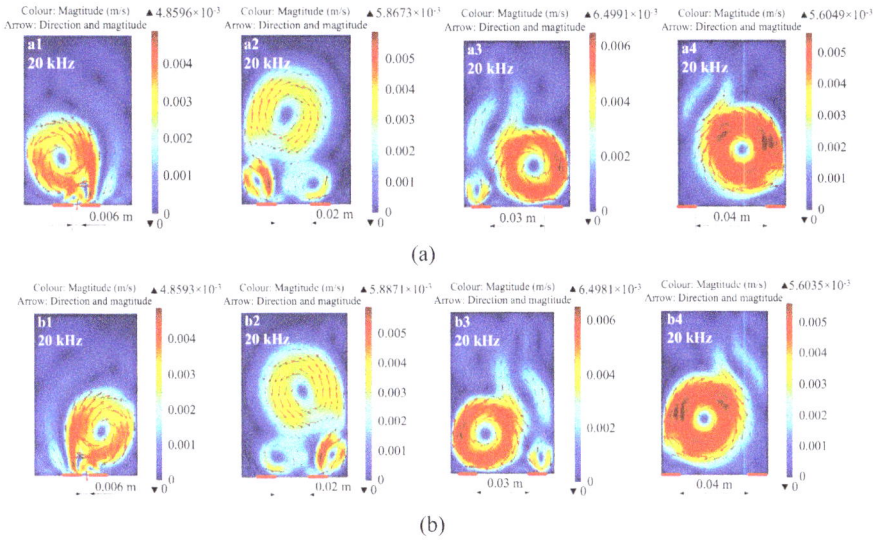

FIGURE 7.16   Patterns of acoustic streaming field generated by two separated vibration sources with constant total length and symmetric spatial distribution about the central axis. (a) $\pi/2$ vibration phase difference. (b) $-\pi/2$ vibration phase difference. Reproduced from Ref. 4 with permission from Elsevier.

known that using the out-of-phase vibration source can eliminate the two eddies in the corners.

Defining the phase of the left vibration source minus that of the right one as the phase difference, the acoustic streaming field versus the separation between the two vibration sources was computed for the phase difference of $\pm\pi/2$, and the results are shown in Fig. 7.16. Unlike the acoustic streaming fields when the phase difference is 0 and π (see Fig. 7.15), the acoustic streaming fields are no more symmetric about the y-axis in this case. Also, it is seen that the acoustic streaming fields for the phase difference of $\pm\pi/2$ are reverse with each other, which means that a phase reversal of the vibration displacement or driving voltage of one of the vibration sources flips the acoustic streaming field's pattern.

The above theoretical results show the diversity in the acoustic streaming field's pattern, caused by the frequency, size, topology and phase distribution of its vibration source. The diversity indicates the possibility to implement new manipulating functions for micro/nano objects, such as concentrating and rotating micro/nano objects at multiple locations on a substrate simultaneously.

Investigating the effect of a micro/nano particle in acoustic field on the acoustic streaming becomes an important issue, now that acoustic streaming has been employed in the ultrasonic micro/nano manipulations. The effect of a 0.5-mm-radius particle in the sound field on the acoustic streaming field was investigated by the FEM, and the results are shown in Fig. 7.17. In images a, b and c in Fig. 7.17, the particle is at (0, 0.05 m), (0.005, 0.05 m) and (0.01, 0.05 m), respectively. The relative coordinates of the particle $(x/W_a, y/L_a)$ are (0, 5/9), (1/12, 5/9) and (1/6, 5/9),

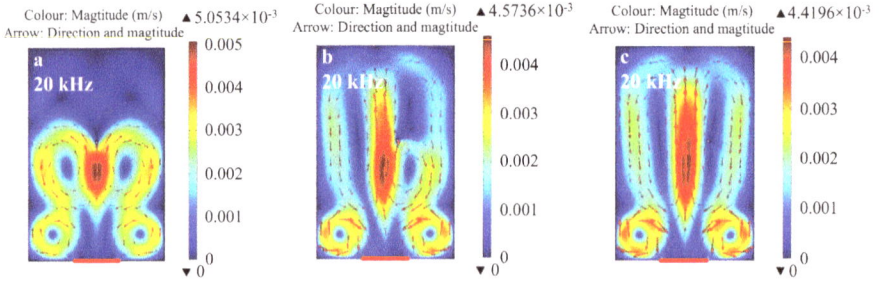

FIGURE 7.17 Patterns of acoustic streaming field with a 0.5-mm-radius particle at (0, 0.05 m), (0.005, 0.05 m) and (0.01, 0.05 m) for images 1, 2 and 3, respectively.

respectively. By comparison with Fig. 7.12(c), it is concluded that existence of the particle remarkably decreases the main eddy length when it is on the central axis, and an additional eddy will be generated when the particle is shifted from the central axis to a location other than the center of a main eddy.

Furthermore, the effect of the particle radius at (0, 0.05 m) on the acoustic streaming field was computed, and the results are shown in Fig. 7.18. The particle radius used in the calculation is 5, 50, 200 and 1000 μm, respectively. Figure 7.18(a) indicates that as the particle becomes large, the main eddies become short while the size of the eddies in the two corners has little change. Figure 7.18(b) shows the acoustic streaming field length versus the particle radius when the particle is at (0, 0.05 m). In the measurement of acoustic streaming field length, the upper boundary of acoustic streaming field is defined by the location where the acoustic streaming velocity is 1/5 of the maximum value. It is seen that when the particle radius is larger than 200 μm (at 20 kHz), the acoustic streaming field length decreases little as the particle becomes large. Moreover, based on our computation, it is known that when the ratio of the particle radius to wavelength is less than 0.03%, the particle on the central axis has little effect on the acoustic streaming field.

Further analyses show that increasing the number of small particles in the acoustic field increases the number of eddies, which is quite useful in practical applications such as ultrasonic mixing. Figure 7.18 also indicates that for a given acoustic field, the pattern and size of the eddies can be tuned by simply inserting a thin rod into a proper location in it.

The effect of temperature change of water in the ultrasonic field on the acoustic streaming was not understood before this work. For the model shown in Fig. 7.11, the effect of water temperature on the acoustic streaming field was investigated, and the results are shown in Figs. 7.19 and 7.20. In the computation, the relationships describing the temperature dependence of the sound speed, density, heat capacity, thermal conductivity and shear viscosity of water at 1 atm, given by COMSOL Multiphysics software, were utilized.

Figure 7.19 shows the pattern of the acoustic streaming field in water at temperature $T = 274$, 318 and 372 K, respectively. It indicates that when the water temperature increases, the length of the main eddies increases and the flowing pattern has little change. Figure 7.20(a) gives the maximum velocity of the main eddies

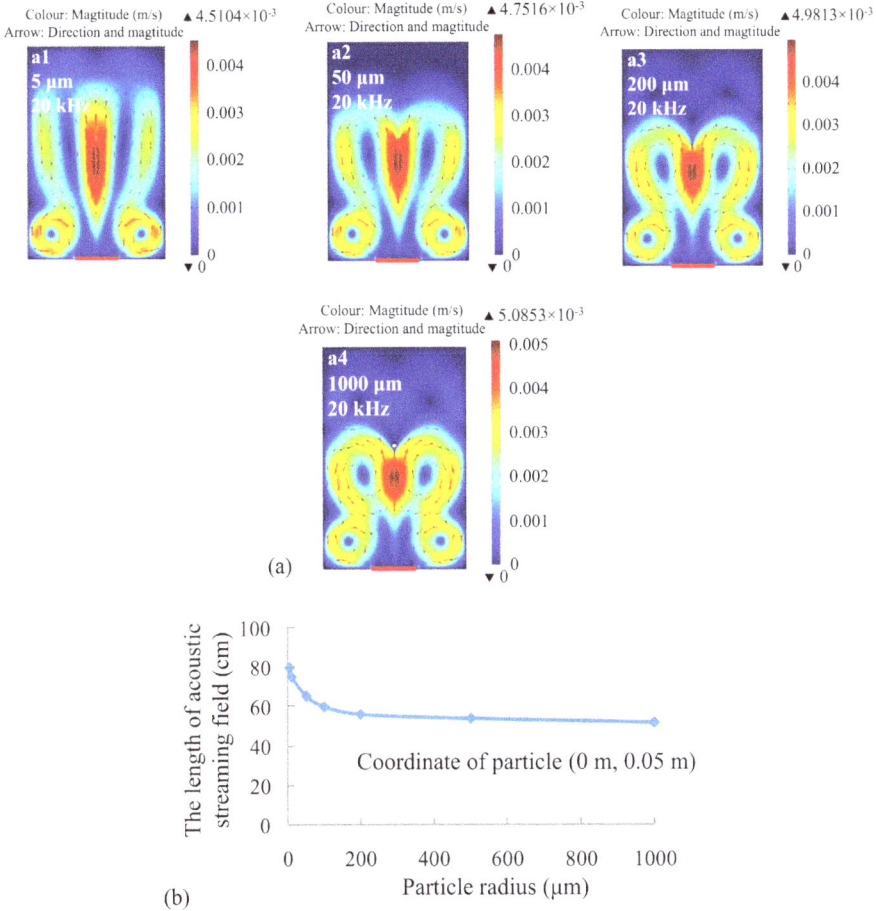

FIGURE 7.18 The effect of the size of a particle at (0 m, 0.05 m) on the acoustic streaming field. (a) Patterns of acoustic streaming field for different particle radii. (b) The length of the acoustic streaming field versus particle radius. Reproduced from Ref. 4 with permission from Elsevier.

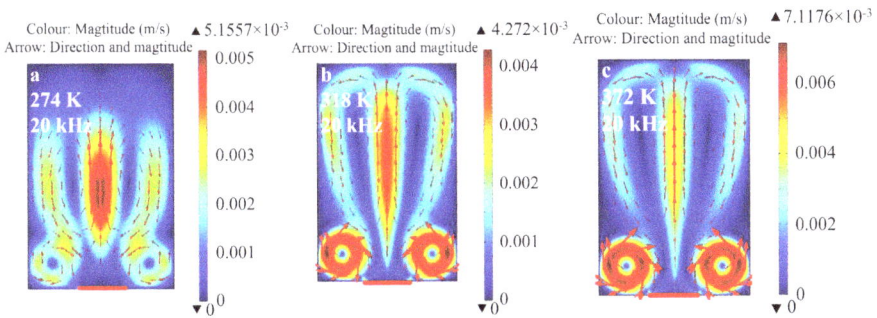

FIGURE 7.19 The acoustic streaming field's patterns in water at different temperatures. Reproduced from Ref. 4 with permission from Elsevier.

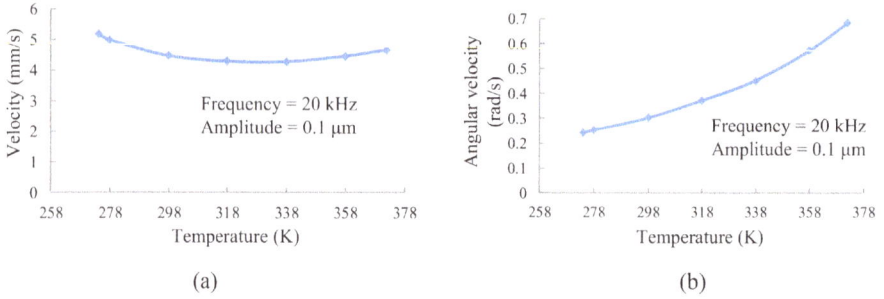

FIGURE 7.20   The acoustic streaming field's velocity in water at different temperatures. (a) Maximum velocity of the main eddies versus temperature. (b) Angular velocity of the corner eddies versus temperature. Reproduced from Ref. 4 with permission from Elsevier.

versus the temperature. It shows that as the temperature increases, the velocity decreases first and then increases. In the temperature range from 1 to 99°C, the acoustic streaming velocity has a fluctuation of 17% and reaches the minimum at about 60°C. Figure 7.20(b) shows the calculated angular velocity of the corner eddies versus the temperature, which shows that the angular velocity increases as the temperature increases.

For the convenience of analyzing Figs. 7.19 and 7.20, values of sound speed c, density ρ, shearing viscosity η and acoustic dissipation factor b of water at different temperatures are listed in Table 7.2. It is seen that as the temperature changes from 274 to 372 K, the acoustic dissipation factor b decreases monotonously. Thus, as the temperature rises, the sound wave can transmit further, which results in longer main eddies, as shown in Fig. 7.19.

From Table 7.2, it is also seen that as the temperature increases from 274 to 332 K, the acoustic dissipation factor b decreases. This causes a decrease of the spatial gradient of Reynolds stress and the maximum velocity of the main eddies, as shown in Fig. 7.20(a). As the temperature increases from 274 to 33 2K, the decrease of the acoustic dissipation factor b is not very remarkable, and neither is the spatial gradient of the Reynolds stress. However, in this temperature range, the shear viscosity decreases as the temperature. This results in an increase of the maximum velocity

TABLE 7.2

Sound Speed, Density, Shear Viscosity and Acoustic Dissipation Factor of Water at Different Temperatures

T (K)	274	298	332	353	372
c (m/s)	1408.7	1494.3	1549.9	1554.2	1543.6
ρ (kg/m³)	1003.8	998.2	985.1	974.0	961.9
η (Pa·s)	0.00174	0.00089	0.00048	0.00036	0.00028
b (Pa·s)	0.00599	0.00308	0.00164	0.00124	0.00098

of the main eddies, as shown in Fig. 7.20(a). The increase of angular velocity of the corner eddies with the temperature increase, as shown in Fig. 7.20(b), is caused by the decrease of the shear viscosity. From Fig. 7.20(b), it is also seen that the curve's slope in the temperature range from 274 to 332 K is less than that from 332 to 372 K. This is because the acoustic dissipation factor b decreases faster with the temperature rise in the range from 274 to 332 K than in the range from 332 to 372 K (see Table 7.2).

7.2.4 CONCLUSIONS

This example demonstrates the FEM-based computation method of acoustic streaming in a 2D rectangular acoustofluidic field with a traveling wave and using water as the acoustic medium, and a diversity of acoustic streaming with the change of the working frequency, vibration excitation source length, and distance and phase difference between two separated symmetric vibration excitation sources. In addition, the computational result indicates that the existence of micro particles has more or less influence on the distribution of acoustic streaming, depending on its position and size. It is also clarified that with an increase of the acoustic medium's temperature, the speed of the main acoustic streaming decreases first and then increases, and the angular velocity of the corner eddies increases monotonously. These conclusions provide useful guidelines for controlling and analyzing the acoustic streaming field in nano handling.

7.3 ACOUSTIC RADIATION FORCE AND STOKES FORCE IN ULTRASONIC TWEEZERS

7.3.1 INTRODUCTION

This example describes how to determine the manipulating force when both of the acoustic radiation force and acoustic streaming-induced Stokes force affect the manipulation performance. In this example [7], the ultrasonic tweezers' manipulation probe is in contact with the droplet-substrate interface and makes an angle with the interface. Micro particles near the irregular gap between the probe's tip and the droplet-substrate interface can be sucked to the probe's tip and trapped. The acoustic radiation force and acoustic streaming-induced Stokes force on the manipulated particles are computed by the FEM.

7.3.2 DEVICE AND USAGE

Figure 7.21(a) shows a schematic diagram of the ultrasonic tweezers, which are composed of a micro manipulating probe (MMP), vibration transmission needle (VTN) and sandwich-type piezoelectric transducer. The MMP is bonded onto the VTN's tip, and the MMP's axis is parallel to the end plate of the transducer. The VTN is bonded onto a corner of the end plate of the transducer. The reason for using the corner as the vibration excitation is that the vibration velocity in the four corners of an end plate is usually bigger. The MMP's tip is inserted into the droplet water and

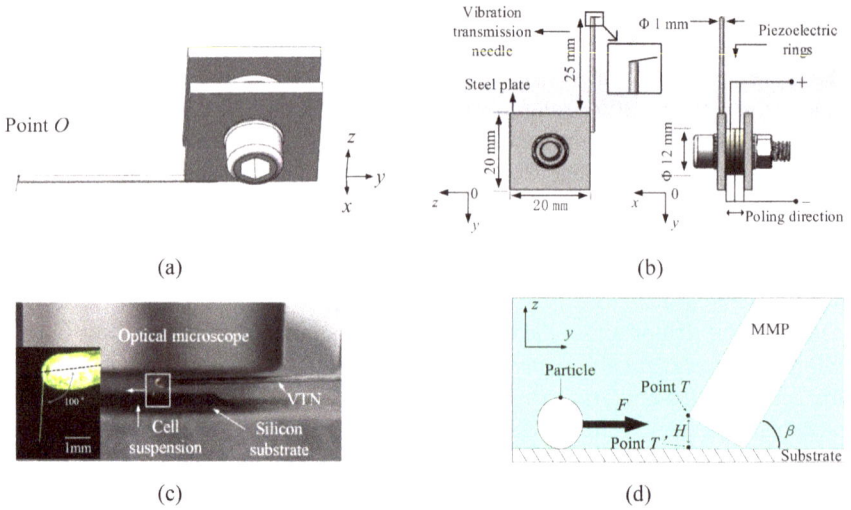

FIGURE 7.21   The device and experimental setup for capture of single biological particles at the interface between a water film and substrate. (a) Schematic diagram of the ultrasonic tweezers. (b) Structure and size of the device. (c) Photograph of the experimental setup. (d) Schematic diagram of the relative positions of the MMP and substrate. Reproduced from Ref. 7 with permission from Elsevier.

in contact with the substrate. Ultrasonic vibration excited by the transducer is transmitted to the MMP via the VTN. In the area around the MMP's tip in vibration, the acoustic radiation force is generated to trap a single micro-cell.

The MMP is made of a glass fiber with a 2.5-mm length and 16-μm diameter. The length and diameter of VTN, which is made of Nickel-plated steel, are 25 and 1 mm, respectively. The angle made by the MMP and VTN is 100°. The sandwich-type piezoelectric transducer consists of four piezoelectric rings made of lead zirconate titanate (P-81, Haiying Enterprise Group Co., Ltd., China), two square-shaped stainless steel end plates, and one nut-screw structure. The nut-screw structure is used to tighten the end plates and piezoelectric rings. The dimensions of the end plates are 20 mm × 20 mm × 2 mm. The thickness, outer and inner diameters of each piezoelectric ring are 1.2 mm, 12 mm and 6 mm, respectively. Material constants of the piezoelectric rings are listed in Table 7.3, and detailed dimensions of the tweezers fabricated and used in this work are illustrated in Fig. 7.21(b).

TABLE 7.3
Material Constants of the Piezoelectric Rings

Piezoelectric Constant d_{33} (C/N)	Electro-mechanical Coupling Factor k_{33}	Mechanical Quality Factor Q_m	Dielectric Dissipation Factor $tan\delta$	Density (kg/m³)
250×10^{-12}	0.63	500	0.6%	7450

The water film was composed of DI (deionized) water and dispersed mirco particles. Yeast cells (Angel Yeast Co. Ltd, China) with a diameter range of 3–7 μm and *Chlorella vulgaris* powder particles with a diameter range of 2–10 μm were used as the experimental samples. The concentration of the yeast cell and *Chlorella vulgaris* powder suspensions was always 0.005 mg/ml.

Figure 7.21(c) shows the experimental setup for the capture. The experiments were carried out under an optical microscopy (VHX-1000E, Keyence, Osaka, Japan). The vibrating MMP was inserted into the water film, the MMP's tip was in contact with the substrate, and the VTN above the water film was parallel to the substrate.

During the operation of the ultrasonic tweezers, single micro particles at the interface could be sucked by and captured to the MMP's tip. The movement of the MMP was achieved by an *X-Y-Z* moving stage, onto which the ultrasonic tweezers were fixed. By moving the MMP, the captured particle was transferred to a desired location at the interface between the water film and substrate.

For the convenience of theoretical analyses and discussion, a Cartesian coordinate system, shown in Fig. 7.21(a), is used for the following analyses and discussion. The *yz* plane of the coordinate system is in the plane formed by the central axes of the VTN and MMP. Also, a diagram describing the relative positions of the MMP and substrate is given in Fig. 7.21(d). Point *T* is the left endpoint of the MMP's cross-section in the *yz* plane, and point *T'* is the orthographic projection of point *T* onto the substrate.

Figure 7.22 shows a process of the capture, transfer and release of a 7-μm-diameter yeast cell at the interface between the water film and silicon substrate, recorded by the optical microscopy. The driving frequency and voltage were 135 kHz and 12 V_{p-p}, respectively. Image *a* shows the state just before the yeast cell was captured. In image *b* to *e*, the yeast cell was captured and transferred. In Image *f*, the driving voltage of ultrasonic tweezers was switched off and the vibration was stopped. In image *g*, the particle was released. The *Chlorella vulgaris* powder particles could also be captured in the experiments. The temperature rise of the vibrating MMP's tip in water film was measured by a thermocouple (GM1312, k-type, Shenzhen Jumaoyuan Science and Technology Co. Ltd, China). It showed that the temperature rise was less than 0.1°C. This means the method proposed in this work is suitable for the capture of biological samples, which are usually quite sensitive to the temperature rise. The temperature rise around the MMP is low for the following two reasons: (I) The MMP is very thin and the acoustic output power radiated from it is not strong; (II) the acoustic streaming around has enhanced the heat dissipation around the MMP.

7.3.3 FEM COMPUTATION FOR PRINCIPLE ANALYSES

The vibration of the ultrasonic tweezers was measured by a laser Doppler vibrometer (PSV-300F, Polytec GmbH, Waldbronn, Germany). Figure 7.23(a) shows the measured vibration trajectory of the MMP's root when the MMP has the capture capability. It shows that the MMP's root vibrates elliptically when it has the capture capability. The measured *x*-, *y*- and *z*-directional vibration velocity components are $1.49 \times 10^{-2} \angle 13.3°$ m/s, $3.83 \times 10^{-2} \angle 149.2°$ m/s, $7.02 \times 10^{-2} \angle 154.5°$ m/s, respectively, in which the angles represent the phase difference between the measured vibration velocity components and the reference voltage. The reference voltage is picked up

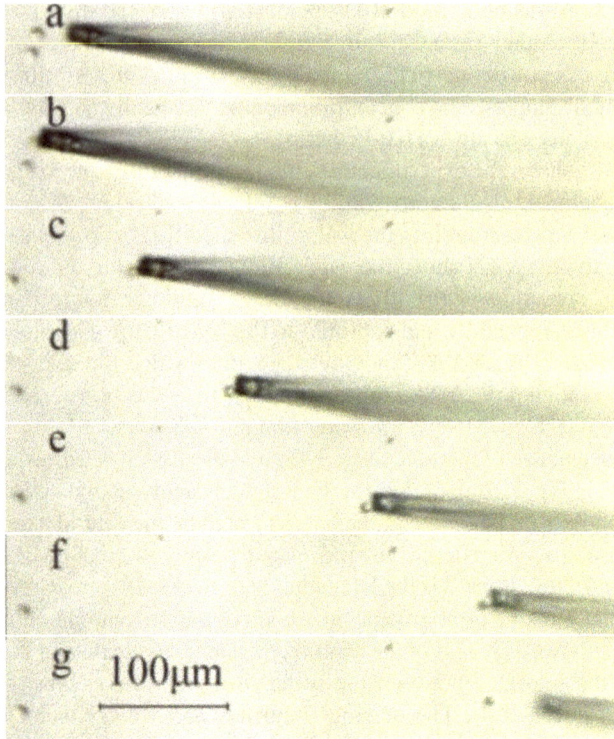

FIGURE 7.22 A sequence of images to show the process of capture, transfer and release of a 7-μm-diameter yeast cell at the interface between a water film and substrate. Reproduced from Ref. 7 with permission from Elsevier.

from the driving voltage of the transducer and applied to the vibrometer. The measured distributions of vibration velocities of the VTN and end plate at 135 kHz and 12 V_{p-p} are given in Fig. 7.23(b, c). Figure 7.23(c) shows that the in-plane vibration velocity of the end plate is larger than the out-of-plane vibration velocity, which means that the piezoelectric rings vibrate in the radial mode.

In this work, the steady-state acoustic field and acoustic radiation force on the micro particles around the MMP's tip were computed by the FEM to analyze the capture mechanism. The computation was implemented by COMSOL Multiphysics 5.2a software. To simplify the computation, only the MMP, water film and substrate surface were included in the FEM model, as shown in Fig. 7.24(a).

The 3D physical model shown in Fig. 7.21(d) was used in the FEM computation, and a meshed FEM model is shown in Fig. 7.24(a). The mesh size of the acoustic field near the ultrasonic needle is smaller than that in the rest region of the acoustic field, in order to decrease the computational error of the acoustic field near the MMP and decrease the computational time. The detailed mesh size values of different regions are as follows. The maximum element size is 0.5 μm (about 0.0045% of the wavelength of the sound field in water at 135 kHz), and the total element number is 1140074 in the area around the MMP's tip. In the region far from the MMP in the

(a)

(b)

(c)

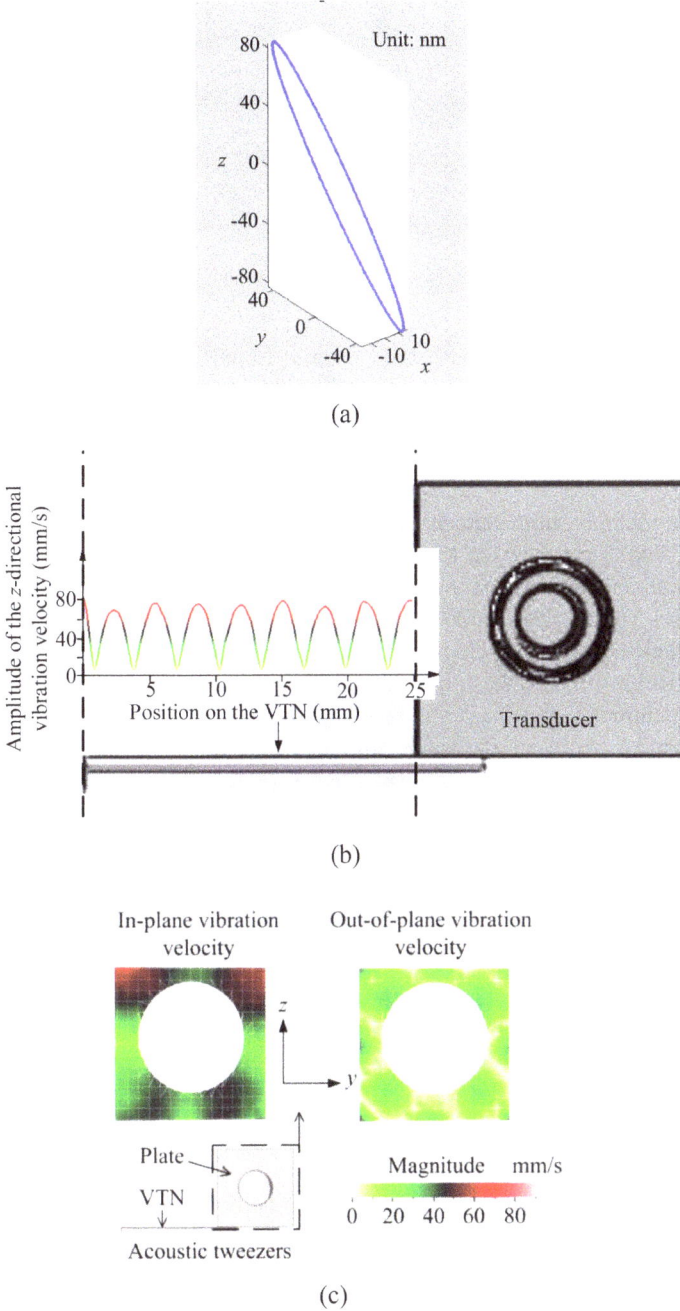

FIGURE 7.23   (a) Vibration trajectory of the MMP's root (point O). (b) Measured distribution of the z-directional vibration velocity of the vibration transmission needle at 135 kHz and 12 $V_{p\text{-}p}$. (c) The in-plane and out-of-plane vibration velocities of the end plate of transducer. Reproduced from Ref. 7 with permission from Elsevier.

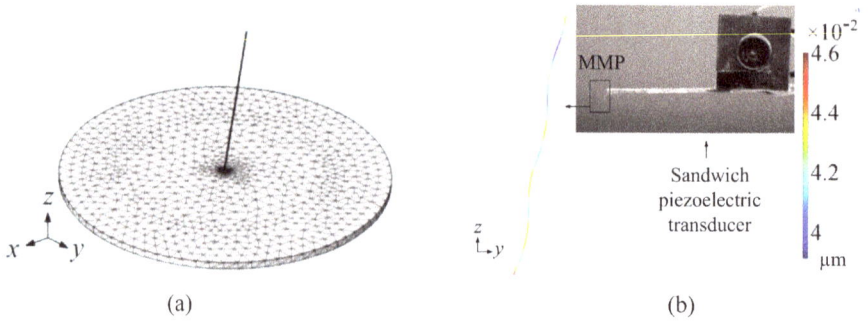

(a)   (b)

FIGURE 7.24 FEM meshed model and computed vibration displacement distribution of the MMP. (a) FEM meshed model for the probe-film system. (b) Computed vibration displacement distribution of the MMP at a driving frequency of 135 kHz and voltage of 12 V_{p-p}. Reproduced from Ref. 7 with permission from Elsevier.

liquid film, the maximum element size is 0.5 mm (about 4.5% of the wavelength of the sound field in water at 135 kHz), and the total element number is 152280. Also, the maximum element size of the needle is 0.2 mm (about 0.56% of the wavelength of the sound field in glass at 135 kHz), and the total element number is 112147. The tetrahedral elements are used in the FEM computation model. It has been proved that the numerical results are mesh-independent and convergent with the above computational conditions.

The boundary between the water film and substrate was acoustically hard. The boundaries between the water film and air were acoustically soft. The acoustic-structure boundaries were used for the interfaces between the water film and MMP. In the computation, sound field in the water film was computed first.

The following wave equation is used to solve the sound field:

$$\rho_f \frac{\partial^2 p}{\partial t^2} = \rho_f c_f^2 \nabla^2 p + b \nabla^2 \frac{\partial p}{\partial t}, \tag{7.18}$$

where p is the sound pressure, ρ_f is the fluid density without sound field, and c_f is the sound speed. The acoustic dissipation factor b is computed by

$$b = \frac{4}{3} \eta + \eta', \tag{7.19}$$

where η and η' are the shear and bulk viscosity coefficients of the acoustic medium, respectively. The vibration velocity u_i (where subscript i represents x, y or z) of the sound field can be calculated by

$$u_i = i \frac{1}{\rho_f \omega} \frac{\partial p}{\partial x_i}, \tag{7.20}$$

where $i = \sqrt{-1}$ and ω is the angular frequency. Then the computed vibration velocity and sound pressure were used to compute the acoustic radiation force.

If we adopt Gor'kov theory, the acoustic radiation force per unit volume (acting on a particle) $\vec{F}$ in the sound field is

$$\vec{F} = -\nabla U, \tag{7.21}$$

where U is the time-averaged force potential per unit particle volume. When the wave number k and the particle radius R satisfy $kR \leq 1$, U is

$$U = -D\langle K_E \rangle + (1-\gamma)\langle P_E \rangle, \tag{7.22}$$

where $\langle K_E \rangle$ and $\langle P_E \rangle$ are the time-averaged kinetic and potential energy densities of the sound field, respectively, D is a parameter determined by the densities of the particle and fluid, and γ is the compressibility ratio between the particle and fluid. D and γ can be calculated by

$$D = \frac{3(\rho_s - \rho_0)}{2\rho_s + \rho_0}, \tag{7.23}$$

$$\gamma = \frac{\rho_0 c_0^2}{\rho_s c_s^2}, \tag{7.24}$$

where ρ_s and ρ_0 are the densities of the particle and fluid, respectively and c_s and c_0 are the sound speed in the particle and fluid, respectively. The time-averaged kinetic energy density is

$$\langle K_E \rangle = \frac{\rho_0 \langle v^2 \rangle}{2}. \tag{7.25}$$

The time-averaged potential energy density is

$$\langle P_E \rangle = \frac{\langle p^2 \rangle}{2\rho_0 c_0^2}, \tag{7.26}$$

where v and p are the vibration velocity and acoustic pressure, respectively.

Parameters used in computation are listed in Tables 7.4 and 7.5. Unless otherwise specified, the measured x-, y- and z- directional vibration velocities at the MMP's

TABLE 7.4

Dimensions of the Probe-Film System Used in the FEM Computation

Probe-Film System	Dimensions
The MMP's length (mm)	2.5
The MMP's radius (µm)	8
The film's thickness (mm)	0.1
The film's radius (mm)	2.5
The angle between the MMP and substrate (degree)	80

TABLE 7.5

Material Constants Used in the FEM Computation

Material	Density (kg/m³)	Velocity of Sound (m/s)	Poison's Ratio	Modulus of Elasticity (Pa)	Shear Viscosity Coefficient (Pa·s)
Glass (MMP)	2200	–	0.3	7.4×10^9	–
Water	1000	1497	–	–	0.0017
Yeast cell	1140	1587.5	–	–	–

root at 135 kHz and 12 $V_{p\text{-}p}$ were used in the FEM analyses. They were (V_x=) $1.49 \times 10^{-2}\angle 13.3°$ m/s, (V_y=) $3.83 \times 10^{-2}\angle 149.2°$ m/s and (Vz=) $7.02 \times 10^{-2}\angle 154.5°$ m/s, respectively. The MMP was in contact with the substrate. Figure 7.24(b) shows the computed vibration displacement distribution of the MMP, which indicates that the MMP vibrates flexurally, and the vibration amplitude at the MMP's tip is larger than that at the MMP's root.

Figure 7.25(a) shows the computed y-directional acoustic radiation force per unit particle volume near the MMP. The white stripe area represents the MMP's cross-section in the yz plane. It can be seen that acoustic radiation force acting on a particle on the left side of MMP is in the +y-direction (toward the MMP's tip). When this +y-directional acoustic radiation force is bigger than the frictional force between the particle and substrate, the particle can be sucked to the MMP's tip. Figure 7.25(b) shows the computed z-directional acoustic radiation force per unit particle volume. It is seen that the z-directional acoustic radiation force per unit particle volume is positive, which means that it decreases the frictional force between the particle and substrate. In addition, due to the trapping force shown in Fig. 7.25 and contact between the trapped particle and MMP's tip make the position of captured particle relative to the MMP's tip stable.

Figure 7.26 shows the computed distribution of the time-averaged force potential per unit volume U in the plane parallel to and 3 μm above the substrate. The white circular area indicates the MMP's cross-section 3 μm above the substrate. It is seen that U decreases as the distance from the MMP decreases, which means the acoustic radiation force per unit particle volume near the MMP points to the MMP. Due to this spatial gradient of the force potential, micro particles near the MMP are pushed to the MMP.

The spatial gradients of the Reynolds stress and mean pressure are the forces which generate the acoustic streaming. The spatial gradient of the Reynolds stress F_j is computed by

$$f_j = -\partial\langle\rho_f u_i u_j\rangle/\partial x_i, \tag{7.27}$$

where u_i and u_j are the vibration velocities of the sound field, repeated suffixes i and j represent x, y and z in the 3D model, and $<>$ represents the time average over one time period. The mean pressure is computed by

$$\bar{p}_2 = \frac{1}{2\rho_f c_f^2}\frac{B}{A}\langle p^2\rangle, \tag{7.28}$$

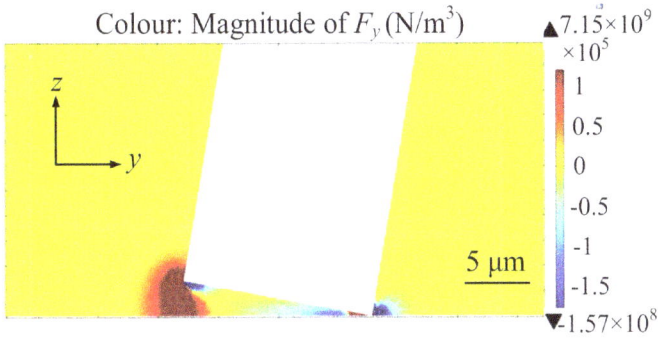

Colour: Magnitude of F_y (N/m^3)

(a)

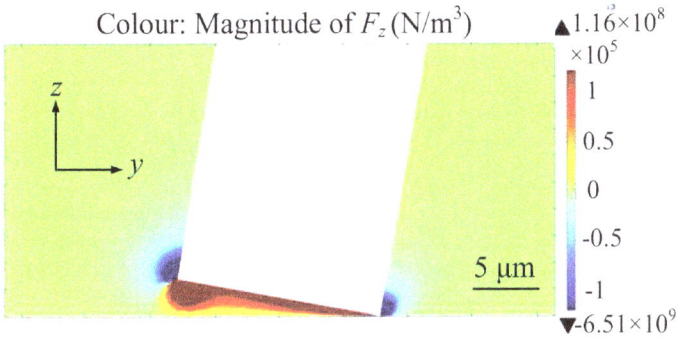

Colour: Magnitude of F_z (N/m^3)

(b)

FIGURE 7.25 Computed y- and z-directional acoustic radiation force per unit particle volume near the MMP in the yz plane ($x = 0$). (a) Computed y-directional acoustic radiation force per unit particle volume. (b) Computed z-directional acoustic radiation force per unit particle volume. The MMP is assumed to be cylindrical for the computations. If the x value is small, the qualitative conclusion derived from this figure has no change. Reproduced from Ref. 7 with permission from Elsevier.

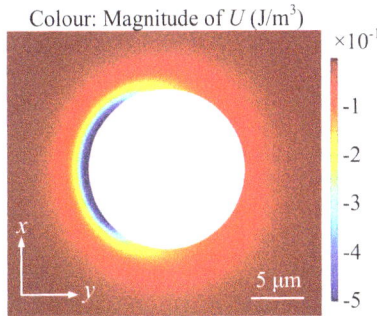

Colour: Magnitude of U (J/m^3)

FIGURE 7.26 Computed distribution of the time-averaged force potential per unit volume U in a plane parallel with and 3 μm above the substrate. Reproduced from Ref. 7 with permission from Elsevier.

where $\dfrac{B}{A}$ is the nonlinear parameter of the fluidic medium. The steady acoustic streaming satisfies the following equation:

$$\rho_f(\bar{u}_i\,\partial\bar{u}_j/\partial x_i) = f_j - \partial\bar{p}_2/\partial x_j + \eta\nabla^2\bar{u}_j, \tag{7.29}$$

where $\bar{u}_i$ is acoustic streaming velocity. The acoustic streaming also satisfies the continuity equation

$$\partial\bar{u}_i/\partial x_i = 0, \tag{7.30}$$

Slip boundary condition was used in the FEM computation of the acoustic streaming. The Stokes force on a particle, caused by the acoustic streaming, was obtained by the following equation:

$$F_d = 6\pi\mu Rv, \tag{7.31}$$

where F_d is the Stokes force, μ is the dynamic viscosity, R is the particle radius and v is the flow velocity relative to the object.

Figure 7.27 shows the estimated Stokes force on a 5-μm-diameter yeast cell on the substrate and near the MMP, in which the white strip is the cross-section of MMP in the yz-plane. It shows that the Stokes force on the cell is in the order of 10^{-10} (N). According to the preceding computation of acoustic radiation force, the acoustic radiation force on the cell is in the order of 10^{-8} (N). Thus, the acoustic radiation force is much larger than the Stokes force for the region near the MMP in this example. Figure 7.27 also shows that the acoustic streaming tends to flush away the particles

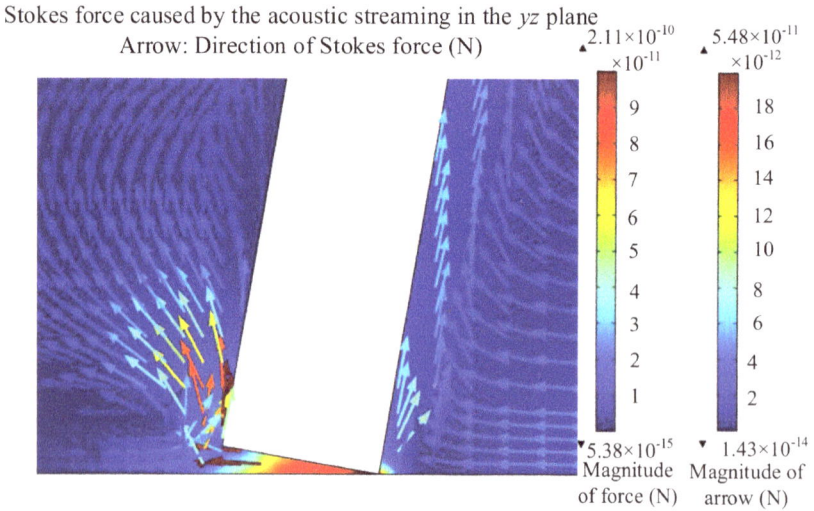

FIGURE 7.27 Stokes force caused by the acoustic streaming in the yz plane. Reproduced from Ref. 7 with permission from Elsevier.

near the MMP. Therefore, it is concluded that it is the acoustic radiation force that captures the cells in the ultrasonic tweezers.

7.3.4 CHARACTERISTICS AND DISCUSSION

In the experiments, it was observed that a micro particle at the interface could be sucked onto the MMP's tip if the distance between the MMP's tip and particle was less than a critical value. This critical value is defined as the maximum capture distance d_m, which is used to measure the capture capability. d_m versus the working frequency at different driving voltages of the transducer was measured for a yeast cell in front of the MMP at the interface between the water film and substrate, and the result is shown in Fig. 7.28(a). The yeast cells used in the measurement has an average diameter of 5 μm. It is seen that there exists a maximum d_m for a given driving voltage. The best capture capability increases as the driving voltage increases, which is caused by the increases of vibration velocity of the transducer in resonance.

In the experiments, it was also observed that the captured micro particle could be transported to a desired location in the water film by moving the ultrasonic tweezers, and the transported particle would fall during the transfer if the ultrasonic conditions were not proper, for example, a too weak MMP vibration. In this work, the transfer success rate is defined by

$$S_r = N_s/N_t, \tag{7.32}$$

where N_t is the total transfer number and N_s is the successful transfer number. The measured success rate versus operating frequency at different driving voltages is shown in Fig. 7.28(b). In the measurement, the moving distance and velocity of the MMP were 300 μm and 30 μm/s, respectively, and the MMP motion was linear and parallel to the substrate. It is seen that a success rate of 100% can be achieved at a driving voltage of 20 V_{p-p} in the operating frequency range from 135 to 135.4 kHz.

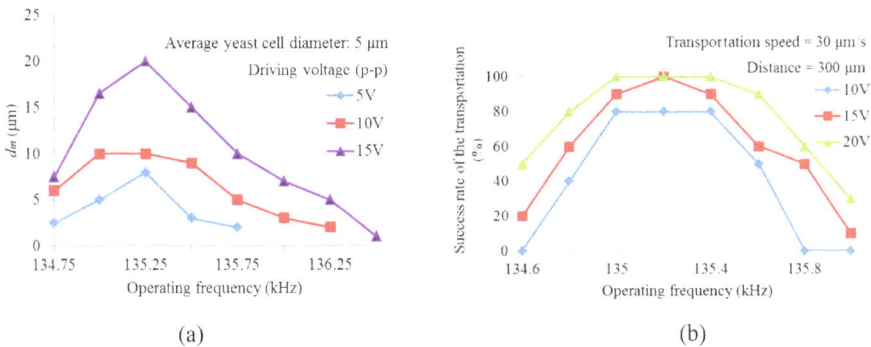

(a)   (b)

FIGURE 7.28   Measured frequency characteristics of the device for single yeast cells with an average diameter of 5 μm. (a) Capture capability d_m versus operating frequency for different driving voltages. (b) Transfer success rate versus operating frequency for different driving voltages. Reproduced from Ref. 7 with permission from Elsevier.

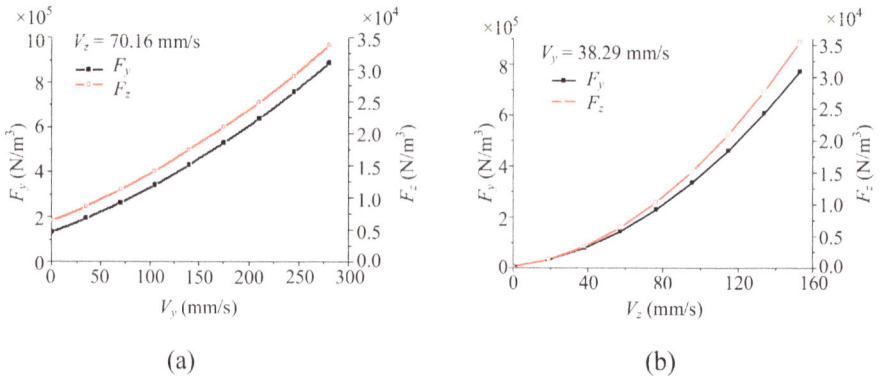

(a)   (b)

FIGURE 7.29 The computed dependency of the acoustic radiation force per unit particle volume at point T' on the vibration velocity at the MMP's root. (a) The dependency of the y- and z-directional acoustic radiation forces per unit particle volume on the y-directional vibration velocity V_y at the MMP's root. (b) The dependency of the y- and z-directional acoustic radiation forces per unit particle volume on the z-directional vibration velocity V_z at the MMP's root. If we change the size and shape of the VTN, the ratio of V_y and V_z can be changed, and the vibration velocity change shown in this figure may happen theoretically. Reproduced from Ref. 7 with permission from Elsevier.

This figure indicates that a strong vibration is beneficial to increase the success rate of the particle transfer.

The computed dependencies of the y- and z-directional acoustic radiation forces per unit particle volume on the y- and z-directional vibration velocities V_y and V_z are shown in Fig. 7.29(a, b). It is seen that dF_z/dV_z is larger than dF_y/dV_y, which indicates that the z-directional vibration velocity of the MMP has a larger effect on the capture than the y-directional vibration velocity. This is because the z-directional vibration of the MMP has a larger effect on the ultrasonic field between the MMP's tip and substrate. Thus, one must pay more attention to the vibration velocity V_z in the control of the capturing force.

The dependency of the y- and z-directional acoustic radiation forces per unit particle volume at point T' on the tilt angle β was computed, and the result is shown in Fig. 7.30(a). In the computation, the driving frequency and voltage were 135 kHz and 12 V_{p-p}, respectively. It is seen that the change of a small tilt angle β affects the acoustic radiation force little, and the acoustic radiation force increases rapidly as the tilt angle β increases when β is large. When β is small, the distance H between point T and T' (see Fig. 7.21) is large and the sound field between the MMP's end surface and substrate is weak, which causes little change of the acoustic radiation force at point T' as β changes. When β is large, H is small and the sound field between the MMP's end surface and substrate is strong. In this case, as β increases, the sound field between the MMP's end surface and the substrate increases rapidly, and so does the acoustic radiation force at point T'. Therefore, to obtain a strong capture capability, one needs to use a large tilt angle β.

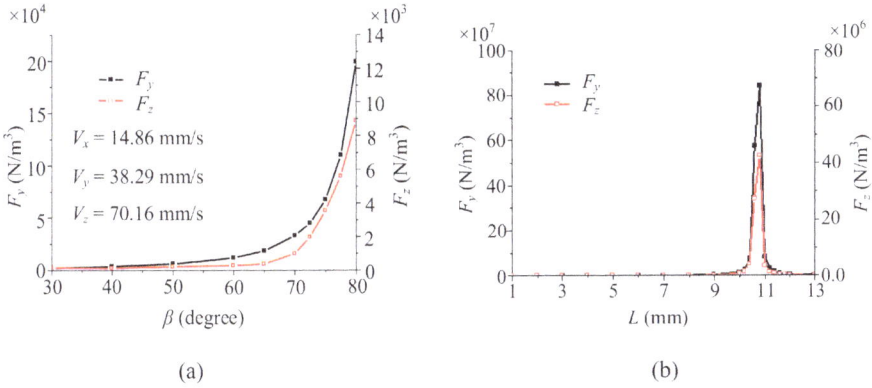

FIGURE 7.30 The dependency of the y- and z-directional acoustic radiation forces per unit particle volume at point T' (a) on the MMP's tilt angle β and (b) on length L at 135 kHz and 12 $V_{p\text{-}p}$. Reproduced from Ref. 7 with permission from Elsevier.

The dependency of the y- and z-directional acoustic radiation forces per unit particle volume at point T' on the MMP's length L was computed, and the result is shown in Fig. 7.30(b). In the computation, the driving frequency and voltage were 135 kHz and 12 $V_{p\text{-}p}$, respectively, and L changed from 1 to 10 mm. It indicates that the MMP resonates at 135 kHz when its length is 10.8 mm, which increases the acoustic radiation force sharply. This provides an effective method to generate a large acoustic radiation force to capture the micro particles.

In the experiments, it was found that it was quite difficult to capture the micro particles if the MMP and substrate were not in contact. To find the reason for this phenomenon, the y- and z-directional acoustic radiation forces per unit particle volume F_y and F_z were computed when the MMP and substrate were not in contact, and the results are shown in Fig. 7.31. In the computation, the distance between the MMP's lower end and substrate was 3 µm. For comparison, the results when the MMP and substrate are in contact are also listed in Fig. 7.31. It is seen that a tiny separation between the MMP and substrate decreases F_y and F_z substantially, which well explains the experimental phenomenon. The decrease of acoustic radiation force is caused by a weak sound field under the MMP, resulting from the increase of the distance between the MMP and substrate.

7.3.5 CONCLUSIONS

This example is to demonstrate the FEM method combined with Gor'kov theory for computing the acoustic radiation force in an irregular ultrasonic field, and the FEM method to compute the Stokes force generated by acoustic streaming in the same ultrasonic field. With the demonstrated computational method, one can compare the acoustic radiation force and the Stokes force quantitatively, and clarify the working principle of a micro/nano handling process.

(a)

(b)

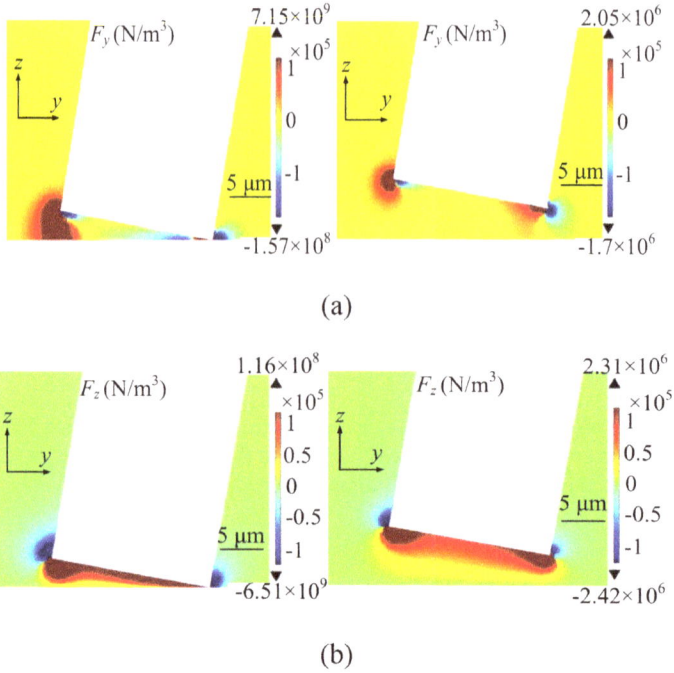

FIGURE 7.31 The change of the y- and z-directional acoustic radiation force distributions when the MMP is lifted up from the substrate by 3 μm. (a) F_y. (b) F_z. The color represents the magnitude of the acoustic radiation force. Reproduced from Ref. 7 with permission from Elsevier.

7.4 REMARKS

This chapter gives three examples of FEM modeling and computation of the ultrasonic fields, acoustic radiation force, acoustic streaming and Stokes force generated by the acoustic streaming in practical ultrasonic devices. In all of the computations, ultrasonic fields are computed first, and then acoustic radiation force and acoustic streaming are computed after post-processing the computational results of ultrasonic fields. In all of the three examples, the ultrasonic fields are not rotational. For an ultrasonic field within the acoustic boundary or coupled with a temperature field, one may utilize the thermo-acoustic module in COMSOL Multiphysics software to compute the acoustic pressure and vibration velocity distributions first. Then, with the acoustic pressure and vibration velocity distributions, the acoustic radiation force and acoustic streaming can be computed.

REFERENCES

1. Y. Liu, J. Hu and C. Zhao, "Dependence of acoustic trapping capability on the orientation and shape of particles," *IEEE Trans. Ultrason. Ferroelectr. Freq. Control*, 57(6), pp. 1443–1450, 2010.
2. Y. Liu and J. Hu, "Trapping of particles by the leakage of a standing wave ultrasonic field," *J. Appl. Phys.*, 106(3), 034903, 2009.

3. J. Hu, *Ultrasonic Micro/Nano Manipulations: Principles and Examples*, (World Scientific, New Jersey, London, Singapore, 2014), pp. 15–19.

4. Q. Tang and J. Hu, "Diversity of acoustic streaming in a rectangular acoustofluidic field," *Ultrasonics*, 58, pp. 27–34, 2015.

5. O. V. Abramov, *High-Intensity Ultrasonics: Theory and Industrial Applications*, (Gordon and Breach Science Publishers, 1998).

6. W. L. Nyborg, *Acoustic Streaming, in Physical Acoustics*, in W. P. Mason, Ed., (Academic Press, New York, 1965), Vol. 2B, pp. 265–331.

7. Q. Liu, Q. Tang and J. Hu, "A new strategy to capture single biological micro particles at the interface between a water film and substrate by ultrasonic tweezers," *Ultrasonics*, 103, 106067, 2020.

8. L. E. Kinsler, A. R. Frey, A. B. Coppens and J. V. Sanders, *Fundamentals of Acoustics*, (Hamilton Press, 1999).

8 Concluding Remarks

Ultrasonic nano/microfabrication, handling, and driving is a branch of nano/micro manipulation technology, which is an emerging field in actuation technology. It has an interdisciplinary nature, with the working principles in physical acoustics, device structures of piezoelectric actuators, potential applications in nano device fabrication, biological sample handling, high-performance sensing, etc.

It is well known that the liquid-borne power ultrasound can be employed in the fabrication of graphene materials. The examples in this book indicate that ultrasonic vibration/field can also be employed in nano rolling/cutting and fabrication of nano strain sensors and flexible gas sensors. The merits of ultrasonic fabrication methods, demonstrated in this book, include room temperature and normal pressure fabrication, simple and low-cost equipment, and use of less or no organic solvent. To implement a real application in the fabrication of commercial products of nano devices, one has to increase the number of devices ultrasonically fabricated in each batch, that is, to solve the scaling-up problem of ultrasonic nano fabrication.

The ultrasonic concentration or enrichment is one kind of ultrasonic handling, and the examples of ultrasonic concentration in this book show that properly controlled ultrasound in a droplet can be employed to concentrate nanoscale samples at the droplet-substrate interface into a linear or spot pattern, and the concentrated materials may be moved by shifting the device. Examples of computation of acoustic streaming field are also given in this book. The acoustic streaming field is one of the important physical fields to implement nano concentration, which has potential applications in high-sensitivity bio-sensing systems, nano device fabrication, etc.

The probe-type ultrasonic nano tweezer is another important device in ultrasonic handling. The examples in this book demonstrate various potential applications of this type of nano tweezers in manipulating nanoscale samples

Application examples of ultrasonic driving of gas molecules and microfluid in this book show that ultrasonic nano/micro driving has potential applications in high-performance sensing systems, chemical batteries and cooling of hot spots. The airborne ultrasound is an effective way to catalyze the reactions at sensing surfaces and at the cathode of a metal-air battery. To widen the application range of this technology, it is necessary to investigate the exact mechanism of the ultrasonic catalysis effect or ultrasonic gas molecule driving process. The investigation work will involve the quantitative analyses of acoustic and thermal fields near the reaction surfaces, which are coupled together, and gas molecular dynamics process near the sonicated reaction surfaces.

Moreover, examples of FEM modeling and computation, described in this book, elaborate the numerical analysis method for analyzing the devices for ultrasonic nano/microfabrication, handling, and driving.

In ultrasonic nano/micro fabrication, handling and driving, the key technique is how to design the ultrasonic devices or transducers for a particular function. At the present

DOI: 10.1201/9781003404705-8

stage, there are still no related systematic design guidelines. The main challenge in such a design is to propose effective topological structures of the ultrasonic devices.

After reading the examples in this book, one may naturally raise the question whether ultrasound can manipulate much smaller substances such as atoms and electrons. Considering the fact that the scale of the objects, which have been actuated by ultrasound in recent 50 years, decreases from macro, micro and nano scale to molecular scale, and reported experimental phenomena of sound/vibration induced luminescence, the answer to this question is yes, as my point of view.

Index

For Product Safety Concerns and Information please contact our EU
representative GPSR@taylorandfrancis.com
Taylor & Francis Verlag GmbH, Kaufingerstraße 24, 80331 München, Germany

www.ingramcontent.com/pod-product-compliance
Lightning Source LLC
Chambersburg PA
CBHW060339220326
41598CB00023B/2751